Le Hai Khoi • Javad Mashreghi

Theory of $\mathcal{N}_p$ Spaces

Birkhäuser

Le Hai Khoi
Department of General Education
University of Science and Technology of Hanoi,
Vietnam Academy of Science and Technology
Hanoi, Vietnam

Javad Mashreghi
Department of Mathematics and Statistics
Université Laval
Québec, QC, Canada

ISSN 1660-8046   ISSN 1660-8054 (electronic)
Frontiers in Mathematics
ISBN 978-3-031-39703-5   ISBN 978-3-031-39704-2 (eBook)
https://doi.org/10.1007/978-3-031-39704-2

This book is published under the imprint Birkhäuser, www.birkhauser-science.com by the registered company
Springer Nature Switzerland AG
The registered company address is: Gewerbestrasse 11, 6330 Cham, Switzerland

Paper in this product is recyclable.

Preface

Banach and Hilbert spaces of analytic functions on the open unit disc $\mathbb{D}$ and the operators acting on such spaces are currently active domains of research. The classical part of this subject was developed in the early days of the preceding century. However, due to numerous applications in science and engineering as well as its dominant role in other branches of mathematics, the topic has gained enormous momentum in recent years. A wide range of mathematicians and applied scientists are presently working on different aspects of the theory.

A list of abstract analytic spaces is long, and there are several excellent books on each space. Since we are mainly dealing with Hardy, Dirichlet, and Bergman spaces in this note, a few references are mentioned below. A panoramic approach to function spaces is provided in [108]. Theory of Hardy Space is discussed in depth in [23, 38, 56, 66]. Some further spaces which live inside the Hardy space H^2 are studied in [31, 32, 35, 36, 67]. Contrary to Hardy spaces, two close cousins, i.e., Bergman and Dirichlet spaces, are not completely settled yet and each topic contains numerous unsolved problems. Theory of Bergman spaces is available in [21, 40, 107], and Dirichlet spaces are discussed in two new texts [4, 25]. The above references also contain scattered information on Bloch, Bloch-type, and Bergman-type spaces.

The Bergman spaces and Bergman-type spaces have been recently the center of some intense studies. This stems from the fact that such spaces play a vital role in both function theoretic and operator theoretic development of function spaces and have a close connection to Bloch spaces, which appear as the images of the bounded functions under the Bergman projection. Bloch spaces by themselves are also important and play the role of the dual spaces of the Bergman spaces. Operators on these spaces, in particular, composition operators on $A^{-p}(\mathbb{D})$, are concrete objects to study the geometric properties of these spaces. See [40] and references therein.

In the family of Bergman and Bergman-type spaces, the class of Q_p-spaces, $p > 0$, on $\mathbb{D}$ have proved to be more important [7, 102]. The Q_p-space consists of functions in $\mathrm{Hol}(\mathbb{D})$ such that

$$\sup_{a \in \mathbb{D}} \int_{\mathbb{D}} |f'(z)|^2 (1 - |\sigma_a(z)|^2)^p \, dA(z) < \infty.$$

Here $\sigma_a(z) = (a - z)/(1 - \bar{a}z)$ is the automorphism of $\mathbb{D}$ that exchanges 0 and a, and dA is Lebesgue area measure on the plane, normalized so that $A(\mathbb{D}) = 1$. It is well-known that Q_p-spaces coincide with the classical Bloch space $\mathcal{B}$ for $p \in (1, \infty)$ and Q_1 is precisely the celebrated BMOA, the space of holomorphic functions on $\mathbb{D}$ with bounded mean oscillation. However, for $p \in (0, 1)$, the Q_p-spaces are all totally new objects.

The initial motivation of this work, the $\mathcal{N}_p$-spaces, comes from the study of the various properties of Q_p-spaces. One notices that if, in the definition of the Q_p-space, $f'(z)$ is replaced by $f(z)$, then we have a new family, the so-called $\mathcal{N}_p$-space on $\mathbb{D}$ [73, 95]. More explicitly, the $\mathcal{N}_p$-space consists of functions in $\mathrm{Hol}(\mathbb{D})$ for which

$$\sup_{a \in \mathbb{D}} \int_{\mathbb{D}} |f(z)|^2 (1 - |\sigma_a(z)|^2)^p \, dA(z) < \infty.$$

As a matter of fact, due to the similarities and the differences between Q_p-spaces and the Bergman-type spaces $\mathcal{B}$ and A^{-1}, it is natural to study the $\mathcal{N}_p$-spaces and the operators acting on them.

Some basic properties of $\mathcal{N}_p$-spaces are as follows. For $p > 1$, the $\mathcal{N}_p$-space coincides with the Begrman-type space A^{-1} consisting of holomorphic functions on the disc for which $\sup |f(z)|(1 - |z|^2) < \infty$. However, for $p \in (0, 1]$, the $\mathcal{N}_p$-spaces are all different function spaces with different topological and algebraic properties. Composition operators as well as weighted composition operators acting on $\mathcal{N}_p$-spaces, or from $\mathcal{N}_p$-spaces into Bergman-type spaces A^{-q}, have different features and form interesting objects to study as concrete operators.

We also treat $\mathcal{N}_p$-spaces and their operators in higher dimensions. Let $\mathbb{B}$ denote the open unit ball in $\mathbb{C}^n$, with $\mathbb{S}$ as its boundary. The space $\mathrm{Hol}(\mathbb{B})$ consists of all holomorphic functions in $\mathbb{B}$ equipped with the compact-open topology. Let φ denote a non-constant holomorphic self-map of $\mathbb{B}$, and let ψ be a holomorphic function on $\mathbb{B}$, which is not identically equal to zero. These functions induce the linear operator $W_{\psi, \varphi} : \mathrm{Hol}(\mathbb{B}) \to \mathrm{Hol}(\mathbb{B})$, which is so-called *weighted composition operator* with symbols ψ and φ, and is defined by

$$W_{\psi, \varphi}(f)(z) = \psi(z) \cdot (f \circ \varphi(z)), \qquad f \in \mathrm{Hol}(\mathbb{B}), \ z \in \mathbb{B}.$$

Note that if ψ is the constant function 1, then $W_{\psi, \varphi} = C_\varphi$ is the classical *composition operator*, while if φ is the identity mapping, then $W_{\psi, \varphi} = M_\psi$ is precisely the *multiplication operator* $W_{\psi, \varphi} = M_\psi$. The literature on composition operators and weighted composition operators acting on spaces of holomorphic functions on $\mathbb{D}$ is simply beyond control. We just refer the readers to the monographs [17, 83] for further information. For composition operators on $A^{-p}(\mathbb{D})$, see [40, 107] and references therein.

This monograph consists of 12 chapters to describe $\mathcal{N}_p$-spaces. At the beginning, we introduce some weighted spaces as well as the classical Hardy space H^2, Bergman space B^2, and Dirichlet space $\mathcal{D}$. These three families of Hilbert function spaces are the most senior function spaces that are needed in studying $\mathcal{N}_p$-spaces. The definitions of these spaces can be generalized, which lead us to some classes of Banach spaces, such as Bloch $\mathcal{B}^p$ spaces and Bergman-type spaces A^{-p}, where $p > 0$. As a counterpart to these families on the open unit disc $\mathbb{D}$, an important part of our studies is about entire functions in the complex plane $\mathbb{C}$, and even their several complex variable version in $\mathbb{C}^n$. A representative family is the Fock space of entire functions which is treated in depth.

Hanoi, Vietnam   Le Hai Khoi
Québec, QC, Canada   Javad Mashreghi

Contents

1.1 Classical Function Spaces

Let $\mathbb{D}$ be the open unit disc in the complex plane $\mathbb{C}$, let $\text{Hol}(\mathbb{D})$ be the space of holomorphic functions on $\mathbb{D}$, and let $dA = dx\,dy$ be the area Lebesgue measure. In studying N_p spaces, several classical function spaces enter our discussion. For reference, we just provide their definitions below. Then we treat some less well-known function spaces in further detail.

The most classical function space is the *Hardy space*

$$H^2 := \left\{ f \in \text{Hol}(\mathbb{D}) : \sup_{0<r<1} \int_0^{2\pi} |f(re^{i\theta})|^2 \, d\theta < \infty \right\}.$$

The norm is naturally defined by

$$\|f\|_{H^2} := \sup_{0<r<1} \left(\frac{1}{2\pi} \int_0^{2\pi} |f(re^{i\theta})|^2 \, d\theta \right)^{\frac{1}{2}},$$

with respect to which H^2 is a Hilbert space. His close cousin is the *Bergman space*

$$B^2 := \left\{ f \in \text{Hol}(\mathbb{D}) : \int_{\mathbb{D}} |f(z)|^2 \, dA(z) < \infty \right\}.$$

In the same manner, the norm is defined by

$$\|f\|_{B^2} := \left(\int_{\mathbb{D}} |f(z)|^2 \, dA(z) \right)^{\frac{1}{2}},$$

© The Author(s), under exclusive license to Springer Nature Switzerland AG 2023
L. H. Khoi, J. Mashreghi, *Theory of N_p Spaces*, Frontiers in Mathematics,
https://doi.org/10.1007/978-3-031-39704-2_1

with respect to which B^2 is a Hilbert space. The next Hilbert space is the *Dirichlet space*

$$\mathcal{D}^2 := \left\{ f \in \text{Hol}(\mathbb{D}) : \int_{\mathbb{D}} |f'(z)|^2 \, dA(z) < \infty \right\}.$$

In this case, since the constant functions are eliminated by the above defining property, the definition of norm is slightly different. Two popular norms that are used in studying Dirichlet spaces are

$$\|f\|^2_{\mathcal{D}^2} := \|f\|^2_{H^2} + \int_{\mathbb{D}} |f'(z)|^2 \, dA(z)$$

and

$$\|f\|^2_{\mathcal{D}^2} := |f(0)|^2 + \int_{\mathbb{D}} |f'(z)|^2 \, dA(z).$$

The above three families of Hilbert spaces are the most senior function spaces that are needed in studying $\mathcal{N}_p$ spaces. The corresponding definitions can be generalized, which lead to some classes of Banach spaces.

The *Bloch space* is defined by

$$\mathcal{B}^p := \left\{ f \in \text{Hol}(\mathbb{D}) : \sup_{z \in \mathbb{D}} |f'(z)|(1 - |z|^2)^p < \infty \right\},$$

where $1 \leq p < \infty$. As in the definition of Dirichlet space, to take care of constant functions, the norm is defined by

$$\|f\|^p_{\mathcal{B}^p} := |f(0)|^p + \sup_{z \in \mathbb{D}} |f'(z)|(1 - |z|^2)^p.$$

The last member of these families is the *Bergman-type space*

$$A^{-p} := \left\{ f \in \text{Hol}(\mathbb{D}) : \sup_{z \in \mathbb{D}} |f(z)|(1 - |z|)^p < \infty \right\},$$

with the norm

$$\|f\|_{A^{-p}} := \sup_{z \in \mathbb{D}} |f(z)|(1 - |z|)^p.$$

All the above families consist of analytic functions on the open unit disc $\mathbb{D}$. An important part of our studies is about entire functions in the complex plane $\mathbb{C}$ and even their several complex variable version in $\mathbb{C}^N$. We introduce one such family here. But, there are more which will be introduced later. The *Fock space* is the family of entire functions

$$\mathcal{F}^2 := \left\{ f \in \mathrm{Hol}(\mathbb{C}) : \int_{\mathbb{C}} |f(z)|^2 e^{-|z|^2} \, dA(z) < \infty \right\}.$$

It is a Hilbert space with the norm

$$\|f\|_{\mathcal{F}^2} := \left(\frac{1}{\pi} \int_{\mathbb{C}} |f(z)|^2 e^{-|z|^2} \, dA(z) \right)^{\frac{1}{2}}.$$

1.2 Weighted Sequence Spaces

It is well-known, by the Cauchy–Hadamard theorem, that a power series

$$f(z) = \sum_{n=0}^{\infty} a_n z^n$$

represents a homomorphic function in $\mathbb{D}$ if and only if

$$\limsup_{n \to \infty} |a_n|^{1/n} \le 1. \tag{1.1}$$

Let us denote by $\ell_{\mathbb{D}}$ the set of all complex sequences $\mathbf{a} = (a_n)_{n \ge 0}$ that satisfy the condition (1.1). Clearly, $\ell_{\mathbb{D}}$ is a complex vector space.

To treat the weighted Hardy spaces, we start with introducing the weighted sequence spaces. Let $\beta = (\beta_n)_{n \ge 0}$ be a sequence of positive real numbers. Then for each sequence $\mathbf{a} = (a_n)_{n \ge 0}$ of complex numbers, we define the quantity

$$\|\mathbf{a}\| := \left(\sum_{n=0}^{\infty} \beta_n^2 |a_n|^2 \right)^{1/2} \in [0, +\infty].$$

The weighted sequence space ℓ_β^2 is simply

$$\ell_\beta^2 := \{ \mathbf{a} : \|\mathbf{a}\| < +\infty \}. \tag{1.2}$$

This set ℓ_β^2 becomes a Hilbert space with the inner product

$$\langle \mathbf{a}, \mathbf{b} \rangle = \sum_{n=0}^{\infty} \beta_n^2 a_n \overline{b_n}, \qquad \mathbf{a} = (a_n) \in \ell_\beta^2, \mathbf{b} = (b_n) \in \ell_\beta^2. \tag{1.3}$$

These spaces have many important applications in studying operators on function spaces. To explore the relation between $\ell_{\mathbb{D}}$ and ℓ_β^2, for a sequence of weights $\beta = (\beta_n)_{n\geq 0}$, we define

$$\beta_* = \liminf_{n\to\infty}(\beta_n)^{1/n} \quad\text{and}\quad \beta^* = \limsup_{n\to\infty}(\beta_n)^{1/n}.$$

Clearly, $\beta_* \leq \beta^*$. The following result reveals the *relationship* between $\ell_{\mathbb{D}}$ and ℓ_β^2.

Theorem 1.2.1 *Let $\beta = (\beta_n)_{n\geq 0}$ be a sequence of complex numbers. Then the following cases happen.*

(i) $\ell_\beta^2 \subsetneq \ell_{\mathbb{D}}$ *if and only if $\beta_* \geq 1$.*
(ii) $\ell_{\mathbb{D}} \subsetneq \ell_\beta^2$ *if and only if $\beta^* < 1$.*
(ii) $\ell_{\mathbb{D}} \setminus \ell_\beta^2 \neq \varnothing$ *and $\ell_\beta^2 \setminus \ell_{\mathbb{D}} \neq \varnothing$ if and only if $\beta_* < 1 \leq \beta^*$.*

Proof (i) Suppose $\ell_\beta^2 \subsetneq \ell_{\mathbb{D}}$. Thus, every sequence in ℓ_β^2 generates a holomorphic function on $\mathbb{D}$. We consider a very explicit sequence in ℓ_β^2. Let $\mathbf{a} = (a_n)$, where

$$a_n = \frac{1}{(n+1)\beta_n}, \qquad (n \geq 0).$$

Clearly, $\mathbf{a} \in \ell_\beta^2$, and so by (1.1), we must have

$$\limsup_{n\to\infty}\left(\frac{1}{(n+1)\beta_n}\right)^{1/n} \leq 1.$$

But,

$$\limsup_{n\to\infty}\left(\frac{1}{(n+1)\beta_n}\right)^{1/n} = \lim_{n\to\infty}(n+1)^{-1/n} \cdot \frac{1}{\liminf_{n\to\infty}\beta_n^{1/n}} = 1 \cdot \beta_*^{-1}.$$

Thus, we obtain $\beta_* \geq 1$.

Conversely, suppose that $\beta_* \geq 1$. For any $\mathbf{a} \in \ell_\beta^2$, since

$$\|\mathbf{a}\| = \left(\sum_{n=0}^{\infty} \beta_n^2 |a_n|^2\right)^{1/2} < \infty,$$

we have $|a_n|\beta_n < 1$ for all n large enough. As a result,

$$\limsup_{n\to\infty} a_n^{1/n} \le \limsup_{n\to\infty} \frac{1}{\beta_n^{1/n}} = \frac{1}{\liminf_{n\to\infty}\beta_n^{1/n}} = \frac{1}{\beta_*} \le 1.$$

Therefore, by the Cauchy–Hadamard theorem, the power series $\sum_{n=0}^{\infty} a_n z^n$ is holomorphic on $\mathbb{D}$. In other words, $\ell_\beta^2 \subset \ell_{\mathbb{D}}$. However, consider the sequence $\mathbf{b} = (b_n)$, where $b_n = \beta_n^{-1}$. Then, since $\beta_* \ge 1$, we have $\mathbf{b} \in \ell_{\mathbb{D}}$, but clearly $\mathbf{b} \notin \ell_\beta^2$. Hence, we showed $\ell_\beta^2 \subsetneq \ell_{\mathbb{D}}$.

(ii) Suppose $\beta^* < r < 1$, for some $r > 0$. Thus, starting from some $N_1 \ge 1$, we have $\beta_n^{1/n} < r$ for all $n > N_1$. For any sequence $\mathbf{a} = (a_n) \in \ell_{\mathbb{D}}$, from (1.1), we can find some $N_2 \in \mathbb{N}$ such that $|a_n|^{1/n} \le 1$ for all $n > N_2$. Let $N = \max\{N_1, N_2\}$. Then

$$\sum_{n=0}^{\infty} |a_n|^2 \beta_n^2 = \sum_{n=0}^{N} |a_n|^2 \beta_n^2 + \sum_{n=N+1}^{\infty} |a_n|^2 \beta_n^2$$

$$\le \sum_{n=0}^{N} |a_n|^2 \beta_n^2 + \sum_{n=N+1}^{\infty} r^{2n}$$

$$= \sum_{n=0}^{N} |a_n|^2 \beta_n^2 + \frac{r^{2N+2}}{1-r^2} < +\infty,$$

which shows that $\mathbf{a} \in \ell_\beta^2$. In short, $\ell_{\mathbb{D}} \subset \ell_\beta^2$. Now consider $\mathbf{b} = (b_n)$, where $b_n = ((n+1)\beta_n)^{-1}$. Then $\mathbf{b} \in \ell_\beta^2$, but

$$\limsup_{n\to\infty} |b_n|^{1/n} \ge \liminf_{n\to\infty} |b_n|^{1/n}$$

$$= \lim_{n\to\infty} (n+1)^{-1/n} \cdot \frac{1}{\limsup_{n\to\infty} \beta_n^{1/n}} = 1 \cdot \beta^{*-1} > 1.$$

Hence, $\mathbf{b} \notin \ell_{\mathbb{D}}$. Therefore, we have $\ell_{\mathbb{D}} \subsetneq \ell_\beta^2$.

Conversely, suppose that $\ell_{\mathbb{D}} \subsetneq \ell_\beta^2$. We show that $\beta^* \ge 1$ leads to a contradiction. If $\beta^* = 1$, then there is a subsequence (β_{n_k}) such that $\lim_{n\to\infty} \beta_{n_k}^{1/n_k} = 1$. Define $\mathbf{a} = (a_n)$ by

$$a_n = \begin{cases} \beta_{n_k}^{-1} & \text{if } n = n_k, \text{ for some } k, \\ 0 & \text{otherwise.} \end{cases}$$

Clearly, $\limsup_{n\to\infty} |a_n|^{1/n} = \lim_{k\to\infty} \beta_{n_k}^{-1/n_k} = 1$, and thus $\mathbf{a} \in \ell_{\mathbb{D}}$, but since $|a_{n_k}|\beta_{n_k} = 1 \ne 0$ for infinitely many n_k, we conclude $\mathbf{a} \notin \ell_\beta^2$, which contradicts $\ell_{\mathbb{D}} \subsetneq \ell_\beta^2$.

If $\beta^* > 1$, then there is a subsequence (β_{n_k}) such that $\beta_{n_k}^{1/n_k} > r > 1$ for some r. Thus, $\beta_{n_k} > 1$ which implies

$$\sum_{n=0}^{\infty} \beta_n^2 = +\infty.$$

Now, consider $\mathbf{a} = (a_n)$, where $a_n \equiv 1$. Then $\mathbf{a} \in \ell_{\mathbb{D}}$. Since $\ell_{\mathbb{D}} \subsetneqq \ell_\beta^2$, we must have $\|\mathbf{a}\|^2 = \sum_{n=0}^{\infty} \beta_n^2 < +\infty$, which is a again a contradiction.

(iii) This part follows from *(i)* and *(ii)*.

$\square$

1.3   Special Families of Weighted Sequence Spaces

For any positive sequence $\beta = (\beta_n)_{n \geq 0}$, we define the number

$$\beta_\rho = \liminf_{n \to \infty} \frac{\log \beta_n}{n \log n}$$

and the function $\mu_\beta : \mathbb{N} \to \mathbb{R}^+$ by

$$\mu_\beta(n) = \frac{n \beta_{n-1}}{\beta_n}, \qquad (n \geq 1).$$

We also introduce the following family of sequences $\beta = (\beta_n)_{n \geq 0}$ that are of great importance in the sequel:

$$\mathbf{A} = \{\beta : \beta_* = +\infty\},$$
$$\mathbf{B} = \{\beta : \beta_\rho > 0\},$$
$$\mathbf{C} = \{\beta : \beta_\rho < +\infty\},$$
$$\mathbf{D} = \{\beta : \beta_n/\beta_{n+1} \downarrow 0\},$$
$$\mathbf{G} = \{\beta : \mu_\beta \text{ is bounded}\},$$
$$\mathbf{H} = \{\beta : \exists \tau = \tau(\beta) > 0 \text{ such that } \mu_\beta(n) \downarrow \tau\}.$$

The interrelation between these classes are addressed in the following result.

Proposition 1.3.1 *The following assertions are true.*

(i) $\mathbf{D} \subsetneq \mathbf{A}$ *but* $\mathbf{C} \not\subseteq \mathbf{A}$.
(ii) $\mathbf{D} \setminus \mathbf{G} \neq \varnothing$ *and* $\mathbf{G} \setminus \mathbf{D} \neq \varnothing$.
(iii) $\mathbf{G} \subsetneq \mathbf{B} \subsetneq \mathbf{A}$, *so* $\mathbf{B} \setminus \mathbf{D} \neq \varnothing$ *and* $\mathbf{D} \setminus \mathbf{B} \neq \varnothing$.
(iv) $\mathbf{H} \subsetneq (\mathbf{A} \cap \mathbf{B} \cap \mathbf{C} \cap \mathbf{D} \cap \mathbf{G}) = \mathbf{C} \cap \mathbf{D} \cap \mathbf{G}$.
(v) $(\mathbf{C} \cap \mathbf{G}) \setminus (\mathbf{D} \cap \mathbf{G}) \neq \varnothing$ *and* $(\mathbf{D} \cap \mathbf{G}) \setminus (\mathbf{C} \cap \mathbf{G}) \neq \varnothing$.

Proof *(i)* To show $\mathbf{D} \subseteq \mathbf{A}$, let $\beta = (\beta_n) \in \mathbf{D}$. Since $\beta_{n+1}/\beta_n \uparrow +\infty$, there is some N_1 such that for $n > N_1$, $\beta_{n+1}/\beta_n > 2$. So $\beta_n - \beta_{n-1} > \beta_{n-1} \geq \beta_{N_1}$ ($n > N_1$). Therefore, $\beta_n \uparrow +\infty$. Choose $N_2 > N_1$ such that $\beta_n > 1$ for $n > N_2$.

For any $S > 0$ arbitrarily large, choose N_3 such that whenever $n \geq N_3$, $\beta_{n+1}/\beta_n > S^2$. Let $N = \max\{N_2, N_3\}$; then $\beta \in \mathbf{A}$ because for $n > 2N$,

$$\beta_n = \beta_N \prod_{p=N}^{n-1} \frac{\beta_{p+1}}{\beta_p} > 1 \cdot S^{2(n-N)} = S^{n+(n-2N)} > S^n.$$

Therefore, $\beta_* = +\infty$. To see why $\mathbf{D} \neq \mathbf{A}$, choose the following sequence $\beta^{(1)}$, which is in $\mathbf{A}$ but not $\mathbf{D}$:

$$\beta_n^{(1)} = \begin{cases} n^n & \text{if } n \text{ is odd,} \\ n^{2n} & \text{if } n \text{ is even.} \end{cases}$$

To show $\mathbf{C} \not\subseteq \mathbf{A}$, let $\beta_n \equiv 1$ for all n. Then clearly $\beta_\rho = 0 < +\infty$, but $\beta_* = 1 < +\infty$.

(ii) To verify $\mathbf{D} \setminus \mathbf{G} \neq \varnothing$, let $\beta_n^{(2)} = \sqrt{n!}$, then $\beta_n^{(2)}/\beta_{n+1}^{(2)} = (\sqrt{n+1})^{-1} \downarrow 0$, but $\mu_\beta^{(2)}(n) = \sqrt{n} \uparrow \infty$. Equivalently, $\beta^{(2)} \in \mathbf{D} \setminus \mathbf{G}$.

To show, choose $\beta^{(3)} = (\beta_n^{(3)})$:

$$\beta_n^{(3)} = \begin{cases} n! & \text{if } n \text{ is odd,} \\ 2n! & \text{if } n \text{ is even.} \end{cases}$$

Then

$$\mu_\beta^{(3)}(n) = \begin{cases} 2 & \text{if } n \text{ is odd,} \\ \frac{1}{2} & \text{if } n \text{ is even,} \end{cases} \quad \text{and} \quad \frac{\beta_n^{(3)}}{\beta_{n+1}^{(3)}} = \begin{cases} \frac{1}{2(n+1)} & \text{if } n \text{ is odd,} \\ \frac{2}{n+1} & \text{if } n \text{ is even.} \end{cases}$$

Therefore, $\beta^{(3)} \in \mathbf{G}$ but $\beta^{(3)} \notin \mathbf{D}$, which implies $\mathbf{G} \setminus \mathbf{D} \neq \varnothing$.

(iii) Let $\beta \in \mathbf{B}$. Then there is $c > 0$ such that

$$\frac{\log \beta_n}{n \log n} > c, \qquad n \ge 1,$$

and thus

$$\lim_{n \to \infty} \sqrt[n]{\beta_n} \ge \lim_{n \to \infty} n^c \to \infty.$$

This observation shows $\mathbf{B} \subseteq \mathbf{A}$. To see, take $\beta_0^{(4)} = \beta_1^{(4)} = \beta_2^{(4)} = 1$ and $\beta_n^{(4)} = (\log n)^n$ for $n \ge 3$. Then $\beta_*^{(4)} = +\infty$, but $\beta_\rho^{(4)} = 0$, so $\beta^{(4)} \in \mathbf{A} \setminus \mathbf{B}$. Note also that $\beta^{(4)} \in \mathbf{D}$, so $\mathbf{D} \not\subseteq \mathbf{B}$.

To prove $\mathbf{G} \subseteq \mathbf{B}$, let $\beta \in \mathbf{G}$. There is $S > 0$ such that $\mu_\beta(n) < S$ for all n, so

$$\beta_n > \frac{n\beta_{n-1}}{S} > \frac{n(n-1)\beta_{n-2}}{S^2} > \cdots > \frac{n!\beta_0}{S^n}.$$

Since $n! \ge n^{n/2}$, we have

$$\frac{\log \beta_n}{n \log n} > \frac{\log \beta_0 + \log n! - n \log S}{n \log n} \ge \frac{\log \beta_0 + \frac{n}{2} \log n - n \log S}{n \log n} \to \frac{1}{2}$$

as $n \to \infty$. Hence, $\beta_\rho > 0$ as we need. Now choose $\beta = \beta^{(1)}$ as above, clearly $\beta_\rho^{(1)} = 1$, but $\mu_\beta^{(1)}(2n+1) = \left(\frac{4n^2}{2n+1}\right)^{2n}$ is not bounded, so $\beta \notin \mathbf{G}$ and $\mathbf{G} \subsetneq \mathbf{B}$. This together with $\mathbf{G} \not\subseteq \mathbf{D}$ also implies $\mathbf{B} \not\subseteq \mathbf{D}$. Therefore, $\mathbf{G} \subsetneq \mathbf{B} \subsetneq \mathbf{A}$; $\mathbf{B} \setminus \mathbf{D} \ne \varnothing$ and $\mathbf{D} \setminus \mathbf{B} \ne \varnothing$.

(iv) Let $\beta \in \mathbf{H}$. Since $(\mu_\beta(n))$ is convergent, it is bounded and clearly $\beta \in \mathbf{G}$. From the hypothesis $\frac{n\beta_{n-1}}{\beta_n} \downarrow \tau > 0$, we have

$$\frac{\beta_{n-1}}{\beta_n} > \frac{(n+1)\beta_n}{n\beta_{n+1}} > \frac{\beta_n}{\beta_{n+1}}$$

and

$$\frac{\beta_n}{\beta_{n+1}} < \frac{n\beta_{n-1}}{(n+1)\beta_n} < \cdots < \frac{\beta_0}{(n+1)\beta_1} \to 0,$$

so $\beta \in \mathbf{D}$. Also, as $n! \le n^n$,

$$\beta_n < \frac{n\beta_{n-1}}{\tau} < \cdots < \frac{n!\beta_0}{\tau^n},$$

which implies

$$\frac{\log \beta_n}{n \log n} < \frac{\log n! + \log \beta_0 - n \log \tau}{n \log n} \le \frac{n \log n + \log \beta_0 - n \log \tau}{n \log n} \to 1,$$

as $n \to \infty$, so $\beta \in \mathbf{C}$. Therefore, $\mathbf{H} \subseteq \mathbf{C} \cap \mathbf{D} \cap \mathbf{G}$.

Finally, choose $\beta_0^{(5)} = 1$, $\beta_n^{(5)} = n^{2n}$, $(n > 0)$. Then it is clear that $\beta^{(5)} \in \mathbf{C} \cap \mathbf{D} \cap \mathbf{G}$, but $\mu_\beta^{(5)}(n) \downarrow 0$, so $\beta \notin \mathbf{H}$. Hence, $\mathbf{H} \subsetneq \mathbf{C} \cap \mathbf{D} \cap \mathbf{G}$.

(v) Consider $\beta = \beta^{(3)}$; then $\beta \in (\mathbf{C} \cap \mathbf{G}) \setminus (\mathbf{D} \cap \mathbf{G})$. Also, for $\beta = \beta^{(5)}$ where $\beta_n^{(5)} = n^{n^2}$, then $\beta \in (\mathbf{D} \cap \mathbf{G}) \setminus (\mathbf{C} \cap \mathbf{G})$.

$\square$

1.4   Weighted Hardy Spaces

Let

$$\mathcal{H}(\beta) = \left\{ f(z) = \sum_{n=0}^{\infty} a_n z^n, \ (a_n) \in \ell_\beta^2 \cap \ell_{\mathbb{D}} \right\}.$$

More explicitly, the weighted Hardy space $\mathcal{H}(\beta)$ consists of analytic functions

$$f(z) = \sum_{n=0}^{\infty} a_n z^n \in \mathrm{Hol}(\mathbb{D})$$

such that

$$\|f\|_{\mathcal{H}(\beta)} := \left(\sum_{n=0}^{\infty} \beta_n^2 |a_n|^2 \right)^{1/2} < +\infty.$$

As in (1.3), with

$$\langle f, g \rangle = \sum_{n=0}^{\infty} \beta_n^2 a_n \overline{b_n},$$

where (a_n) and (b_n) are, respectively, the Taylor coefficients of f and g, we see that $\mathcal{H}(\beta)$ is in fact an inner product function space. If there is no ambiguity, we write $\|f\|$ for $\|f\|_{\mathcal{H}(\beta)}$. Depending on (β_n), the space $\ell_\beta^2 \cap \ell_{\mathbb{D}}$ may not be complete in norm (1.2), so $\mathcal{H}(\beta)$ is not necessarily complete and thus not a Hilbert space. Using Theorem 1.2.1, we obtain the following theorem which characterizes the completeness of $\mathcal{H}(\beta)$ in terms of sequence β.

Theorem 1.4.1 *The space $\mathcal{H}(\beta)$ is a Hilbert space of holomorphic functions on the open unit disc $\mathbb{D}$ if and only if*

$$\beta_* = \liminf_{n \to \infty} \beta_n^{1/n} \geq 1. \tag{1.4}$$

The function space $\mathcal{H}(\beta)$ induced by the weight β that satisfies condition (1.4) is known as a weighted Hardy space. The name *weighted Hardy space* comes from the observation that if $\beta_n \equiv 1$, then $\mathcal{H}(\beta)$ becomes the classical Hardy space $H^2(\mathbb{D})$. Two further interesting cases are as follows. If

$$\beta_n = \frac{1}{\sqrt{n+1}}, \qquad (n \geq 0),$$

we obtain the Bergman space. If

$$\beta_n = \sqrt{n+1}, \qquad (n \geq 0),$$

we end up with the Dirichlet space.

A practically useful class of the weight sequences consists of those β satisfying

$$\sum_{n=0}^{\infty} \frac{1}{\beta_n^2} < +\infty. \tag{1.5}$$

It is clear that condition (1.5) is stronger than condition (1.4). Any space $\mathcal{H}(\beta)$ with β satisfying condition (1.5) is called a small weighted Hardy space. An important property of these spaces is that functions in $\mathcal{H}(\beta)$ extend continuously to the closed unit disc $\overline{\mathbb{D}} = \{z \in \mathbb{C} : |z| \leq 1\}$. Here is yet another more extreme case which consists of entire functions.

Theorem 1.4.2 *The space $\mathcal{H}(\beta)$ is a Hilbert space of entire functions if and only if $\beta_* = +\infty$.*

In particular, since in this case $\beta_* > 1$ and (1.5) is satisfied, $\mathcal{H}(\beta)$ is a small weighted Hardy space. If

$$\beta_n = \sqrt{n!}, \qquad (n \geq 0),$$

we obtain the classical Fock space $\mathcal{F}^2(\mathbb{C})$ of entire functions.

1.5   Special Families of Weighted Hardy Spaces

Using Proposition 1.3.1, we define some special types of Hilbert spaces of entire functions. Depending on the weights β, we denote several subclasses $\mathcal{H}(\beta_E)$ as follows:

$$
\begin{cases}
\mathcal{H}(\beta_\rho), & \text{if } \beta \in \mathbf{B}, \\[2mm]
\mathcal{H}(\beta_\rho^+), & \text{if } \beta \in \mathbf{B} \cap \mathbf{C}, \\[2mm]
\mathcal{H}(\beta_\rho^+, T), & \text{if } \beta \in \mathbf{C} \cap \mathbf{G}, \\[2mm]
\mathcal{E}_\gamma^2(T), & \text{if } \beta \in \mathbf{D} \cap \mathbf{G}, \\[2mm]
\mathcal{E}_\gamma^2(T, \tau), & \text{if } \beta \in \mathbf{H}.
\end{cases}
$$

Note that $\mathcal{E}_\gamma^2(T, \tau)$ is the smallest subclass of $\mathcal{H}(\beta_E)$ we consider in this section.

One of the main characteristics of an entire function is its growth order. Recall that an entire function with power series representation $f(z) = \sum a_n z^n$ has finite order ρ if and only if

$$
\rho = \limsup_{n \to \infty} \frac{n \log n}{-\log |a_n|} < +\infty.
$$

We consider the following two properties:

(P_1) Every function $f \in \mathcal{H}(\beta_E)$ has a finite order.
(P_2) There exists a function $g \in \mathcal{H}(\beta_E)$ which has a nonzero order.

In the next two theorems, we characterize the above two properties in $\mathcal{H}(\beta_E)$ spaces.

Theorem 1.5.1 *A Hilbert space $\mathcal{H}(\beta_E)$ has property (P_1) if and only if it is induced by $\beta \in \mathbf{B}$.*

Proof *Necessity*: Assume that every function in $\mathcal{H}(\beta_E)$ has a finite order, but $\beta_\rho = 0$. Thus, there exists an increasing sequence $(n_k)_{k \geq 1}$ such that

$$
\frac{\log \beta_{n_k}}{n_k \log n_k} < \frac{1}{k}, \qquad (k \geq 1),
$$

or equivalently

$$\log \beta_{n_k} < \frac{n_k \log n_k}{k}, \qquad (k \geq 1).$$

Choose the function $f(z) = \sum_{n=0}^{\infty} a_n z^n$ with Taylor coefficients $(a_n)_{n \geq 0}$ as follows:

$$a_n = \begin{cases} \frac{1}{n\beta_n} & \text{if } n = n_k, \\ 0 & \text{otherwise.} \end{cases} \qquad (1.6)$$

Clearly, $\|f\|^2 \leq \pi^2/6 < +\infty$, and since $\lim_{k \to \infty} (\beta_{n_k})^{1/n_k} = +\infty$,

$$\limsup_{n \to \infty} |a_n|^{1/n} = \limsup_{k \to \infty} \frac{1}{n_k^{1/n_k} \beta_{n_k}^{1/n_k}} = \limsup_{k \to \infty} \frac{1}{\beta_{n_k}^{1/n_k}} = \lim_{k \to \infty} \frac{1}{\beta_{n_k}^{1/n_k}} = 0,$$

which means $f \in \mathcal{H}(\beta_E)$. But then,

$$\limsup_{n \to \infty} \frac{n \log n}{-\log |a_n|} = \limsup_{k \to \infty} \frac{n_k \log n_k}{\log(n_k \beta_{n_k})} \geq \limsup_{k \to \infty} \frac{n_k \log n_k}{\log n_k + \frac{n_k \log n_k}{k}} = +\infty,$$

a contradiction to f being of finite order.

Sufficiency Since $\beta \in \mathbf{B} \subsetneq \mathbf{A}$ (Theorem 1.3.1(iii)), it induces a $\mathcal{H}(\beta_E)$ space. Assume $\beta_\rho > 0$, but there is some $f(z) = \sum_{n=0}^{\infty} a_n z^n$ not having finite order in $\mathcal{H}(\beta_E)$. We can thus choose an increasing sequence $(n_k)_{k \geq 0}$ such that $|a_{n_k}| < 1$ and

$$\frac{n_k \log n_k}{-\log |a_{n_k}|} > k, \qquad (k \geq 1),$$

or equivalently

$$|a_{n_k}| > n_k^{-n_k/k}, \qquad (k \geq 1).$$

From the hypothesis $\beta_\rho > 0$, there are numbers $c > 0$ and $N \in \mathbb{N}$ such that

$$\frac{\log \beta_n}{n \log n} > c, \qquad (n > N).$$

Thus,

$$\beta_n > n^{cn}, \qquad (n > N).$$

Take the integer K such that $1/K < c$. Then

$$|a_{n_k}|\beta_{n_k} > n_k^{cn_k - n_k/k} = n_k^{n_k(c - 1/k)} > 1, \qquad (k \geq K).$$

This contradicts $\|f\| < +\infty$. $\qquad\qquad\qquad\qquad\qquad\qquad\qquad\qquad\qquad\qquad\square$

Theorem 1.5.2 *A Hilbert space $\mathcal{H}(\beta_E)$ has property (P_2) if and only if it is induced by $\beta \in \mathbf{A} \cap \mathbf{C}$.*

Proof Since β induces a $\mathcal{H}(\beta_E)$ space if and only if $\beta \in \mathbf{A}$, we only need to prove that $\mathcal{H}(\beta_E)$ has property (P_2) if and only if $\beta \in \mathbf{C}$.

Necessity: Suppose $\beta_\rho = +\infty$. For any $(a_n) \in \ell_\beta^2$ that induces the power series of an element $f \in \mathcal{H}(\beta_E)$, there is an integer N such that $|a_n|\beta_n < 1$ for all $n > N$. Hence,

$$\limsup_{n\to\infty} \frac{n \log n}{-\log|a_n|} \leq \limsup_{n\to\infty} \frac{n \log n}{\log \beta_n} = 0,$$

which shows that all $f \in \mathcal{H}(\beta_E)$ have order 0.

Sufficiency Suppose $\beta_\rho < +\infty$. Then there exist $c > 0$ and an increasing sequence $(n_k)_{k\geq 1}$ such that

$$\frac{\log \beta_{n_k}}{n_k \log n_k} < c.$$

Choose $(a_n)_{n\geq 1}$ as in (1.6). Thus, $f \in \mathcal{H}(\beta_E)$ and, moreover,

$$\limsup_{n\to\infty} \frac{n \log n}{-\log|a_n|} = \limsup_{k\to\infty} \frac{n_k \log n_k}{\log n_k \beta_{n_k}} \geq \limsup_{k\to\infty} \frac{n_k \log n_k}{(cn_k + 1)\log n_k} = \frac{1}{c} > 0,$$

so the power series induced by this sequence is of nonzero order. $\qquad\qquad\qquad\square$

As a consequence of Theorems 1.5.1 and 1.5.2, we have the following corollary.

Corollary 1.5.3 *For a Hilbert space $\mathcal{H}(\beta_E)$ induced by β, precisely one of the following alternative cases happens.*

(i) *Every function $f \in \mathcal{H}(\beta_E)$ has order 0 if and only if $\beta_\rho = +\infty$, i.e., $\beta \in \mathbf{B} \setminus \mathbf{C}$.*

(ii) *There exists a function $f \in \mathcal{H}(\beta_E)$ that does not have finite order if and only if $\beta_\rho = 0$, i.e., $\beta \in \mathbf{C} \setminus \mathbf{B}$.*

(iii) *Every $f \in \mathcal{H}(\beta_E)$ has finite order and there exists $g \in \mathcal{H}(\beta_E)$ having positive order if and only if $\beta_\rho \in (0, +\infty)$, i.e., $\beta \in \mathbf{B} \cap \mathbf{C}$.*

By Corollary 1.5.3, a space $\mathcal{H}(\beta_\rho)$ is a Hilbert space of entire functions of finite orders, while a space $\mathcal{H}(\beta_\rho^+)$ is a space $\mathcal{H}(\beta_\rho)$ having at least one function of a positive order. Therefore, $\mathcal{H}(\beta_\rho)$ is a special case of $\mathcal{H}(\beta_E)$, $\mathcal{H}(\beta_\rho^+)$ is a special case of $\mathcal{H}(\beta_\rho)$, and $\mathcal{H}(\beta_\rho^+, T)$ is a special case of $\mathcal{H}(\beta_\rho^+)$.

It is clear that any space $\mathcal{E}_\gamma^2(T, \tau)$ is a space $\mathcal{H}(\beta_\rho^+)$ (Proposition 1.3.1 (iv)) and any space F_α^2 is a space $\mathcal{H}(\beta_\rho^+)$ with $\beta_n = \sqrt{\alpha^{-n} n!}$. Hence, both $\mathcal{E}_\gamma^2(T, \tau)$ and F_α^2 belong to the same class $\mathcal{H}(\beta_\rho^+)$. However, while $\mathcal{E}_\gamma^2(T, \tau)$ also belongs to the subclass $\mathcal{H}(\beta_\rho^+, T)$, it is not true for F_α^2 (as $\mu_\beta(n) = \sqrt{\alpha n} \uparrow +\infty$).

1.6   Hilbert Spaces of Entire Functions

Denote by m_N the ordinary Lebesgue measure on $\mathbb{R}^N$ (i.e., the unit cube has measure 1). We denote the unit ball of $\mathbb{R}^N$ and its boundary, the closed unit sphere, respectively, by $\mathbf{B}_N$ and $\mathbf{S}_{N-1}$. Calculation of the volume of $\mathbf{B}_N$ and the surface area of $\mathbf{S}_{N-1}$ is a standard exercise in measure theory. It is well-known that they are, respectively, given by

$$V_N = \frac{\pi^{N/2}}{\Gamma(1 + N/2)} \tag{1.7}$$

and

$$A_{N-1} = \frac{2\pi^{N/2}}{\Gamma(N/2)}. \tag{1.8}$$

The quantities (1.7) and (1.8) are used to define two normalized measures: the Lebesgue normalized measure on $\mathbb{R}^N$,

$$\eta_N := \frac{m_N}{V_N},$$

with respect to which the volume of $\mathbf{B}_N$ is one, and the area normalized measure on $\mathbf{S}_{N-1}$,

$$\sigma_{N-1} := \frac{m_N}{A_{N-1}},$$

with respect to which the surface area of $\mathbf{S}_{N-1}$ is one. It is well-known that all these measures are rotation invariant. We just need the second normalized measure in our calculation.

From now on, we assume that N is fixed. Hence, for simplicity of notations, we denote the unit ball and the closed unit sphere of $\mathbb{C}^N$, respectively, by $\mathbf{B}$ and $\mathbf{S}$. More explicitly, we write $\mathbf{B} := \mathbf{B}_{2N}$ and $\mathbf{S} := \mathbf{S}_{2N-1}$. Moreover, we put

$$\sigma := \sigma_{2N-1}.$$

The space of all square integrable functions on the closed unit sphere is denoted by $L^2 = L^2(\mathbf{S}, \sigma) = L^2(\mathbf{S})$. This is a Hilbert space with inner product

$$\langle f, g \rangle_{L^2} = \int_{\mathbf{S}} f(z)\overline{g(z)} <, d\sigma(z), \qquad f, g \in L^2(\mathbf{S}).$$

For every multi-index $\alpha = (\alpha_1, \ldots, \alpha_N)$ whose coordinates are nonnegative integers and $z = (z_1, \ldots, z_n) \in \mathbb{C}^N$, we define

$$|\alpha| = \alpha_1 + \cdots + \alpha_N,$$

$$\alpha! = \alpha_1! \times \cdots \times \alpha_N!,$$

$$z^\alpha = z_1^{\alpha_1} \times \cdots \times z_N^{\alpha_N}.$$

By direct verification, two different multi-indices α and γ and the monomials z^α and z^β are orthogonal in $L^2(\mathbf{S})$. We now compute $\|z^\alpha\|_{L^2}^2$.

Lemma 1.6.1 *For every multi-index* $\alpha = (\alpha_1, \ldots, \alpha_N)$, *we have*

$$v_\alpha := \|z^\alpha\|_{L^2}^2 = \int_{\mathbf{S}} |z^\alpha|^2 \, d\sigma(z) = \frac{(N-1)!\,\alpha!}{(N-1+|\alpha|)!}.$$

Proof Let

$$I := \int_{\mathbb{C}^N} |z^\alpha|^2 e^{-|z|^2} \, dm_{2N}(z).$$

We compute I by two methods and then compare the results to deduce the required identity.

On the one hand,

$$I = \int_{\mathbb{C}^N} \left(\prod_{j=1}^N |z_j|^{2\alpha_j} e^{-|z_j|^2} \right) dm_{2N}(z).$$

Recalling that, for any positive integer k,

$$\frac{1}{\pi} \int_{\mathbb{C}} |\lambda|^{2k} e^{-|\lambda|^2} \, dm_2(\lambda) = k!,$$

we easily see

$$I = \prod_{j=1}^{N} \int_{\mathbb{C}} |\lambda|^{2\alpha_j} e^{-|\lambda|^2} \, dm_2(\lambda) = \pi^N \alpha!.$$

On the other hand, the formula for changing to polar coordinates is

$$\int_{\mathbb{R}^N} f(z) \, dm_N(z) = \int_0^\infty r^{N-1} \left(\int_S f(rz) \, dm_N(z) \right) dr.$$

Applying this to compute I, we get

$$\begin{aligned}
I &= \int_{\mathbb{C}^N} |z^\alpha|^2 e^{-|z|^2} \, dm_{2N}(z) \\
&= \int_0^\infty r^{2N-1} e^{-r^2} \left(\int_S r^{2|\alpha|} |z^\alpha|^2 \, dm_{2N}(z) \right) dr \\
&= \int_0^\infty r^{2|\alpha|+2N-1} e^{-r^2} \, dr \times \int_S |z^\alpha|^2 \, dm_{2N}(z) \\
&= \int_0^\infty r^{2|\alpha|+2N-1} e^{-r^2} \, dr \times A_{2N-1} \int_S |z^\alpha|^2 \, d\sigma(z) \\
&= \frac{\pi^N (|\alpha| + N - 1)!}{(N-1)!} \int_S |z^\alpha|^2 \, d\sigma(z).
\end{aligned}$$

Hence, by the above two formulas for I,

$$\int_S |z^\alpha|^2 \, d\sigma(z) = \frac{(N-1)! \, \alpha!}{(N-1+|\alpha|)!}.$$

$\square$

Since two different multi-indices α and γ and the monomials z^α and z^β are orthogonal in L^2 and the norm of a monomial is given in Lemma 1.6.1, for a polynomial $p(z) = \sum_\alpha c_\alpha z^\alpha$, we have

$$\|p_n\|_{L^2}^2 = \sum_\alpha |c_\alpha|^2 v_\alpha. \tag{1.9}$$

For an integer n, write

$$v_n := \binom{N-1+n}{n}^{-1} = \frac{(N-1)! \, n!}{(N-1+n)!}, \qquad (n \geq 1).$$

Now, it is easy to see that

$$v_{(n,0,\dots,0)} = v_n, \qquad v_{(n,1,0,\dots)} = \frac{v_{n+1}}{n+1}, \tag{1.10}$$

and, more generally,

$$\frac{v_\alpha}{\alpha!} = \frac{v_{|\alpha|}}{|\alpha|!}. \tag{1.11}$$

Thus, we have the following expansion:

$$\sum_{|\alpha|=n} \frac{|z^\alpha|^2}{v_\alpha} = \frac{1}{v_n} \sum_{|\alpha|=n} \frac{n!}{\alpha!} |z^\alpha|^2$$

$$= \frac{1}{v_n} \sum_{|\alpha|=n} \frac{n!}{\alpha_1! \cdots \alpha_N!} |z_1|^{2\alpha_1} \cdots |z_N|^{2\alpha_N}$$

$$= \frac{1}{v_n} \left(|z_1|^2 + \cdots + |z_N|^2 \right)^n = \frac{|z|^{2n}}{v_n}. \tag{1.12}$$

1.7 Hilbert Spaces of Formal Power Series

A polynomial p in $\mathbb{C}^N$ is homogeneous of degree n if

$$p(tz) = t^n p(z)$$

for all $t \in \mathbb{C}$ and $z \in \mathbb{C}^N$. As a degenerate case, the zero polynomial is homogeneous of degree n, for all integers $n \geq 0$. Let $\beta = (\beta_n)$ be a sequence of positive real numbers. For each integer $n \geq 0$, let $\mathcal{P}_n(\beta)$ be the Hilbert space of all homogeneous polynomials of degree n equipped with the inner product

$$\langle p, q \rangle_\beta = \beta_n^2 \langle p, q \rangle_{L^2},$$

which is simply a scaled version of the inner product of p and q in $L^2(\mathbf{S}, d\sigma)$. If $p(z) = \sum_{|\alpha|=n} c_\alpha z^\alpha$, then (1.9) gives

$$\|p\|_\beta = \beta_n \left(\sum_{|\alpha|=n} |c_\alpha|^2 \|z^\alpha\|_{L^2}^2 \right)^{1/2} = \left(\sum_{|\alpha|=n} v_\alpha \beta_{|\alpha|}^2 |c_\alpha|^2 \right)^{1/2}.$$

We then define $\mathcal{H}(\beta)$ to be the Hilbert space direct sum

$$\mathcal{H}(\beta) = \bigoplus_{n=0}^{\infty} \mathcal{P}_n(\beta). \tag{1.13}$$

The space $\mathcal{H}(\beta)$ can be considered as the space of formal (convergence is not an issue) homogeneous expansions $f = \sum_{n=0}^{\infty} p_n$ with the norm

$$\|f\|_{\mathcal{H}(\beta)} = \left(\sum_{n=0}^{\infty} \|p_n\|_{\beta}^2\right)^{1/2} = \left(\sum_{n=0}^{\infty} \beta_n^2 \|p_n\|_{L^2}^2\right)^{1/2} < \infty,$$

where $p_n \in \mathcal{P}_n(\beta)$. Writing each p_n as a linear combination of monomials $\sum_{|\alpha|=n} c_\alpha z^\alpha$, we see that f can equally be considered as a formal power series $f(z) = \sum_\alpha c_\alpha z^\alpha$ with norm

$$\|f\|_{\mathcal{H}(\beta)} = \left(\sum_\alpha v_\alpha \beta_{|\alpha|}^2 |c_\alpha|^2\right)^{1/2}. \tag{1.14}$$

The inner product in $\mathcal{H}(\beta)$ is then given by

$$\langle f, g \rangle_{\mathcal{H}(\beta)} = \sum_\alpha v_\alpha \, \beta_{|\alpha|}^2 \, c_\alpha \bar{d}_\alpha,$$

where

$$f = \sum_\alpha c_\alpha z^\alpha \quad \text{and} \quad g = \sum_\alpha d_\alpha z^\alpha.$$

It follows that the set of multiples of monomials $\{(\beta_{|\alpha|}\sqrt{v_\alpha})^{-1}z^\alpha : \alpha \in \mathbb{N}^N\}$ forms an orthonormal basis for $\mathcal{H}(\beta)$. We shall refer to it as the *standard orthonormal basis*.

The space $\mathcal{H}(\beta)$ is actually a weighted Hardy space as introduced in Sect. 1.4. Hence, the notation is fine despite the fact that β has different meanings. With appropriate choices of the weight, $\mathcal{H}(\beta)$ can be identified with the classical Hardy, Bergman, and weighted Bergman spaces of holomorphic functions over the unit ball. On the other hand, with certain choices of β, the formal power series of an element may diverge. In the following theorem, we obtain a characterization of the weight for which all elements of $\mathcal{H}(\beta)$ are entire functions on $\mathbb{C}^N$.

Lemma 1.7.1 *Let $\beta = (\beta_n)$ be a sequence of positive numbers and $\mathcal{H}(\beta)$ be the associated Hilbert space of formal power series. Then all elements of $\mathcal{H}(\beta)$ are entire functions over $\mathbb{C}^N$ if and only if*

$$\lim_{n \to \infty} \beta_n^{1/n} = \infty. \tag{1.15}$$

Proof *Necessity*: Assume that all elements of $\mathcal{H}(\beta)$ are entire functions on $\mathbb{C}^N$. Consider the formal power series

$$f(z) = \sum_{n=1}^{\infty} \frac{z_1^n}{n \, \beta_n}.$$

By (1.10),

$$\|z_1^n\|_{L^2}^2 = v_{n,0,\dots,0} = v_n = \left(\frac{(N-1)! \, n!}{(N-1+n)!} \right)^2 \le 1.$$

Thus, by (1.14), we have

$$\|f\|_{\mathcal{H}(\beta)}^2 = \sum_{n=1}^{\infty} v_n \, \beta_n^2 \frac{1}{n^2 \, \beta_n^2} \le \sum_{n=1}^{\infty} \frac{1}{n^2} < \infty.$$

Therefore, f belongs to $\mathcal{H}(\beta)$, and hence, it is an entire function on $\mathbb{C}^N$. Therefore, we must have

$$\lim_{n \to \infty} \left(\frac{1}{n \, \beta_n} \right)^{1/n} = 0.$$

Since $\lim_{n \to \infty} n^{1/n} = 1$, we conclude that (1.15) holds.

Sufficiency Suppose that (1.15) holds. Let $f = \sum_{\alpha} c_{\alpha} z^{\alpha}$ be a formal power series belonging to $\mathcal{H}(\beta)$. Then, for any $z \in \mathbb{C}^N$, by the Cauchy–Schwarz inequality and (1.12),

$$\sum_{\alpha} |c_{\alpha} z^{\alpha}| \le \left(\sum_{\alpha} |c_{\alpha}|^2 \, v_{\alpha} \, \beta_{|\alpha|}^2 \right)^{1/2} \left(\sum_{\alpha} \frac{|z^{\alpha}|^2}{v_{\alpha} \, \beta_{|\alpha|}^2} \right)^{1/2}$$

$$= \|f\|_{\mathcal{H}(\beta)} \left(\sum_{n=0}^{\infty} \beta_n^{-2} \sum_{|\alpha|=n} \frac{|z^{\alpha}|^2}{v_{\alpha}} \right)^{1/2}$$

$$= \|f\|_{\mathcal{H}(\beta)} \left(\sum_{n=0}^{\infty} v_n^{-1} \beta_n^{-2} |z|^{2n} \right)^{1/2}. \tag{1.16}$$

Since $\lim_{n \to \infty} \beta_n^{-1/n} = 0$ and $\lim_{n \to \infty} v_n^{-1/n} = 1$, the last series above converges for all $z \in \mathbb{C}^N$. Hence, the formal power series of f absolutely converges for all $z \in \mathbb{C}^N$ and so it defines an entire function.

$\square$

Example 1.7.2 Let $\omega > 0$. For

$$\beta_n = \omega^{-n} \left(\frac{(N-1+n)!}{(N-1)!} \right)^{1/2},$$

it is easy to verify that the condition (1.15) holds. Then the space $\mathcal{H}(\beta)$ is identified with the Fock space $\mathcal{F}_\omega^2$ discussed in Sect. 1.4.

Example 1.7.3 Consider $N = 1$. Let β be any weight sequence for which the sequence of ratios $\frac{n\beta_{n-1}}{\beta_n}$ is bounded and that

$$\liminf_{n \to \infty} \frac{\log \beta_n}{n \log n} < +\infty.$$

From Proposition 1.3.1, it follows that condition (1.15) is fulfilled. Every element f in the induced space $\mathcal{H}(\beta)$ has finite growth order ρ_f, and at least one element $g \in \mathcal{H}(\beta)$ has positive growth order. Moreover, every translation operator $T_u : f(z) \mapsto f(z+u)$ $(u \in \mathbb{C})$ is bounded on $\mathcal{H}(\beta)$. This space is denoted by $\mathcal{H}(\beta_\rho^+, T)$.

Let $\mathcal{H}$ be a Hilbert space whose elements are functions from the set X to $\mathbb{C}$. Then $\mathcal{H}$ is called a *reproducing kernel Hilbert space* (RKHS) if for every $y \in X$, the evaluation functional

$$\begin{aligned} \delta_y : X &\longrightarrow \mathbb{C} \\ f &\longmapsto f(y) \end{aligned}$$

is bounded. Hence, by the Riesz representation theorem, there exists a unique element $K_y \in \mathcal{H}$ such that

$$f(y) = \langle f, K_y \rangle_\mathcal{H}, \qquad (f \in \mathcal{H}). \tag{1.17}$$

We call K_y the *reproducing kernel at the point* y. The function $K : X \times X \to \mathbb{C}$ defined by

$$K(x, y) := K_y(x), \qquad (x, y \in X),$$

is called the *reproducing kernel for* $\mathcal{H}$. Since $K_y \in$, the identity (1.17) immediately implies

$$K(x, y) = \langle K_y, K_x \rangle_\mathcal{H}, \qquad (x, y \in X),$$

If $\{e_j : j \in J\}$ is an orthonormal basis for $\mathcal{H}$, then

$$K_y = \sum_{j \in J} c_j e_j,$$

where the coefficients c_j are given by

$$c_j = \langle K_y, e_j \rangle = \overline{\langle e_j, K_y \rangle} = \overline{e_j(y)}.$$

Hence, we have the representation

$$K_y = \sum_{j \in J} \overline{e_j(y)} \, e_j,$$

in which the series is convergent in the norm of $\mathcal{H}$. Furthermore, since the evaluation functionals δ_x are continuous on $\mathcal{H}$, if we apply this functional to this representation, we see that

$$K(x, y) = \sum_{j \in J} e_j(x)\overline{e_j(y)},$$

where the convergence is pointwise for each $x, y \in X$. To tie the space introduced in this section to concepts of Sect. 1.4, we show that if all elements of $\mathcal{H}(\beta)$ are entire functions, then $\mathcal{H}(\beta)$ is a reproducing kernel Hilbert space.

Theorem 1.7.4 *Let $\beta = (\beta_n)$ be a positive sequence satisfying*

$$\lim_{n \to \infty} \beta_n^{1/n} = \infty.$$

Then the space $\mathcal{H}(\beta)$ is a reproducing kernel Hilbert space over $\mathbb{C}^N$ with the reproducing kernel $K : \mathbb{C}^N \times \mathbb{C}^N \to \mathbb{C}$ given by

$$K(z, w) = \sum_{n=0}^{\infty} \frac{\langle z, w \rangle^n}{v_n \beta_n^2}.$$

The convergence is uniform on compact subsets of $\mathbb{C}^N \times \mathbb{C}^N$.

Proof From (1.16) in the proof of Lemma 1.7.1, we see that for any $z \in \mathbb{C}^N$, there exists a constant $M_z > 0$ such that

$$|f(z)| \le M_z \|f\|_{\mathcal{H}(\beta)}, \qquad f \in \mathcal{H}(\beta).$$

Hence, each evaluation functional δ_z is bounded, which shows that $\mathcal{H}(\beta)$ is an RKHS. To obtain an expression for K, recall that the set of monomials $\{(\beta_{|\alpha|}\sqrt{\nu_\alpha})^{-1}z^\alpha : \alpha \in \mathbb{N}^N\}$ form an orthonormal basis of $\mathcal{H}(\beta)$. Using the multinomial expansion and (1.11), we get

$$K(z, w) = \sum_\alpha \frac{z^\alpha \bar{w}^\alpha}{\nu_\alpha \beta_{|\alpha|}^2} = \sum_{n=0}^\infty \frac{1}{\nu_n \beta_n^2} \sum_{|\alpha|=n} \frac{n!}{\alpha!} z^\alpha \bar{w}^\alpha = \sum_{n=0}^\infty \frac{\langle z, w \rangle^n}{\nu_n \beta_n^2}.$$

Note that since $\lim_{n\to\infty}(\nu_n\beta_n^2)^{-1/n} = 0$, the series converges uniformly on compact subsets of $\mathbb{C}^N \times \mathbb{C}^N$. $\square$

In the rest of the text, unless otherwise stated, whenever we write β or $\mathcal{H}(\beta)$, we implicitly assume that β satisfies the condition (1.15) in Theorem 1.7.1, so that all elements of $\mathcal{H}(\beta)$ are entire functions.

1.8   Some Estimations

We end this chapter with some integral estimations which will be used in studying operators on function spaces. The notation $A \asymp B$ means that there are two constants c_1 and c_2, independent of certain variables and parameters, such that

$$c_1 B \leq A \leq c_2 B,$$

uniformly with respect to all variables and parameters involved. If there is ambiguity, we explicitly mention the independence status of c_1 and c_2.

Lemma 1.8.1 *Let $a \in \mathbb{D}$, and let $b > -1$. Then*

$$\int_\mathbb{D} \frac{(1-|z|^2)^b}{|1-\bar{a}z|^{1+b+c}} \, dA(z) \asymp \begin{cases} 1, & c < 0, \\ \log(1-|a|^2)^{-1}, & c = 0, \\ (1-|a|^2)^{-c}, & c > 0. \end{cases}$$

For fixed b and c, the constants in $\asymp$ are uniforms with respect to $a \in \mathbb{D}$, in particular as $|a| \to 1^-$.

Proof Write

$$I_{b,c}(a) = \int_\mathbb{D} \frac{(1-|z|^2)^b}{|1-\bar{a}z|^{2+b+c}} \, dA(z), \qquad (a \in \mathbb{D}).$$

Since $b > -1$, the integral $I_{b,c}(z)$ converges for every $a \in \mathbb{D}$. Put

$$k = \frac{1 + b + c}{2}.$$

Note that whenever $k \le 0$, we have $c < 0$ and $I_{b,c}(a)$ is bounded. So, in the following, we consider the case where $k > 0$.

The power series representation

$$\frac{1}{(1 - \bar{a}z)^k} = \sum_{n=0}^{\infty} \frac{\Gamma(n + k)}{n!\Gamma(k)} (\bar{a}z)^n$$

is a useful tool in the following. The measure $(1 - |z|^2)^b \, dA(z)$ is orthogonal to all harmonics $e^{in\theta}$, $n \ne 0$. Hence, by expanding

$$\frac{1}{|1 - \bar{a}z|^{2k}} = \frac{1}{(1 - \bar{a}z)^k} \frac{1}{(1 - a\bar{z})^k},$$

we see that

$$\begin{aligned}
I_{b,c}(a) &= \int_{\mathbb{D}} \frac{(1 - |z|^2)^b}{|1 - \bar{a}z|^{2k}} \, dA(z) \\
&= \sum_{n=0}^{\infty} \left(\frac{\Gamma(n + k)}{n!\,\Gamma(k)} \right)^2 |a|^{2n} \int_{\mathbb{D}} (1 - |z|^2)^b |z|^{2n} \, dA(z) \\
&= \sum_{n=0}^{\infty} \left(\frac{\Gamma(n + k)}{n!\,\Gamma(k)} \right)^2 |a|^{2n} \times 2 \int_0^1 (1 - r^2)^b r^{2n} \, r\,dr \\
&= \sum_{n=0}^{\infty} \left(\frac{\Gamma(n + k)}{n!\,\Gamma(k)} \right)^2 |a|^{2n} \times \frac{\Gamma(b + 1)\Gamma(n + 1)}{\Gamma(b + n + 2)} \\
&= \frac{\Gamma(b + 1)}{\Gamma(k)^2} \sum_{n=0}^{\infty} \frac{\Gamma(n + k)^2}{n!\,\Gamma(b + n + 2)} |a|^{2n}.
\end{aligned}$$

By Stirling's formula,

$$\frac{\Gamma(n + k)^2}{n!\,\Gamma(b + n + 2)} \sim n^{c-1}, \qquad (n \to \infty).$$

Thus,

$$I_{b,c}(a) \asymp \sum_{n=0}^{\infty} \frac{|a|^{2n}}{(n + 1)^{1-c}}.$$

Note that the constants involved in $\asymp$ depend on b and c, but uniformly hold with respect to a for a fixed b and c. There are now three cases to consider for c as follows.

(i) $c < 0$: the series

$$\sum_{n=0}^{\infty} \frac{|a|^{2n}}{(n+1)^{1-c}}$$

defines a bounded function on $\mathbb{D}$, and hence, $I_{b,c}(a)$ is bounded on $\mathbb{D}$.
(ii) $c = 0$: in this case,

$$I_{b,c}(a) \sim \sum_{n=0}^{\infty} \frac{|a|^{2n}}{n+1} \sim \log \frac{1}{1-|a|^2}, \qquad (|a| \to 1^-).$$

(iii) $c > 0$: we have

$$I_{b,c}(a) \sim \sum_{n=0}^{\infty} (n+1)^{c-1}|a|^{2n} \sim \frac{1}{1-|a|^c}, \qquad (|a| \to 1^-),$$

since, again by Stirling's formula,

$$(n+1)^{c-1} \sim \frac{\Gamma(n+c)}{n!\,\Gamma(c)}$$

and

$$\sum_{n=0}^{\infty} \frac{\Gamma(n+c)}{n!\,\Gamma(c)} |a|^{2n} = \frac{1}{1-|a|^c}.$$

$\square$

Lemma 1.8.2 *Let $a \in \mathbb{D}$. Then*

$$\int_0^{2\pi} \frac{d\theta}{|1 - ae^{-i\theta}|^{1+c}} \asymp \begin{cases} 1, & c < 0, \\ \log(1-|a|^2)^{-1}, & c = 0, \\ (1-|a|^2)^{-c}, & c > 0. \end{cases}$$

For a fixed c, the constants in $\asymp$ are uniforms with respect to $a \in \mathbb{D}$, in particular as $|a| \to 1^-$.

Proof The proof is similar to the proof of Lemma 1.8.1, and hence, it is omitted. $\square$

Notes on Chapter 1

Most of definitions and results of this chapter are taken from [19,20]. The sequence spaces ℓ_β^2 have many important applications in studying operators on function spaces; see, e.g., [17,83]. The *weighted Hardy spaces* $\mathcal{H}(\beta)$ are treated extensively in the past two decades. In particular, properties of composition operators on $\mathcal{H}(\beta)$ spaces are investigated in [17], with an extension to the case of unit ball in several complex variables. See also [84]. For more details on entire functions and their growth, e.g., order as well as a type, we refer to [58]. The proof of Theorem 1.2.1 is quite analogous to the case of entire functions in [19], and part of it is shown in [76]. Lemma 1.6.1 is discussed in [79], while it was also stated in some other works [8,28,30]. Hilbert spaces $\mathcal{H}(\beta)$ of formal power series are considered in [17, Chapter 2] as weighted Hardy spaces. The family of function spaces introduced in Example 1.7.3 and composition operators acting on them are studied in [19]. The theory of reproducing kernel Hilbert spaces was investigated in great details by Aronszajn in [5] in the 1950s. See also the new book [76]. Lemma 1.8.1 can be found in [40, Theorem 1.7] as well as in [108, Lemma 3.10], while Lemma 1.8.2 can be found in [40, Theorem 1.7]. Both lemmas are frequently used in estimations of operators on function spaces.

2.1 The Nevanlinna Counting Function

Let $\varphi: \mathbb{D} \to \mathbb{D}$ be a holomorphic self-map of the open unit disc $\mathbb{D}$. Then the *Nevanlinna counting function* of φ is defined by

$$N_\varphi(w) = \sum_{z \in \varphi^{-1}\{w\}} \log \frac{1}{|z|}, \qquad (w \in \mathbb{D} \setminus \{\varphi(0)\}), \tag{2.1}$$

where $\varphi^{-1}\{w\}$ denotes the sequence of inverse images of w under φ, each point being repeated according to its multiplicity. For a more detailed study of the counting function, let $w \in \mathbb{C} \setminus \{\varphi(0)\}$, and let $\{z_j(w) : j \geq 1\}$ denote the sequence of the preimages $\varphi^{-1}\{w\}$, arranged increasing order of absolute values, each point repeated according to its multiplicity. Also, for $0 \leq r < 1$, let $n(r, w) = n_\varphi(r, w)$ denote the number of these points in the disc $r\mathbb{D}$. Then we define the *partial counting functions* of φ by

$$N_\varphi(r, w) = \sum_{j=1}^{n(r,w)} \log \frac{r}{|z_j|}. \tag{2.2}$$

With this notation, the original Nevanlinna counting function is precisely $N_\varphi(1, w)$, i.e.,

$$N_\varphi(w) = N_\varphi(1, w) = \sum_j \log \frac{1}{|z_j(w)|}.$$

With the convention that, for $0 \leq r \leq 1$, $N_\varphi(r, w) = 0$, whenever $w \notin \varphi(r\mathbb{D})$, and so the counting functions can be regarded as defined on the entire complex plane. Note that for each fixed complex number w, the partial counting function $N_\varphi(r, w)$ increases with r. Hence, by the monotone convergence theorem,

$$\lim_{r \to -1} N_\varphi(r, w) = N_\varphi(w).$$

Another useful reformulation of (2.2) is

$$N_\varphi(r, w) = \int_0^r \log(\frac{r}{t}) \, dn(t, w), \tag{2.3}$$

and in particular, (2.1) becomes

$$N_\varphi(w) = \int_0^1 \log(\frac{1}{t}) \, dn(t, w). \tag{2.4}$$

2.2   Littlewood's Inequality

We start with a simple, but very useful, estimation of $N_\varphi(w)$.

Theorem 2.2.1 (Littlewood)   *Let* $\varphi : \mathbb{D} \to \mathbb{D}$ *be analytic, and let* N_φ *be the Nevanlinna counting function* (2.1). *Then, for each* $w \in \mathbb{D} \setminus \{\varphi(0)\}$, *we have*

$$N_\varphi(w) \leq \log \left| \frac{1 - \overline{\varphi(0)}w}{\varphi(0) - w} \right|. \tag{2.5}$$

Regarding the case of equality, the following are equivalent.

(i) *Equality holds for some* w.
(ii) *Equality holds for all* $w \in \mathbb{D}$, *except for an exceptional set of logarithmic capacity zero.*
(iii) φ *is an inner function.*

Proof Jensen identity is the main ingredient of proof. Let $f : \mathbb{D} \to \mathbb{C}$ be analytic with $f(0) \neq 0$, and let $0 \leq r < 1$. Using the notations introduced in Sect. 2.1, the identity says

$$\frac{1}{2\pi} \int_0^{2\pi} \log |f(re^{i\theta})| \, d\theta = \log |f(0)| + \sum_{j=1}^{n(r,w)} \log \frac{r}{|z_j|}.$$

We rewrite the identity as

$$N_\varphi(r, 0) = \frac{1}{2\pi} \int_0^{2\pi} \log |f(re^{i\theta})|\, d\theta - \log |f(0)|. \qquad (2.6)$$

Put

$$\varphi_w(z) = \frac{w - \varphi(z)}{1 - \overline{w}\varphi(z)} \qquad (z \in \mathbb{D}).$$

Then φ_w is a holomorphic self-map on $\mathbb{D}$, and $\varphi^{-1}\{w\}$ is precisely the zero set of φ_w. Applying the representation (2.6) with $f = \varphi_w$, where $w \neq \varphi(0)$, we obtain

$$N_\varphi(r, w) = N_{\varphi_w}(r, 0) = \frac{1}{2\pi} \int_0^{2\pi} \log |\varphi_w(re^{i\theta})|\, d\theta - \log |\varphi_w(0)|. \qquad (2.7)$$

Since $|\varphi_w| \leq 1$ on $\mathbb{D}$, the integral on the right-hand side of (2.7) is negative, and thus, for all $w \in \mathbb{D} \setminus \{\varphi(0)\}$, we have

$$N_\varphi(r, w) \leq -\log |\varphi_w(0)|.$$

Letting $r \to 1$, the Littlewood inequality (2.5) immediately follows.

To prove the equivalence of (i), (ii), and (iii), first note that, by (2.7), equality holds in (2.5) if and only if

$$\lim_{r \to 1} \int_0^{2\pi} \log |\varphi_w(re^{i\theta})|\, d\theta = 0, \qquad (2.8)$$

which is a characterization of Blaschke products. It is also elementary that φ is inner if and only so is φ_w. Thus, $(i) \implies (iii)$. Now, a classical theorem of Frostman says that if φ is an inner function, then φ_w is a Blaschke product for every $w \in \mathbb{D}$, except for a set of logarithmic capacity zero, i.e., $(iii) \implies (ii)$. That $(ii) \implies (i)$ is trivial. $\square$

2.3   A Change of Variable Formula

We used dA to denote the normalized planar Lebesgue measure so that the area of the unit disc $\mathbb{D}$ is 1. Set

$$dA_1(z) = \log\left(\frac{1}{|z|}\right) dA(z), \qquad (z \in \mathbb{D}).$$

Theorem 2.3.1 *Let g be a positive measurable function on $\mathbb{D}$, and let $\varphi : \mathbb{D} \to \mathbb{D}$ be analytic. Then*

$$\int_{\mathbb{D}} (g \circ \varphi)|\varphi'|^2 \, dA_1 = \int_{\mathbb{D}} g N_\varphi \, dA,$$

where N_φ is the Nevanlinna counting function (2.1).

Proof The singular points of φ are the zeros of φ', i.e., the set

$$\mathcal{E} := \{z \in \mathbb{D} : \varphi'(z) = 0\}$$

Let $\mathbb{D}' := \mathbb{D} \setminus \mathcal{E}$. Then φ is a local homeomorphism on the open set $\mathbb{D}'$. Thus, there exists a countable collection $\{R_j\}$ of closed polar rectangles whose interiors are disjoint and their union is $\mathbb{D}'$, and more importantly, φ is one-to-one on each R_j.

Let us denote by ψ_j the inverse of the restriction of φ to R_j, so that ψ_j is a one-to-one map taking $\varphi(R_j)$ back onto R_j. By the usual change of variable formula applied on R_j, with $z = \psi_j(w)$, we have

$$\int_{R_j} (g \circ \varphi)(z) \, |\varphi'(z)|^2 \, dA_1(z) = \int_{\varphi(R_j)} g(w) \log \left(\frac{1}{|\psi_j(w)|} \right) dA(w). \qquad (2.9)$$

Hence, if χ_j denotes the characteristic function of the set $\varphi(R_j)$, then

$$\int_{\mathbb{D}} (g \circ \varphi)|\varphi'|^2 \, dA_1 = \int_{\mathbb{D}} g(w) \left\{ \sum_j \chi_j(w) \left(\frac{1}{|\psi_j(w)|} \right) \right\} dA(w). \qquad (2.10)$$

This is the desired formula, since, by definition, the term in curly brackets on the right-hand side of the equation above is precisely $N_\varphi(w)$. □

In the above proof, we are slightly cheating. The points on the boundaries of R_j are counting more than once when we add up the integrals in the left of (2.9) to obtain the left of (2.10). At the same token, to write

$$\sum_j \chi_j(w) \left(\frac{1}{|\psi_j(w)|} \right) = N_\varphi(w),$$

there are w's which are counted more than enough (images of the boundaries of R_j) and possibly even w's which are not counted enough (image of singular points). However, this does not create any problem since the collection of all such exceptional points form sets of (two-dimensional Lebesgue) measure zero on both sides of (2.10).

2.4   The Generalized Nevanlinna Counting Function

For a holomorphic function $\varphi : \mathbb{D} \to \mathbb{D}$, $0 \le r < 1$, and $\gamma \ge 0$, the *generalized Nevanlinna partial counting function* is defined by

$$N_{\varphi,\gamma}(r, w) := \sum_{j=1}^{n(r,w)} \log\left(\frac{r}{|z_j(w)|}\right)^{\gamma}, \qquad (w \in \mathbb{D} \setminus \{\varphi(0)\}),$$

and thus, the *generalized Nevanlinna counting function* is given by

$$N_{\varphi,\gamma}(w) = \sum_{j\ge 1}\left[\log\frac{1}{|z_j(w)|}\right]^{\gamma}, \qquad (w \in \mathbb{D} \setminus \{\varphi(0)\}),$$

where $\{z_j(w)\}$ denotes the sequence of φ-preimages of w, multiplicity counted. In other words, $N_{\varphi,\gamma}(w) = N_{\varphi,\gamma}(1, w)$. Note also that $N_{\varphi,0}(r, w)$ is the *multiplicity function* $n(r, w)$ of φ and $N_{\varphi,1}$ is the classical Nevanlinna counting function. The representation (2.3) and (2.4) can also be, respectively, generalized as

$$N_{\varphi,\gamma}(r, w) = \int_0^r \left(\log\frac{r}{t}\right)^{\gamma} dn(t, w) \tag{2.11}$$

and

$$N_{\varphi,\gamma}(w) = \int_0^1 \left(\log\frac{1}{t}\right)^{\gamma} dn(t, w). \tag{2.12}$$

In the light of above formulas, we define the measure dA_{γ} on $\mathbb{D}$ by

$$dA_{\gamma}(w) = \left[\log\frac{1}{|w|}\right]^{\gamma} dA(w).$$

The importance of $N_{\varphi,\gamma}$ stems from the following result, which generalizes the change of variable formula.

Theorem 2.4.1 *Let g be a positive measurable function on $\mathbb{D}$, and let $\varphi : \mathbb{D} \to \mathbb{D}$ be analytic. Then*

$$\int_{\mathbb{D}} (g \circ \varphi)|\varphi'|^2 dA_{\gamma} = c(\gamma) \int_{\mathbb{D}} g N_{\varphi,\gamma} dA,$$

where $c(\gamma) = \frac{2^{\gamma}}{\Gamma(\gamma+1)}$.

Proof The proof is almost identical with the original case $\gamma = 1$, i.e., Theorem 2.3.1, and hence, it is omitted. □

We also have the following extension of Littlewood's inequality (Theorem 2.2.1).

Theorem 2.4.2 *Let* $\varphi : \mathbb{D} \to \mathbb{D}$ *be analytic, and let* $\gamma \geq 1$. *Then we have*

$$N_{\varphi,\gamma}(w) \leq \left(\log \left| \frac{1 - \overline{\varphi(0)}w}{\varphi(0) - w} \right| \right)^{\gamma}, \qquad (w \in \mathbb{D} \setminus \{\varphi(0)\}).$$

Proof The case $\gamma = 1$ is the original Littlewood's inequality, which was studied in Sect. 2.2. For each $\gamma > 1$ and for any positive sequence of numbers (a_j), we have

$$\sum a_j^{\gamma} \leq \left(\sum a_j \right)^{\gamma}.$$

Hence, we immediately deduce

$$N_{\varphi,\gamma}(w) \leq N_{\varphi,1}^{\gamma}(w) \qquad (w \in \mathbb{D}).$$

Now, the desired result easily follows from the case $\gamma = 1$. □

2.5   Interrelation Between Nevanlinna Counting Functions

We have the following representation of the Nevanlinna counting function $N_{\varphi,\gamma}$ in terms of $N_\varphi = N_{\varphi,1}$.

Lemma 2.5.1 *Let* $\varphi : \mathbb{D} \to \mathbb{D}$ *be analytic, and let* $\gamma \geq 1$. *Then, for* $w \neq \varphi(0)$, *we have*

$$N_{\varphi,\gamma}(r, w) = \gamma(\gamma - 1) \int_0^r t^{-1} N_\varphi(t, w) \left(\log \frac{r}{t} \right)^{\gamma-2} dt.$$

Proof We integrate by parts twice. In each case, the integrated terms sum up to zero, because the condition $w \neq \varphi(0)$ guarantees that $n(t, w)$ and $N_\varphi(t, w)$ both vanish for all t sufficiently close to zero. By (2.11), the first integration by parts yields

$$N_{\varphi,\gamma}(r, w) = \int_0^r \left(\log \frac{r}{t} \right)^{\gamma} dn(t, w)$$

$$= \gamma \int_0^r t^{-1} n(t, w) \left(\log \frac{r}{t} \right)^{\gamma-1} dt.$$

In particular, in the special case $\gamma = 1$, we obtain

$$N_\varphi(r, w) = \int_0^r t^{-1} n(t, w) \, dt.$$

In other words, we can write

$$t^{-1} n(t, w) \, dt = dN_\varphi(t, w).$$

Thus, applying the second integration by parts, we deduce

$$\int_0^r t^{-1} n(t, w) \left(\log \frac{r}{t} \right)^{\gamma-1} dt = (\gamma - 1) \int_0^r t^{-1} N_\varphi(t, w) \left(\log \frac{r}{t} \right)^{\gamma-2} dt,$$

from which the desired result follows.   □

The following result, which is very useful in applications, shows that the counting functions satisfy a specific mean value property.

Lemma 2.5.2 *Let φ be a holomorphic self-map of $\mathbb{D}$, and let $\gamma > 0$. If $\varphi(0) \neq 0$ and $0 < r < |\varphi(0)|$, then*

$$N_{\varphi,\gamma}(0) \leq \frac{1}{r^2} \int_{r\mathbb{D}} N_{\varphi,\gamma}(z) \, dA(z).$$

Proof First, we prove the lemma for the classical case $N_\varphi(w)$. As usual, for $0 \leq r < 1$, we write $\varphi_r(z) = \varphi(rz)$, $z \in \mathbb{D}$. Set $d\sigma = d\theta/2\pi$, and let μ be the Borel probability measure defined on $\mathbb{C}$ (but supported only on $\varphi_r(\partial\mathbb{D})$) by $d\mu = \varphi_r^{-1} d\sigma$. Then for each fixed complex number $w \neq \varphi(0)$, Jensen's formula applied to $f = \varphi - w$ is rewritten as

$$N_\varphi(r, w) + \log |\varphi(0) - w| = \int \log |\zeta - w| \, d\mu(\zeta).$$

The left-hand side of the above equation, being the logarithmic potential of a compactly supported probability measure, is subharmonic in the complex plane. Thus, $N_\varphi(r, w)$ is a subharmonic function of w on $\mathbb{C} \setminus \{\varphi(0)\}$. Consequently, $N_\varphi(r, w)$ is subharmonic, and since $N_\varphi(r, w) \uparrow N_\varphi(w)$ as $r \uparrow 1$, the monotone convergence theorem insures that $N_\varphi(w)$ inherits from $N_\varphi(r, w)$ the subharmonicity and the submean inequality holds.

Now, we return to the general case. By Lemma 2.5.1, $N_{\varphi,\gamma}(r, w)$ inherits the continuity and subharmonicity of $N_\varphi(r, w)$ on $\mathbb{D} \setminus \{\varphi(0)\}$ for each $0 \leq r < 1$. By similar arguments as for the classical case and using again the monotone convergence theorem, the desired result follows.   □

2.6   The Littlewood–Paley Formula

The norm in Hardy space H^p was defined as the supremum, even as the limit, of certain line integrals in Sect. 1.1. In the following result, we provide another formula in which the area integral is used.

Theorem 2.6.1 (Littlewood–Paley) *Let $f \in H^p(\mathbb{D})$, $0 < p < \infty$, with $f \not\equiv 0$ and $f(0) = 0$. Then we have*

$$\frac{1}{2\pi} \int_0^{2\pi} |f(e^{i\theta})|^p \, d\theta = \frac{p^2}{2} \int_{\mathbb{D}} |f(z)|^{p-2} |f'(z)|^2 \log(1/|z|) \, dA(z).$$

Proof Let ρ be so small that on the closed disc $\{|z| \leq \rho\}$, there are no zeros of f, except of course the origin. Let $1 > \varrho > \rho$ be such that on the circle $\{|z| = \varrho\}$, f has no zeros. Hence, on the annulus $\{\rho < |z| < \varrho\}$, there are a finite number of the zeros of f, say $z_1, z_2, \cdots, z_n$. Let ε be so small that all the discs $\{|z - z_k| \leq \varepsilon\}$, $1 \leq k \leq n$, are entirely in the annulus $\{\rho < |z| < \varrho\}$. Finally, let

$$\Omega = \{\rho < |z| < \varrho\} \setminus \bigcup_{k=1}^{n} \{|z - z_k| \leq \varepsilon\}.$$

In the following, we shall eventually let $\varepsilon \to 0$, and then $\rho \to 0$, and $\varrho \to 1$ through a sequence $(\varrho_n)_{n \geq 1}$, so that there is no zero on the circles $\{|z| = \varrho_n\}$.

The function $W(z) = |f(z)|^p$ is infinitely differentiable in a neighborhood of $\bar{\Omega}$. Hence, by Green's theorem, we have

$$\iint_{\Omega} \log(\varrho/|z|) \, \nabla^2 W(z) \, dx \, dy = \int_{\partial \Omega} \left(\log(\varrho/|z|) \frac{\partial W}{\partial n} - W \frac{\partial \log(\varrho/|z|)}{\partial n} \right) d\ell.$$

First of all, a simple calculation shows that

$$\nabla^2 W(z) = p^2 \, |f(z)|^{p-2} \, |f'(z)|^2.$$

Hence, the Green formula becomes

$$p^2 \iint_{\Omega} \log(\varrho/|z|) \, |f(z)|^{p-2} \, |f'(z)|^2 \, dx \, dy = I_{\varrho} - I_{\rho} - \sum_{k=1}^{n} I_k, \qquad (2.13)$$

where the integrals on the right side are explained below. On the boundary $\{|z| = \varrho\}$, we have

$$I_\varrho = \int_0^{2\pi} \left(\log(1) \, \frac{\partial W}{\partial r}(\varrho e^{i\theta}) + W(\varrho e^{i\theta}) \frac{1}{\varrho} \right) \varrho \, d\theta$$

$$= \int_0^{2\pi} |f(\varrho e^{i\theta})|^p \, d\theta$$

By a well-known result in the theory of H^p spaces, we know that

$$I_\varrho = \int_0^{2\pi} |f(\varrho e^{i\theta})|^p \, d\theta \longrightarrow \int_0^{2\pi} |f(e^{i\theta})|^p \, d\theta, \qquad (2.14)$$

as $\varrho \to 1$. On the boundary $\{ |z| = \rho \}$, we have

$$I_\rho = \int_0^{2\pi} \left(\log(\varrho/\rho) \, \frac{\partial W}{\partial r}(\rho e^{i\theta}) + W(\rho e^{i\theta}) \frac{1}{\rho} \right) \rho \, d\theta$$

$$= \int_0^{2\pi} |f(\rho e^{i\theta})|^p \, d\theta + \rho \log(\varrho/\rho) \int_0^{2\pi} \frac{\partial |f|^p}{\partial r}(\rho e^{i\theta}) \, d\theta.$$

Since f is continuous at the origin, then

$$\int_0^{2\pi} |f(\rho e^{i\theta})|^p \, d\theta \longrightarrow 2\pi \, |f(0)|^p = 0,$$

as $\rho \to 0$. On the other hand, since

$$\frac{\partial |f|^p}{\partial r} = p \, (u u_r + v v_r) \, |f|^{p-2},$$

by the Cauchy–Schwarz inequality,

$$\left| \int_0^{2\pi} \frac{\partial |f|^p}{\partial r}(\rho e^{i\theta}) \, d\theta \right| \leq p \int_0^{2\pi} |f(\rho e^{i\theta})|^{p-1} |f'(\rho e^{i\theta})| \, d\theta.$$

If f has a zero of order $n_0 \geq 1$ at the origin, then $|f(\rho e^{i\theta})| \asymp \rho^{n_0}$, and $|f'(\rho e^{i\theta})| \asymp \rho^{n_0-1}$, as $\rho \to 0$. Hence,

$$\int_0^{2\pi} |f(\rho e^{i\theta})|^{p-1} |f'(\rho e^{i\theta})| \, d\theta \asymp \rho^{p n_0 - 1},$$

which gives

$$\left| \rho \log(\varrho/\rho) \int_0^{2\pi} \frac{\partial |f|^p}{\partial r}(\rho e^{i\theta}) \, d\theta \right| \leq C \, \rho^{p n_0} \, |\log \rho|.$$

Thus, we also have

$$\rho \log(\varrho/\rho) \int_0^{2\pi} \frac{\partial |f|^p}{\partial r}(\rho e^{i\theta}) \, d\theta \longrightarrow 0,$$

as $\rho \to 0$. Therefore,

$$I_\rho \to 0, \tag{2.15}$$

as $\rho \to 0$. Finally, on the boundary $\{ |z - z_k| = \varepsilon \}$, we have

$$I_k = \int_0^{2\pi} \left(\log(\varrho/|z_k + \varepsilon e^{i\theta}|) \, \frac{\partial |f|^p}{\partial n}(z_k + \varepsilon e^{i\theta}) \right.$$
$$\left. - |f(z_k + \varepsilon e^{i\theta})|^p \, \frac{\partial \log(\varrho/|z|)}{\partial n}(z_k + \varepsilon e^{i\theta}) \right) \varepsilon \, d\theta.$$

Since

$$\left| \frac{\partial |f|^p}{\partial n} \right| \le |\nabla |f|^p| \le p |f|^{p-1} |f'|,$$

if f has a zero of order $n_k \ge 1$ at the z_k, then we have

$$|I_k| \le C \, \varepsilon^{p n_k}.$$

Note that the constant C depends on ϱ. However, for a fixed ϱ, we have

$$I_k \to 0, \tag{2.16}$$

as $\varepsilon \to 0$. Now, first let $\varepsilon \to 0$. Hence, by the monotone convergence theorem, and by (2.16), the Green formula (2.13) becomes

$$p^2 \iint_{\rho < |z| < \varrho} \log(\varrho/|z|) \, |f(z)|^{p-2} \, |f'(z)|^2 \, dx \, dy = I_\varrho - I_\rho.$$

Then, let $\rho \to 0$, and finally let $\varrho \to 1$. Thus, again by the monotone convergence theorem, and by (2.14) and (2.15), we get

$$p^2 \iint_{|z| < 1} \log(1/|z|) \, |f(z)|^{p-2} \, |f'(z)|^2 \, dx \, dy = \int_0^{2\pi} |f(e^{i\theta})|^p \, d\theta.$$

$$\square$$

Since

$$\|f\|_{H^2}^2 = \|f - f(0)\|_{H^2}^2 + |f(0)|^2, \tag{2.17}$$

an interesting special case of Theorem 2.6.1 is the following.

Corollary 2.6.2 *Let* $f \in H^2(\mathbb{D})$. *Then we have*

$$\|f\|_{H^2}^2 = |f(0)|^2 + 2 \int_{\mathbb{D}} |f'(z)|^2 \, \log(1/|z|) \, dA(z).$$

However, for $p \neq 2$, the simple but crucial identity (2.17) does not hold, and we need to either assume $f(0) = 0$ or work with $f - f(0)$. At the same token, using the notation of Sect. 2.3, we can rewrite the formula in Theorem 2.6.1 as

$$\|f\|_{H^p}^p = \frac{p^2}{2} \int_{\mathbb{D}} |f(z)|^{p-2} |f'(z)|^2 \, dA_1(z), \qquad (f(0) = 0).$$

If φ is an analytic self-map of $\mathbb{D}$ and we apply the above formula to

$$g(z) := f(\varphi(z)) - f(\varphi(0)),$$

we obtain

$$\|f \circ \varphi - a_0\|_{H^p}^p = \frac{p^2}{2} \int_{\mathbb{D}} |(f \circ \varphi)(z) - a_0|^{p-2} |f'(\varphi(z))|^2 |\varphi'(z)|^2 \, dA_1(z),$$

where $a_0 = f(\varphi(0))$. Now, if we apply the change of variable formula discussed in Theorem 2.3.1, we immediately obtain the following result.

Corollary 2.6.3 *Let* $f \in H^p$, *and let* $\varphi : \mathbb{D} \to \mathbb{D}$ *be analytic. Then*

$$\|f \circ \varphi - a_0\|_{H^p}^p = \frac{p^2}{2} \int_{\mathbb{D}} |f(w) - a_0|^{p-2} |f'(w)|^2 N_\varphi(w) \, dA(w),$$

where $a_0 = f(\varphi(0))$.

Applying Corollary 2.6.3 with $\varphi(z) = rz$, we obtain a further special case which is required in some applications.

Corollary 2.6.4 *Let* $f \in H^p$, *with* $f(0) = 0$, *and let* $r < 1$. *Then*

$$\frac{1}{2\pi} \int_0^{2\pi} |f(re^{i\theta}|^p \, d\theta = \frac{p^2}{2} \int_{r\mathbb{D}} |f(w)|^{p-2} |f'(w)|^2 \log \frac{r}{|w|} \, dA(w).$$

Under the assumption of Corollary 2.6.3, with $p \geq 2$, we also deduce the estimation

$$\|f \circ \varphi\|_{H^p}^p \asymp |f(\varphi(0))|^p + \int_{\mathbb{D}} |f(w)|^{p-2} |f'(w)|^2 N_\varphi(w) \, dA(w). \tag{2.18}$$

Notes on Chapter 2

Materials in Sect. 2.1 are taken from [82,86]. Moreover, further detail can be found in [83, Chapter 10]. The Littlewood inequality was proved in [62]. Lemma 2.4.1 is from [82]. Lemma 2.5.2 was first proved for the classical Nevanlinna counting function in Section 6 of [26]. The version for the generalized Nevanlinna counting function is from [82]. The Frostman theorem can be found, for example, in [37]. For a treatment of subharmonic functions and the submean inequality used in Lemma 2.5.2, see Chapter 4 of [66]. The Littlewood–Paley (Theorem 2.6.1) formula is discussed in many standard textbooks. For example, see [114] and Lemma 1.2 in [65]. Corollary 2.6.3 can be considered as a version of the Hardy–Stein identity [88]. See also [108] and [87].

In this chapter, the $\mathcal{N}_p$-spaces of holomorphic functions on the open unit disc $\mathbb{D}$ are studied. This family of spaces is also known as Bergman-type or Beurling-type spaces. The Bergman spaces and Bergman-type spaces have attracted a great deal of attention and have been considered by many mathematicians. These spaces, on the one hand, play an important role in both function theoretic and operator theoretic development of function spaces and, on the other hand, have a closed connection to Bloch spaces which appear as the images of the bounded functions under the Bergman projections. Bloch spaces also play the role of the dual spaces of the Bergman spaces. Also among the Bergman and Bergman-type spaces, the so-called $\mathcal{Q}_p$-spaces are of great interest. Adopting the intrinsic relation between the classical Dirichlet space and the Bergman space, the appearance of $\mathcal{N}_p$-spaces is also related to $\mathcal{Q}_p$ spaces. Hence, some basic Banach space properties of the $\mathcal{N}_p$-spaces are discussed below. In particular, it will be showed that, for $p \in (0, 1)$, the $\mathcal{N}_p$-spaces are all different topological vector spaces with independent interest.

3.1 Comparison with $\mathcal{Q}_p$

Let

$$\sigma_a(z) = \frac{a - z}{1 - \bar{a}z}, \qquad (z \in \mathbb{D}),$$

be the automorphism of disc that interchanges 0 and a. The identity

$$1 - |\sigma_a(z)|^2 = \frac{(1 - |a|^2)(1 - |z|^2)}{|1 - \bar{a}z|^2}$$

is easily verified and henceforth will be used frequently without any further reference. To provide a motivation for introducing $\mathcal{N}_p$-spaces, let us look at its cousin the class of Q_p-spaces, $0 < p < \infty$, on the open unit disc $\mathbb{D}$. Fixing p, the space Q_p consists of functions in $\text{Hol}(\mathbb{D})$ such that

$$\sup_{a \in \mathbb{D}} \int_{\mathbb{D}} |f'(z)|^2 (1 - |\sigma_a(z)|^2)^p \, dA(z) < \infty,$$

where dA represents the Lebesgue area measure on the plane, normalized so that $A(\mathbb{D}) = 1$. It is known that:

(i) For $p \in (1, \infty)$, Q_p-spaces coincide with the classical Bloch space $\mathcal{B}$
(ii) Q_1 is equal to BMOA, the space of holomorphic functions on $\mathbb{D}$ with bounded mean oscillation
(iii) For $p \in (0, 1)$, the Q_p-spaces are all different Fréchet topological vector spaces of independent interest

Adopting the relation between the classical Dirichlet space and the Bergman space, if in the definition of the Q_p-space we replace $f'(z)$ by $f(z)$, then we obtain the so-called $\mathcal{N}_p$-space in the unit disc $\mathbb{D}$. Hence, more explicitly, the space $\mathcal{N}_p$ consists of functions in $\text{Hol}(\mathbb{D})$ for which

$$\sup_{a \in \mathbb{D}} \int_{\mathbb{D}} |f(z)|^2 (1 - |\sigma_a(z)|^2)^p \, dA(z) < \infty.$$

Recall that A^{-1} consists of holomorphic functions on $\mathbb{D}$ for which

$$\sup_{z \in \mathbb{D}} |f(z)|(1 - |z|^2) < \infty.$$

See Sect. 1.1. Parallel to the properties that we mentioned above for Q_p-spaces, we will see that:

(i) For $p > 1$, the $\mathcal{N}_p$-spaces coincide with the Bergman-type space A^{-1}
(ii) For $p \in (0, 1]$, the $\mathcal{N}_p$-spaces are all different topological vector spaces
(iii) The class of $\mathcal{N}_p$-spaces is larger than the class of Q_p-spaces

In the rest of this chapter, we establish some elementary properties of $\mathcal{N}_p$-spaces.

3.2   The Test Function k_w

It is clear that analytic polynomials and more generally rational functions with poles outside the closed unit disc are in $\mathcal{N}_p$. However, there is one special subclass, which is of independent interest and is used in several applications as a test function. For $w \in \mathbb{D}$, we define the *test function*

$$k_w(z) = \frac{1 - |w|^2}{(1 - \overline{w}z)^2}, \qquad (z \in \mathbb{D}).$$

The following result for test functions in $\mathcal{N}_p$ is rather simple but important in what follows.

Lemma 3.2.1 *For each $w \in \mathbb{D}$, $k_w \in \mathcal{N}_p$ and, moreover,*

$$\|k_w\|_{\mathcal{N}_p} \le 1.$$

Proof It is obvious that $k_w \in \mathrm{Hol}(\mathbb{D})$. Furthermore, by Theorem 2.3.1,

$$\|k_w\|_{\mathcal{N}_p}^2 = \sup_{a \in \mathbb{D}} \int_{\mathbb{D}} |\sigma_w'(z)|^2 (1 - |\sigma_a(z)|^2)^p \, dA(z)$$

$$\le \int_{\mathbb{D}} |\sigma_w'(z)|^2 \, dA(z)$$

$$= \int_{\mathbb{D}} N_{\sigma_w}(z) \, dA(z)$$

$$= \int_{\mathbb{D}} dA(z) = A(\mathbb{D}) = 1.$$

In the last step, we used the fact that σ_w is an automorphism of $\mathbb{D}$ and thus

$$N_{\sigma_w}(z) = 1, \qquad (z \in \mathbb{D}).$$

$\square$

3.3   An Equivalent Norm for A_γ^p

We denoted the normalized Lebesgue measure on $\mathbb{D}$ by dA and, for $\gamma > -1$, define the measure dA_γ on $\mathbb{D}$ by

$$dA_\gamma(w) = [\log(1/|w|)]^\gamma \, dA(w).$$

Then, for $0 < p < \infty$ and $\gamma > -1$, we define the *weighted Bergman space* A_γ^p by

$$A_\gamma^p = \left\{ f \in \mathcal{O}(\mathbb{D}) : \|f\|_{A_\gamma^p} := \int_{\mathbb{D}} |f(w)|^p \, dA_\gamma(w) < \infty \right\}.$$

Due to some observations, we will see that this definition can be extended to include $\gamma = -1$. The formula in Corollary 2.6.4 leads to a very useful equivalent norm for A_γ^p.

Lemma 3.3.1 *Let $0 < p < \infty$. Then*

$$\|f\|_{A_\gamma^p}^p \asymp |f(0)|^p + \int_{\mathbb{D}} |f(w)|^{p-2} |f'(w)|^2 \, dA_{\gamma+2}(w).$$

Proof Without loss of generality, we may assume that $f(0) = 0$. By Corollary 2.6.4, we have

$$\frac{1}{2\pi} \int_0^{2\pi} |f(re^{i\theta})|^p \, d\theta = \frac{p^2}{2} \int_{r\mathbb{D}} |f(w)|^{p-2} |f'(w)|^2 \log \frac{r}{|w|} \, dA(w).$$

We multiply both sides by $2r \left(\log \frac{1}{r} \right)^\gamma$, integrate with respect to r from 0 to 1, and then apply Fubini's theorem to obtain

$$\int_{\mathbb{D}} |f(w)|^p \, dA_\gamma(w)$$

$$= \frac{p^2}{2} \int_0^1 \left(\int_{\mathbb{D}} \chi_{r\mathbb{D}}(w) |f(w)|^{p-2} |f'(w)|^2 \log \frac{r}{|w|} \, dA(w) \right) 2r \left(\log \frac{1}{r} \right)^\gamma dr$$

$$= \frac{p^2}{2} \int_{\mathbb{D}} |f(w)|^{p-2} |f'(w)|^2 \left(\int_{|w|}^1 2r \left(\log \frac{1}{r} \right)^\gamma \log \frac{r}{|w|} \, dr \right) dA(w).$$

For $1/2 \le |w| \le 1$, the inner integral with respect to r is comparable to $(\log 1/r)^{\gamma+2}$. For $0 \le w \le 1/2$, there are a finite number of singularities, and the integrand is integrable. Thus, the inner integral can be replaced by $(\log 1/r)^{\gamma+2}$ without changing the space of functions, for which it is finite, and an equivalent norm results. □

By Lemma 3.3.1 and formula in Corollary 2.6.3 with $\varphi(z) \equiv z$, it is natural to extend the definition of A_γ^p and define A_{-1}^p to be the Hardy space H^p. Then we can write

$$\|f\|_{A_\gamma^p}^p \asymp |f(0)|^p + \int_{\mathbb{D}} |f(w)|^{p-2} |f'(w)|^2 \, dA_{\gamma+2}(w), \qquad (\gamma \ge -1), \qquad (3.1)$$

and so we are able to provide a unified treatment of the Hardy spaces and the weighted Bergman spaces.

Theorem 3.3.2 *Let φ be a holomorphic self-map of $\mathbb{D}$, and let f be holomorphic on $\mathbb{D}$. Then, for $\gamma \geq -1$, we have*

$$\|f \circ \varphi\|_{A_\gamma^p}^p \asymp |f(\varphi(0))|^p + \int_{\mathbb{D}} |f|^{p-2}|f'|^2 N_{\varphi,\gamma+2}\, dA.$$

Proof This is an immediate consequence of the norm estimate (3.1) and the change of variable formula, which was studied in Lemma 2.4.1. □

3.4 $\mathcal{N}_p$ Versus A^{-1}

Some basic Banach space or Fréchet space properties of the $\mathcal{N}_p$-spaces are discussed below. In particular, it will be showed that, for $p \in (0, 1)$, the $\mathcal{N}_p$-spaces are all different topological vector spaces with independent interest. It will also be justified that the $\mathcal{N}_p$-spaces form a much bigger class of functions than the $\mathcal{Q}_p$-spaces.

Proposition 3.4.1 *(i) For $p \in (0, \infty)$, we have $\| \cdot \|_{A^{-1}} \lesssim \| \cdot \|_{\mathcal{N}_p}$. That is, the inclusion $\mathcal{N}_p \hookrightarrow A^{-1}$ is well-defined and bounded.*
(ii) For $p \in (1, \infty)$, we have $\| \cdot \|_{A^{-1}} \asymp \| \cdot \|_{\mathcal{N}_p}$. That is, we have the Banach space equivalence $\mathcal{N}_p = A^{-1}$.

Proof To prove part (i), first note that, for each $w \in \mathbb{D}$,

$$f(w) = f \circ \sigma_w(0).$$

Then, by subharmonicity of $|f \circ \sigma_w(\cdot)|^2$, we have

$$|f(w)|^2 = |(f \circ \sigma_w)(0)|^2$$
$$\leq \text{const} \cdot \int_{\{z:|\sigma_0(z)|<1/\sqrt{2}\}} |f \circ \sigma_w(z)|^2\, dA(z).$$

Applying the change variable $u = \sigma_w(z)$ implies

$$|f(w)|^2 \leq \text{const} \cdot \int_{\{z:|\sigma_w(z)|<1/\sqrt{2}\}} |f(u)|^2 \cdot \frac{(1-|u|^2)^2}{|1-\langle w, u\rangle|^4}\, dV(u)$$
$$= \text{const} \cdot \int_{\{z:|\sigma_w(z)|<1/\sqrt{2}\}} |f(u)|^2 \cdot \frac{(1-|\sigma_w(u)|^2)^2}{(1-|w|^2)^2}\, dV(u).$$

Here, we also used the identity

$$1 - |\sigma_w(u)|^2 = \frac{(1 - |w|^2)(1 - |u|^2)}{|1 - \langle w, u \rangle|^2}.$$

Consequently, for each $p > 0$, we have

$$(1 - |w|^2)^2 |f(w)|^2 = \text{const} \cdot \int_{\{z:|\sigma_w(z)|<1/\sqrt{2}\}} |f(u)|^2 (1 - |\sigma_w(u)|^2)^2 dV(u)$$

$$\leq \text{const} \cdot \int_{\{z:|\sigma_w(z)|<1/\sqrt{2}\}} |f(u)|^2 (1 - |\sigma_w(u)|^2)^p dV(u)$$

$$\leq \text{const} \cdot \int_{\mathbb{D}} |f(z)|^2 (1 - |\sigma_w(z)|^2)^p dV(z).$$

In the second line, we could replace 2 by p, since $1 - |\sigma_w(u)|^2 \in (\frac{1}{2}, 1)$. Finally, by taking the supremum over all $w \in \mathbb{D}$, the desired result is obtained.

To establish part (ii), we note that the inequality $\| \cdot \|_{A^{-1}} \leq \text{const} \cdot \| \cdot \|_{\mathcal{N}_p}$ is already showed in part (i). Hence, we just need to verify the other direction. We have

$$\|f\|_p^2 = \sup_{a \in \mathbb{D}} \int_{\mathbb{D}} |f(z)|^2 (1 - |\sigma_a(z)|^2)^p \, dA(z)$$

$$= \sup_{a \in \mathbb{D}} \int_{\mathbb{D}} |f(z)|^2 \cdot \frac{(1 - |z|^2)^p (1 - |a|^2)^p}{|1 - \langle a, z \rangle|^{2p}} \, dA(z)$$

$$= \sup_{a \in \mathbb{D}} (1 - |a|^2)^p \int_{\mathbb{D}} |f(z)|^2 (1 - |z|^2)^2 \cdot \frac{(1 - |z|^2)^{p-2}}{|1 - \langle a, z \rangle|^{2p}} \, dA(z)$$

$$\leq \|f\|_{A^{-1}}^2 \cdot \sup_{a \in \mathbb{D}} (1 - |a|^2)^p \int_{\mathbb{D}} \frac{(1 - |z|^2)^{p-2}}{|1 - \langle a, z \rangle|^{2p}} \, dA(z).$$

Since $p > 1$, by Lemma 1.8.1, for any $a \in \mathbb{D}$, we have

$$\int_{\mathbb{D}} \frac{(1 - |z|^2)^{p-2}}{|1 - \langle a, z \rangle|^{2p}} \, dA(z) = \int_{\mathbb{D}} \frac{(1 - |z|^2)^{p-2}}{|1 - \langle a, z \rangle|^{2+(p-2)+p}} \, dA(z)$$

$$= \text{const} \cdot (1 - |a|^2)^{-p}.$$

Combining both inequalities yields

$$\|f\|_p^2 \leq \text{const} \|f\|_{A^{-1}}^2 \sup_{a \in \mathbb{D}} (1 - |a|^2)^p \cdot \frac{1}{(1 - |a|^2)^p}$$

$$= \text{const} \|f\|_{A^{-1}}^2.$$

$\square$

Proposition 3.4.2 *Let $0 < p < \infty$. Then the $\mathcal{N}_p$-space, endowed with the norm $\| \cdot \|_{\mathcal{N}_p}$, is a functional Banach space, and the norm topology of $\mathcal{N}_p$ is finer than the compact-open topology.*

Proof It is easy to see that $\| \cdot \|_{\mathcal{N}_p} \geq 0$ and $\|f\|_{\mathcal{N}_p} = 0 \Longleftrightarrow f = 0$. Moreover, $\lambda \| \cdot \|_{\mathcal{N}_p} = |\lambda| \| \cdot \|_{\mathcal{N}_p}$. To prove that $\| \cdot \|_p$ is a norm, it remains to verify the triangle inequality. Fix $a \in \mathbb{D}$. By the ordinary triangle inequality and by Hölder's inequality, we have

$$\int_{\mathbb{D}} |f(z) + g(z)|^2 (1 - |\sigma_a(z)|^2)^p \, dA(z)$$

$$\leq \|f\|_p^2 + \|g\|_p^2 + 2 \int_{\mathbb{D}} |f(z)g(z)|(1 - |\sigma_a(z)|^2)^p \, dA(z)$$

$$\leq \|f\|_p^2 + \|g\|_p^2 + 2 \left(\int_{\mathbb{D}} |f|^2 (1 - |\sigma_a(z)|^2)^p \, dA(z) \right)^{1/2}$$

$$\times \left(\int_{\mathbb{D}} |g|^2 (1 - |\sigma_a(z)|^2)^p \, dA(z) \right)^{1/2}$$

$$= \|f\|_p^2 + \|g\|_p^2 + 2\|f\|_p \|g\|_p = (\|f\|_p + \|g\|_p)^2.$$

Hence, taking the supremum with respect to $a \in \mathbb{D}$ gives

$$\|f + g\|_p^2 = \sup_{a \in \mathbb{D}} \int_{\mathbb{D}} |f(z) + g(z)|^2 (1 - |\sigma_a(z)|^2)^p \, dA(z)$$

$$\leq \|f\|_p^2 + \|g\|_p^2 + 2\|f\|_p \|g\|_p = (\|f\|_p + \|g\|_p)^2.$$

Thus, $\| \cdot \|_p$ is a norm. Since the norm topology of $A^{-1}(\mathbb{D})$ is clearly stronger than the compact-open topology, the second statement follows from the fact

$$\|f\|_{A^{-1}} \leq C\|f\|_p, \qquad (f \in \mathcal{N}_p(\mathbb{D})),$$

which means $\mathcal{N}_p(\mathbb{D}) \hookrightarrow A^{-1}(\mathbb{D})$.

Next, we prove the completeness of $\mathcal{N}_p(\mathbb{D})$. Let $\{f_m\}$ be a Cauchy sequence in $\mathcal{N}_p(\mathbb{D})$ with respect to the norm $\| \cdot \|_p$. From this, it follows that $\{f_m\}$ is a Cauchy sequence in the space $\text{Hol}(\mathbb{D})$, and hence it converges to some $f \in \text{Hol}(\mathbb{D})$. We need to show that $f \in \mathcal{N}_p(\mathbb{D})$. Indeed, there exists a $\ell_0 \in \mathbb{N}$ such that for all $m, \ell \geq \ell_0$, we have $\|f_m - f_\ell\|_p \leq 1$. Fix an arbitrary $a \in \mathbb{D}$. Then, by Fatou's lemma, we have

$$\int_{\mathbb{D}} |f(z) - f_{\ell_0}(z)|^2 (1 - |\sigma_a(z)|^2)^p \, dA(z)$$

$$\leq \lim_{\ell \to \infty} \int_{\mathbb{D}} |f_\ell(z) - f_{\ell_0}(z)|^2 (1 - |\sigma_a(z)|^2)^p \, dA(z)$$

$$\leq \lim_{\ell \to \infty} \|f_\ell - f_{\ell_o}\|_p^2 \leq 1,$$

which implies

$$\|f - f_{\ell_o}\|_p = \sup_{a \in \mathbb{D}} \int_{\mathbb{D}} |f(z) - f_{\ell_o}(z)|^2 (1 - |\sigma_a(z)|^2)^p \, dA(z) \leq 1,$$

and hence $\|f\|_p \leq 1 + \|f_{\ell_o}\|_p < \infty$.

To show that $f_n \to f$ in the norm, we note that the first part shows a pointwise convergence $f_n \to f$. Since the norm topology is stronger than the pointwise convergence topology, applying Fatou's lemma gives the desired result.

Finally, we can check that $\mathcal{N}_p(\mathbb{D})$ is a functional Banach space. This means that we show that for each $z \in \mathbb{D}$, the point evaluation $f \mapsto f(z)$ is continuous on $\mathcal{N}_p(\mathbb{D})$. We have

$$|f(z)| \leq \frac{2^{p+1}}{3^{p/2}(1 - |z|^2)} \|f\|_p, \qquad (f \in \mathcal{N}_p(\mathbb{D})),$$

which means that the point evaluation is bounded on $\mathcal{N}_p(\mathbb{D})$, and hence it is continuous.

$\square$

3.5 $\mathcal{N}_p$ Elements from the Hadamard Gap Class

The holomorphic function

$$f(z) = \sum_{k=0}^{\infty} a_k z^{n_k},$$

with $n_k \in \mathbb{N}$ for all $k \in \mathbb{N}$, is said to belong to the *Hadamard gap class* (also known as lacunary series) if there exists a constant $c > 1$ such that

$$n_{k+1}/n_k \geq c$$

for all $k \geq 1$. To distinguish elements of $\mathcal{N}_p$ which stem from the Hadamard gap class, we need the following simple but technical result.

Theorem 3.5.1 *Let*

$$f(x) = \sum_{n=1}^{\infty} a_n x^n, \qquad (0 \leq x \leq 1),$$

with $a_n \geq 0$. Let

$$I_n = \{k : 2^n \leq k < 2^{n+1}, k \in \mathbb{N}\}$$

and assume that $t_n = \sum_{k \in I_n} a_k$, $n \geq 0$. Then, for each $c > 0$ and $p > 0$, there exists a constant K depending only on c and p, such that

$$\frac{1}{K} \sum_{n=0}^{\infty} 2^{-nc} t_n^p \leq \int_0^1 (1-x)^{c-1} f(x)^p \, dx \leq K \sum_{n=0}^{\infty} 2^{-nc} t_n^p.$$

Theorem 3.5.1 can be directly proved by using Jensen's inequality. Moreover, it is an immediate consequence of the following more general result.

Theorem 3.5.2 *Let F be a real-valued nonnegative function satisfying the following conditions.*

(i) F is defined on the interval $[0, \infty)$, and $F(0) = 0$.

(ii) For some $p \geq q > 0$, the function $\dfrac{F(t)}{t^p}$ is nonincreasing, while the function $\dfrac{F(t)}{t^q}$ is nondecreasing.

Let $(a_k)_{k \geq 1}$ be a sequence of nonnegative real numbers and $c > 0$. Then

$$\int_0^1 F\left(\sum_{k=1}^{\infty} a_k r^k (1-r)^{c-1} \, dr\right) < \infty \tag{3.2}$$

if and only if

$$\sum_{n=0}^{\infty} 2^{-2nc} F\left(\sum_{I_n} a_k\right) < \infty. \tag{3.3}$$

Proof First, for a sequence $(t_n)_{n \geq 1}$ of nonnegative numbers, we prove that the following inequalities hold.

$$\theta^p F(t) \leq F(\theta t) \leq \theta^q F(t), \text{ for } 0 \leq \theta \leq 1 \text{ and } t \geq 0. \tag{3.4}$$

$$F\left(\sum_{n=1}^{\infty} t_n\right) \leq \left(\sum_{n=1}^{\infty} F(t_n)^{1/p}\right)^p. \tag{3.5}$$

$$F\left(\sum_{n=1}^{\infty} t_n\right) \leq \sum_{n=1}^{\infty} F(t_n), \text{ for } 0 < p \leq 1. \tag{3.6}$$

Indeed, (3.4) follows from the assumption that $\dfrac{F(t)}{t^p}$ is nonincreasing, while $\dfrac{F(t)}{t^q}$ is nondecreasing.

For (3.5) we note, by (3.4), that F is continuous, and so we may assume that $t_n = 0$ for $n \geq 3$. Furthermore, since $F(t)^{1/p}/t$ is nonincreasing, $F(t)^{1/p}$ is subadditive from which (3.5) follows. Inequality (3.6) is derived from (3.5) and elementary inequalities.

Next, we prove the following inequality.

$$\sum_{n=0}^{\infty} 2^{nc} r^{2^n} \leq 2\Gamma(c)|\log r|^{-c}, \ 0 < c \leq 1, \ 0 \leq r < 1. \tag{3.7}$$

To do so, we have

$$2\Gamma(c)|\log r|^{-c} = 2^{1-c} \int_0^{\infty} t^{c-1} r^{t/2} \, dt \geq 2^{1-c} \sum_{k=1}^{\infty} k^{c-1} r^{k/2}$$

$$\geq 2^{1-c} \sum_{n=0}^{\infty} 2^n 2^{(n+1)(c-1)} r^{2^n}.$$

Now we are ready to prove the theorem.

- *Necessity.* Suppose that (3.2) holds. For every $n \geq 0$, we put $t_n = \sum_{I_n} a_k$ and $r_n = 1 - 2^{-n}$. Then we have

$$\int_0^1 F\left(\sum_{n=1}^{\infty} a_n r^n\right)(1-r)^{c-1} \, dr \geq \int_{1/2}^1 F\left(\sum_{k=0}^{\infty} t_k r^{2^{k+1}-1}\right)(1-r)^{c-1} \, dr$$

$$\geq \sum_{n=0}^{\infty} F\left(\sum_{k=0}^{n} t_k r_{n+1}^{2^{k+1}-1}\right) \int_{r_{n+1}}^{r_{n+2}} (1-r)^{c-1} \, dr$$

$$\geq \sum_{n=0}^{\infty} F\left(e^{-1} \sum_{k=0}^{n} t_k\right) 2^{-(n+2)c} \log 2$$

$$\geq K_1 \sum_{n=0}^{\infty} F\left(\sum_{k=0}^{n} t_k\right) 2^{-nc},$$

where $K_1 = e^{-p} 2^{-2c} \log 2$. Hence,

$$\int_0^1 F\left(\sum_{n=1}^{\infty} a_n r^n\right)(1-r)^{c-1} \, dr \geq K_1 \sum_{n=0}^{\infty} 2^{-nc} F(S_{2^n}).$$

- *Sufficiency.* Suppose that (3.3) holds. We consider three cases for p.

- For case $p > 1$, due to inequality (3.5), we can take $F(t) = t^p$. Put $c = \min\{1, c/p\}$ and $s(r) = \sum\limits_{n=0}^{\infty} 2^{nc} r^{2n}$ with $0 \leq r < 1$. By Jensen's inequality, we have

$$\left(\sum_{k=1}^{\infty} a_k r^k\right)^p \leq \left(\sum_{n=0}^{\infty} t_n r^{2n}\right)^p \leq s(r)^{p-1} \sum_{n=0}^{\infty} 2^{nc} r^{2n} 2^{-ncp} |t_n|^p.$$

Combining inequality (3.7), the inequality

$$r(1-r)^{c-1} \leq |\log r|^{c-1}, \ 0 \leq r < 1,$$

and the inequality

$$\int_0^1 |\log r|^{c-1} r^{2n-1} \, dr = \Gamma(c) 2^{-nc},$$

yields the desired result.
    Indeed, a direct computation shows that

$$\int_0^1 \left(\sum_{k=1}^{\infty} a_k r^k\right)^p (1-r)^{c-1} \, dr \leq K_2 \sum_{n=0}^{\infty} 2^{-nc} \left(\sum_{I_n} a_k\right)^p,$$

where (in the case $p \geq c$) $K_2 = 2^{p-1} \Gamma(c)^p$.

- For case $p = 1$, the proof is similar, but it is much simpler, because the function $s(r)$ is not needed.
- For case $0 < p < 1$, the proof is similar to $p = 1$, with the use of inequality (3.6).

$\square$

Theorem 3.5.3 *Let $f(z) = \sum_{k=0}^{\infty} b_k z^{n_k}$ be in the Hadamard gap class, and let $p \in (0, 1]$. Then $f \in \mathcal{N}_p$ if and only if*

$$\sum_{k=0}^{\infty} \frac{1}{2^{k(1+p)}} \left(\sum_{2^k \leq n_j < 2^{k+1}} |b_j|^2\right) < \infty.$$

Proof Suppose $f \in \mathcal{N}_p$. By Parseval and Stirling formulas, we have

$$\|f\|_{\mathcal{N}_p}^2 \gtrsim \int_0^1 \int_0^{2\pi} \left|\sum_{k=0}^{\infty} b_k r^{n_k} (e^{i\theta})^{n_k}\right|^2 d\theta (1-r)^p r \, dr$$

$$\asymp \sum_{k=0}^{\infty} |b_k|^2 \int_0^1 r^{2n_k+1}(1-r)^p \, dp$$

$$= \sum_{k=0}^{\infty} \frac{\Gamma(2n_k+2)\Gamma(p+1)}{\Gamma(2n_k+p+3)} |b_k|^2$$

$$\gtrsim \sum_{k=0}^{\infty} n_k^{-(1+p)} |b_k|^2.$$

This estimation implies that

$$\sum_{k=0}^{\infty} \frac{1}{2^{k(1+p)}} \left(\sum_{2^k \le n_j < 2^{k+1}} |b_j|^2 \right) \gtrsim \sum_{k=0}^{\infty} \sum_{2^k \le n_j < 2^{k+1}} n_j^{-(1+p)} |b_j|^2$$

$$\gtrsim \|f\|_{\mathcal{N}_p}^2 < \infty.$$

Conversely, suppose that

$$\sum_{k=0}^{\infty} \frac{1}{2^{k(1+p)}} \left(\sum_{2^k \le n_j < 2^{k+1}} |b_j|^2 \right) < \infty.$$

By Hölder inequality, we have

$$\|f\|_{\mathcal{N}_p}^2 \lesssim \sup_{a \in \mathbb{D}}(1-|a|^2)^p \int_0^1 \left(\sum_{k=0}^{\infty} |b_k| r^{n_k} \right)^2 (1-r)^p \left(\int_0^{2\pi} \frac{1}{|1-\bar{a}re^{i\theta}|} \, d\theta \right)^p dr$$

$$= \sup_{a \in \mathbb{D}}(1-|a|^2)^p \int_0^1 \left(\sum_{k=0}^{\infty} |b_k| r^{n_k} \right)^2 (1-r)^p \left(\frac{2\pi}{1-|a|^2 r^2} \right)^p dr$$

$$\lesssim \int_0^1 \left(\sum_{k=0}^{\infty} |b_k| r^{n_k} \right)^2 (1-r)^p \, dr.$$

By Theorem 3.5.1, if $\alpha, \beta > 0$, and $\alpha_k \ge 0$ for all $k \in \mathbb{N}$, then

$$\int_0^1 \left(\sum_{k=0}^{\infty} a_k r^k \right)^{\beta} (1-r)^{\alpha-1} \asymp \sum_{k=0}^{\infty} 2^{-k\alpha} \left(\sum_{2^k \le n_j < 2^{k+1}} |b_j| \right)^2,$$

where the constants depend only on α and β. Applying this estimate to the last estimation above, we obtain

$$\|f\|_{\mathcal{N}_p}^2 \lesssim \sum_{k=0}^{\infty} \frac{1}{2^{k(p+1)}} \left(\sum_{2^k \leq n_j < 2^{k+1}} |b_j| \right)^2.$$

Since f is in the Hadamard gap class, there exists a constant $c > 1$ such that $n_{j+1} \leq cn_j$ for all $j \in \mathbb{N}$. Then the maximum number of n_j's between 2^k and 2^{k+1} for any $k \in \mathbb{N}$ is the integer part of $\log_c 2 + 1$. This together with the fact

$$(a_1 + \cdots + a_n)^2 \leq n(a_1^2 + \cdots + a_n^2)$$

implies

$$\|f\|_{\mathcal{N}_p}^2 \lesssim \sum_{k=0}^{\infty} \frac{1}{2^{k(p+1)}} \left(\sum_{2^k \leq n_j < 2^{k+1}} |b_j| \right)^2 < \infty.$$

□

Theorem 3.5.4 *Let* $f(z) = \sum_{k=0}^{\infty} b_k z^{n_k}$ *be in the Hadamard gap class, and let* $p \in (1, \infty)$. *Then* $f \in \mathcal{N}_p$ *if and only if*

$$\sup_{k \geq 0} \frac{|b_k|}{n_k} < \infty.$$

Proof Suppose $f \in \mathcal{N}_p = A^{-1}$. By Cauchy integral formula, we have

$$|b_k| \lesssim \int_{r\mathbb{T}} \frac{|f(z)|}{|z|^{n_k+1}} |dz| \lesssim \frac{\|f\|_{A^{-1}}}{r^{n_k}(1-r)}, \qquad (0 < r < 1).$$

Without loss of generality, we may assume that $n_0 \geq 2$. Then choosing the optimal radius $r = 1 - \frac{1}{n_k}$, we obtain

$$\sup_k \frac{|b_k|}{n_k} \lesssim \sup_k \frac{\|f\|_{A^{-1}}}{(1 - \frac{1}{n_k})^{n_k}} \leq \frac{\|f\|_{A^{-1}}}{(1 - \frac{1}{2})^2} \lesssim \|f\|_{A^{-1}}.$$

Conversely, suppose that $\sup_k \frac{|b_k|}{n_k} < \infty$. We have

$$|f(z)| \leq \sum_{k=0}^{\infty} |b_k||z|^{n_k} \lesssim \sum_{k=0}^{\infty} n_k |z|^{n_k}.$$

Then

$$\frac{|f(z)|}{1-|z|} \lesssim \left(\sum_{k=0}^{\infty} n_k |z|^{n_k}\right)\left(\sum_{k=0}^{\infty} |z|^k\right)$$

$$= \sum_{k=0}^{\infty}\sum_{j=0}^{k} n_j |z|^{n_j+k-j}$$

$$= \sum_{k=0}^{\infty}\left(\sum_{n_j\le k} n_j\right)|z|^k.$$

Since f is in the Hadamard gap class, there exists a constant $c > 1$ such that $n_{j+1} \ge cn_j$ for all $j \ge 1$. A straightforward calculation shows that

$$\frac{1}{k}\sum_{n_j\le k} n_j \le \frac{c}{c-1}.$$

Consequently,

$$\frac{|f(z)|}{1-|z|} \lesssim \sum_{k=0}^{\infty} k|z|^k \le \frac{1}{(1-|z|)^2},$$

from which it follows that $f \in \mathcal{N}_p = A^{-1}$.   □

Corollary 3.5.5 *For $0 < p_1 < p_2 \le 1$, we have*

$$\mathcal{B} \subsetneq \mathcal{N}_{p_1} \subsetneq \mathcal{N}_{p_2} \subsetneq A^{-1}.$$

Proof For $f \in \mathcal{B}$, we have the well-known estimation

$$|f(z)| \lesssim \log\left(\frac{2}{1-|z|}\right)\|f\|_{\mathcal{B}}, \quad z \in \mathbb{D}.$$

It follows that $\|\cdot\|_{\mathcal{N}_p} \lesssim \|\cdot\|_{\mathcal{B}}$ for all $p \in (0,\infty)$. The other inclusions are obvious from the definition of the $\mathcal{N}_p$-spaces and Proposition 3.4.1 (i). So it remains to show that the inclusions are strict.

For this, consider the following three test functions:

$$f_1(z) = \sum_{k=0}^{\infty} 2^k z^{2^k}, \quad f_2(z) = \sum_{k=0}^{\infty} 2^{\frac{k(1+p_1)}{2}} z^{2^k}, \quad f_3(z) = \sum_{k=0}^{\infty} \left(\frac{3}{2}\right)^{\frac{k}{2}} z^{2^k}.$$

By Theorems 3.5.3 and 3.5.4 and the fact

$$f \in \mathcal{B} \quad \text{if and only if} \quad \sup_k |a_k| < \infty, \tag{3.8}$$

we deduce that

$$f_1 \in A^{-1} \setminus \mathcal{N}_{p_2}, \quad f_2 \in \mathcal{N}_{p_2} \setminus \mathcal{N}_{p_1}, \quad f_3 \in \mathcal{N}_{p_1} \setminus \mathcal{B}.$$

$\square$

Notes on Chapter 3

The $\mathcal{Q}_p$-spaces were introduced by Aulaskari, Xiao, and Zhao [7]. The background on these spaces can be found in the book [102]. The $\mathcal{Q}_p$-spaces have been studied intensively for more than two decades, and due to both the similarities and the differences between $\mathcal{B}$ and A^{-1}, it is natural to also study the $\mathcal{N}_p$-spaces. In fact, $\mathcal{N}_1$ was already of significant help in [61], where the authors studied the spectra of composition operators on BMOA. The $\mathcal{N}_p$-spaces were also informally used in [74] to obtain a characterization of a special branch of weighted composition operators with closed range on A^{-q}. These Banach spaces have appeared in many works, e.g., see [40] and the related references therein. The generalized Nevanlinna counting function is introduced by Shapiro in [82]. In this work, the author used the classical and the generalized Nevanlinna counting functions to study composition operators on H^2 and on weighted Bergman spaces. Lemma 3.3.1, Theorem 3.3.2, and Lemma 2.5.2 are proved in [82, 86], where one can find more detailed information. Lemma 1.8.1 is taken from [110], while Lemma 3.5.1 is proved in [68], and Theorem 3.5.2 is given in [68]. The estimate 3.8 is provided in [102]. Recently, in [55], a simplification to the formula in Theorem (3.5.3) has been discovered. Namely, the condition can be replaced by

$$\sum_{k=0}^{\infty} \frac{|b_k|^2}{n_k^{p+1}} < \infty.$$

The α-Bloch Spaces

4

In this chapter, we mainly study the α-Bloch spaces, some operators acting on them, and their dual. We also introduce the space H_α^∞. These spaces play a decisive role in studying the weighted composition operators on $\mathcal{N}_p$ spaces.

4.1 The Spaces $\mathcal{B}_\alpha$ and $\mathcal{B}_{\alpha,0}$

Let $\alpha \in (0, \infty)$. Recall the definition of Bloch space $\mathcal{B}_\alpha$, which we also denoted by $\mathcal{B}^p$, from Sect. 1.1. To be more precise, for $\alpha \in (0, \infty)$, we call

$$\mathcal{B}_\alpha = \left\{ f \in \mathrm{Hol}(\mathbb{D}) : \sup_{z \in \mathbb{D}} (1 - |z|^2)^\alpha |f'(z)| < \infty \right\}$$

the α-*Bloch space*. We also put

$$\mathcal{B}_{\alpha,0} = \left\{ f \in \mathcal{B}_\alpha : \lim_{|z| \to 1} (1 - |z|^2)^\alpha |f'(z)| = 0 \right\}$$

and call it the *little α-Bloch space*. The classical Bloch space is $\mathcal{B} := \mathcal{B}_1$. The following are well-known basic properties of the α-Bloch spaces.

Proposition 4.1.1 *(i) The α-Block space $\mathcal{B}_\alpha$ is a Banach space with the norm*

$$\|f\|_\alpha = |f(0)| + \sup_{z \in \mathbb{D}} (1 - |z|^2)^\alpha |f'(z)|.$$

© The Author(s), under exclusive license to Springer Nature Switzerland AG 2023
L. H. Khoi, J. Mashreghi, *Theory of N_p Spaces*, Frontiers in Mathematics,
https://doi.org/10.1007/978-3-031-39704-2_4

(ii) *The little α-Bloch space $\mathcal{B}_{\alpha,0}$ is the closure of the set of polynomials in the norm topology of $\mathcal{B}_\alpha$. In particular, $\mathcal{B}_{\alpha,0}$ itself is a separable Banach space.*

(iii) *For any $\alpha > -1$ and $z \in \mathbb{D}$, if f is a holomorphic function on $\mathbb{D}$ with*

$$\int_\mathbb{D} (1 - |z|^2)^\alpha |f(z)| \, dA(z) < \infty,$$

then

$$f(z) = (\alpha + 1) \int_\mathbb{D} \frac{(1 - |w|^2)^\alpha f(w)}{(1 - z\overline{w})^{2+\alpha}} \, dA(w).$$

(iv) *For each $\alpha > 0$, $z \in \mathbb{D}$, and $f \in \mathcal{B}_\alpha$, we have*

$$f(z) = f(0) + \int_\mathbb{D} \frac{(1 - |w|^2)^\alpha f'(w)}{\overline{w}(1 - z\overline{w})^{1+\alpha}} \, dA(w).$$

(v) *For each nonnegative integer n and each compact set K contained in $\mathbb{D}$, there exists a constant $C > 0$ (depending on only α), such that*

$$\sup_{z \in K} |f^{(n)}(z)| \le C \|f\|_\alpha$$

for all $f \in \mathcal{B}_\alpha$.

From basic properties of Bloch spaces, several interesting characterization of Bloch spaces can be deduced. These properties are also needed in the sequel.

Proposition 4.1.2 *Let $\alpha > 1$. Then:*

(i) *$f \in \mathcal{B}_\alpha$ if and only if $(1 - |z|^2)^{\alpha-1} f(z)$ is bounded on $\mathbb{D}$.*
(ii) *$f \in \mathcal{B}_{\alpha,0}$ if and only if $(1 - |z|^2)^{\alpha-1} f(z) \to 0$ as $|z| \to 1^-$.*

Proof (i) Suppose $f \in \mathcal{B}_\alpha$. By Proposition 4.1.1 (iv), we have

$$f(z) = f(0) + \int_\mathbb{D} \frac{(1 - |w|^2)^\alpha f'(w)}{\overline{w}(1 - z\overline{w})^{\alpha+1}} \, dA(w), \qquad (z \in \mathbb{D}).$$

Hence,

$$|f(z) - f(0)| \le \|f\|_\alpha \int_\mathbb{D} \frac{dA(w}{|w||1 - z\overline{w}|^{\alpha+1}}, \qquad (z \in \mathbb{D}).$$

Since the factor $|w|$ in the denominator does not change the growth rate of the integral for z near the boundary, by Lemma 1.8.1, there exists a constant $C > 0$ such that

$$|f(z) - f(0)| \le C\|f\|_\alpha(1 - |z|^2)^{-(\alpha-1)}, \qquad (z \in \mathbb{D}),$$

which shows that $(1 - |z|^2)^{\alpha-1}f(z)$ is bounded on $\mathbb{D}$.

Conversely, suppose that $(1 - |z|^2)^{\alpha-1}f(z) \le M$ on $\mathbb{D}$ for some constant $M > 0$. By Proposition 4.1.1 (iii), we have

$$f(z) = \alpha \int_\mathbb{D} \frac{(1 - |w|^2)^{\alpha-1}f(w)}{(1 - z\overline{w})^{\alpha+1}} \, dA(w), \qquad (z \in \mathbb{D}).$$

Differentiating under the integral sign, we obtain

$$f'(z) = \alpha(\alpha + 1) \int_\mathbb{D} \frac{\overline{w}(1 - |w|^2)^{\alpha-1}f(w)}{(1 - z\overline{w})^{\alpha+2}} \, dA(w), \qquad (z \in \mathbb{D}).$$

Again by Lemma 1.8.1, there exists a constant $C > 0$ such that

$$|f'(z)| \le \alpha(\alpha + 1)M \int_\mathbb{D} \frac{dA(w)}{(1 - z\overline{w})^{\alpha+2}} \le CM(1 - |z|^2)^{-\alpha}, \qquad (z \in \mathbb{D}),$$

which shows that $f \in \mathcal{B}_\alpha$.

(ii) Analyzing the proof of (i), we see that for $\alpha > 1$, the norm $\| \cdot \|_\alpha$ on $\mathcal{B}_\alpha$ is actually equivalent to the norm

$$\|f\| = \sup_{z \in \mathbb{D}}(1 - |z|^2)^{\alpha-1}|f(z)|.$$

Note that, on the one hand, $\mathcal{B}_{\alpha,0}$ is the closure of the polynomials in $\mathcal{B}_\alpha$ under the norm $\| \cdot \|_\alpha$ and, on the other hand, the closure of the polynomials in $\mathcal{B}_\alpha$ under the norm $\| \cdot \|$ consists of holomorphic functions f with $(1 - |z|^2)^{\alpha-1}f(z) \to 0$ as $|z| \to 1^-$. Then we can conclude that $f \in \mathcal{B}_{\alpha,0}$ if and only if $(1-|z|^2)^{\alpha-1}f(z) \to 0$ as $|z| \to 1^-$.

The proposition is proved completely.

$\square$

Proposition 4.1.3 *Let $\alpha > 0$ and $n > 1$. Then:*

(i) A holomorphic function f on $\mathbb{D}$ is in $\mathcal{B}_\alpha$ if and only if

$$\sup_{z\in\mathbb{D}}\{(1-|z|^2)^{\alpha+n-1}|f^{(n)}(z)|\} < \infty.$$

(ii) A holomorphic function f on $\mathbb{D}$ is in $\mathcal{B}_\alpha$ if and only is

$$\sup_{z\in\mathbb{D}}\{(1-|z|^2)^{\alpha+n-1}|f^{(n)}(z)|\} < \infty.$$

Proof - Suppose $f \in \mathcal{B}_\alpha$. Then by Proposition 4.1.1 (iv),

$$f(z) = f(0) + \int_\mathbb{D} \frac{(1-|w|^2)^\alpha f'(w)}{\overline{w}(1-z\overline{w})^{1+\alpha}} dA(w), \qquad (z \in \mathbb{D}).$$

Differentiating under the integral sign n times and applying Lemma 1.8.1, we obtain

$$\sup_{z\in\mathbb{D}}\{1-|z|^2)^{\alpha+n-1}|f^{(n)}(z)|\} < \infty.$$

- Conversely, since subtracting a polynomial from f neither alters the assumption nor does the conclusion, without loss of generality, we may assume that the first $n+1$ Taylor coefficients of f all vanish. Then the function

$$h(z) = \frac{(\alpha+2)(1-|z|^2)^{\alpha+n-1} f^{(n)}(z)}{(\alpha+n)\cdot(\alpha+2)\overline{z}^n}$$

is bounded on $\mathbb{D}$. By Proposition 4.1.1 (iii),

$$f^{(n)}(z) = (\alpha+n) \int_\mathbb{D} \frac{(1-|w|^2)^{\alpha+n-1} f^{(n)}(w)}{(1-z\overline{w})^{\alpha+n+1}} dA(w), \qquad (z \in \mathbb{D}).$$

Integrating from 0 to z yields

$$f^{(n-1)}(z) - f^{(n-1)}(0) = \int_\mathbb{D} \frac{(1-|w|^2)^{\alpha+n-1} f^{(n)}(w)}{\overline{w}} \left[\frac{1}{(1-z\overline{w})^{\alpha+n}} - 1\right] dA(w).$$

Since $f^{(n-1)}(0) = 0$ and

$$\int_\mathbb{D} \frac{(1-|w|^2)^{\alpha+n-1} f^{(n)}(w)}{\overline{w}} dA(w) = 0,$$

we obtain

$$f^{(n-1)}(z) = \int_{\mathbb{D}} \frac{(1-|w|^2)^{\alpha+n-1} f^{(n)}(w)}{\overline{w}(1-z\overline{w})^{\alpha+n}} \, dA(w), \qquad (z \in \mathbb{D}).$$

Repeating the argument above $n-1$ times yields

$$f'(z) = \int_{\mathbb{D}} \frac{h(w)}{(1-z\overline{w})^{\alpha+2}} \, dA(w), \qquad (z \in \mathbb{D}).$$

Now, by Lemma 1.8.1, there exists a constant $C > 0$ such that

$$|f'(z)| \le \|h\|_\infty \int_{\mathbb{D}} \frac{dA(w)}{(1-z\overline{w})^{\alpha+2}} \le C\|h\|_\infty (1-|z|^2)^{-\alpha}, \qquad (z \in \mathbb{D}),$$

which implies that $f \in \mathcal{B}_\alpha$.

Now analyzing the proof above shows that the norm $\|\cdot\|_\infty$ on $\mathcal{B}_\alpha$ is equivalent to the norm

$$\|f\| = |f(0)| + |f'(0)| + \cdot + |f^{(n-1)}(0)| + \sup_{z \in \mathbb{D}}\{(1-|z|^2)^{\alpha+n+1}|f^{(n)}(z)|\}.$$

Then the closure of the polynomials in $\mathcal{B}_\alpha$ under the norm above consists of holomorphic functions f with $(1-|z|^2)^{\alpha+n+1}|f^{(n)}(z)| \to 0$ as $|z| \to 1^-$. This implies that a holomorphic function f is in $\mathcal{B}_{\alpha,0}$ if and only if

$$(1-|z|^2)^{\alpha+n+1}|f^{(n)}(z)| \to 0$$

as $|z| \to 1^-$. $\square$

4.2   Some Operators on α-Bloch Spaces

Let $L^1(dA)$ denote the Bergman space of holomorphic functions f on $\mathbb{D}$ such that

$$\|f\| = \int_{\mathbb{D}} |f| \, dA < \infty.$$

As is well-known, the space $L^1(dA)$ is a Banach space with the norm above. Our purpose is to establish a duality relationship between $L^1(dA)$ and $\mathcal{B}_\alpha, \mathcal{B}_{\alpha,0}$. To do so, in then sequel, we use the following operators:

$$R_\alpha(f(z)) = \alpha \int_{\mathbb{D}} \frac{f(w)}{(1-z\overline{w})^{1+\alpha}} \, dA(w),$$

$$S_\alpha(f(z)) = 3(1 - |z|^2)^2 \int_{\mathbb{D}} \frac{f(w)}{(1 - z\overline{w})^4} (1 - |w|^2)^{\alpha-1} dA(w),$$

where $z \in \mathbb{D}$. Note that R_α is not a projection, as it does not give holomorphic functions unless $\alpha = 1$.

We use also the following subspaces of $L^\infty(\mathbb{D})$: the space $C_0(\mathbb{D})$ of complex-valued continuous functions on $\mathbb{D}$ which vanish on the boundary, the space $C(\overline{\mathbb{D}})$ of complex-valued continuous functions on the closed unit disc $\overline{\mathbb{D}}$, and the space $HC(\mathbb{D})$ of bounded complex-valued continuous functions on $\mathbb{D}$. The operators R_α and S_α help to find several useful relationships between spaces $L^\infty(\mathbb{D})$, $C_0(\mathbb{D})$, $C(\overline{\mathbb{D}})$, $HC(\mathbb{D})$, $\mathcal{B}_\alpha$, and $\mathcal{B}_{\alpha,0}$.

Proposition 4.2.1 *For each $\alpha > 0$, we have:*

(i) *R_α is a bounded linear operator from $L^\infty(\mathbb{D})$ onto $\mathcal{B}_\alpha$; R_α also maps $HC(\mathbb{D})$ onto $\mathcal{B}_\alpha$.*

(ii) *R_α is a bounded linear operator from $C(\overline{\mathbb{D}})$ onto $\mathcal{B}_{\alpha,0}$; R_α also maps $C_0(\mathbb{D})$ onto $\mathcal{B}_{\alpha,0}$.*

(iii) *There exists a constant $C > 0$ such that*

$$C^{-1}\|f\|_\alpha \le \inf\{\|g\|_\infty : f = R_\alpha g, g \in L^\infty(\mathbb{D})\} \le C\|f\|_\alpha$$

for all $f \in \mathcal{B}_\alpha$ and

$$C^{-1}\|f\|_\alpha \le \inf\{\|g\|_\infty : f = R_\alpha g, g \in C_0(\mathbb{D})\} \le C\|f\|_\alpha$$

for all $f \in \mathcal{B}_{\alpha,0}$.

Proof (i) First, take $g \in L^\infty(\mathbb{D})$ and put $f = R_\alpha g$. We have

$$f(z) = \alpha \int_{\mathbb{D}} \frac{g(w)}{(1 - z\overline{w})^{1+\alpha}} dA(w),$$

and hence

$$f'(z) = \alpha(\alpha + 1) \int_{\mathbb{D}} \frac{\overline{w} g(w)}{(1 - z\overline{w})^{2+\alpha}} dA(w), \qquad (z \in \mathbb{D}).$$

By Lemma 1.8.1, there exists a constant $C > 0$ such that

$$|f'(z)| \le \alpha(\alpha + 1)\|g\|_\infty \int_{\mathbb{D}} \frac{dA(w)}{(1 - z\overline{w})^{2+\alpha}} \le C\|g\|_\infty (1 - |z|^2)^{-\alpha}, \qquad (z \in \mathbb{D}).$$

Also it is clear that $|f(0)| \le \alpha\|g\|_\infty$. These estimates give

$$\|f\|_\alpha \le (C + \alpha)\|g\|_\infty, \quad f \in L^\infty(\mathbb{D}),$$

which shows that R_α maps $L^\infty(\mathbb{D})$ boundedly into $\mathcal{B}_\alpha$.

Next, we prove that R_α maps $HC(\mathbb{D})$ onto $\mathcal{B}_\alpha$. We notice that for any nonnegative integer n,

$$\alpha \int_{\mathbb{D}} \frac{(1 - |w|^2)^2 w^n}{(1 - z\overline{w})^{1+\alpha}} \, dA(w) = \frac{2\alpha(\alpha + 1) \cdots (\alpha + n)}{(n + 3)!}, \quad (z \in \mathbb{D}),$$

which shows that for a polynomial p, then always there exists $g \in C_0(\mathbb{D})$ such that $p = R_\alpha g$.

So if $f \in \mathcal{B}_\alpha$, we can write

$$f(z) = f(0) + f'(0)z + \frac{f''(0)}{2!}z^2 + f_1(z)$$

with f_1 still in $\mathcal{B}_\alpha$. Then we can find a function $g \in C_0(\mathbb{D})$ satisfying

$$f(0) + f'(0)z + \frac{f''(0)}{2!}z^2 = R_\alpha g(z).$$

By Proposition 4.1.1 (iv), we also have $f_1(z) = R_\alpha g_1(z)$, where

$$g_1(z) = \frac{(1 - |z|^2)^\alpha f_1'(z)}{\alpha \overline{z}},$$

which is in $HC(\mathbb{D})$. Thus, $f = R_\alpha(g + g_1)$ and so R_α maps $HC(\mathbb{D})$ onto $\mathcal{B}_\alpha$.

(ii) The "onto" part follows from the proof of part (i). So we prove that R_α maps $C(\overline{\mathbb{D}})$ into $\mathcal{B}_{\alpha,0}$. By the Stone–Weierstrass approximation theorem, each function from $C(\overline{\mathbb{D}})$ can be uniformly approximated by a finite linear combination of functions of the form $z^n \overline{z}^m$ $(m, n \ge 0)$. Since, by part (i), R_α maps $L^\infty(\mathbb{D})$ boundedly into $\mathcal{B}_\alpha$ and $\mathcal{B}_{\alpha,0}$ is closed in $\mathcal{B}_\alpha$, it suffices to show that R_α maps each function $z^n \overline{z}^m$ into $\mathcal{B}_{\alpha,0}$. But it is clear since an easy calculation using polar coordinates shows that R_α maps each function of the form $z^n \overline{z}^m$ to a polynomial.

(iii) This follows from the consideration of the quotient norm and the open mapping theorem.

$\square$

Proposition 4.2.2 *For each $\alpha > 0$, we have:*

(i) *S_α is a bounded linear operator from $\mathcal{B}_\alpha$ into $L^\infty(\mathbb{D})$ and also is a bounded linear operator from $\mathcal{B}_{\alpha,0}$ into $C_0(\mathbb{D})$.*

(ii) *There exists a constant $C > 0$ (depending only on α) such that*

$$C^{-1}\|f\|_\alpha \le \|S_\alpha f\|_\infty \le C\|f\|_\alpha$$

for all $f \in \mathcal{B}_\alpha$.

(iii) *In particular, for a holomorphic function f on $\mathbb{D}$, $f \in \mathcal{B}_\alpha$ if and only if $S_\alpha f \in L^\infty(\mathbb{D})$ and $f \in \mathcal{B}_{\alpha,0}$ if and only if $S_\alpha f \in C_0(\mathbb{D})$.*

Proof For $f \in \mathcal{B}_\alpha$, there exists $g \in L^\infty(\mathbb{D})$ such that $f = R_\alpha g$. By Proposition 4.1.1 (iii) and Fubini's theorem, we have

$$S_\alpha f(z) = 3\alpha \int_{\mathbb{D}} \frac{(1-|z|^2)^2}{(1-z\overline{w})^4}(1-|w|^2)^{\alpha-1}\, dA(w) \int_{\mathbb{D}} \frac{g(u)\, dA(u)}{(1-w\overline{u})^{1+\alpha}}$$

$$= 3(1-|z|^2)^2 \int_{\mathbb{D}} g(u)\, dA(u)\, \alpha \int_{\mathbb{D}} \frac{(1-|w|^2)^{\alpha-1}\, dA(w)}{(1-z\overline{w})^4(1-w\overline{u})^{1+\alpha}}$$

$$= 3(1-|z|^2)^2 \int_{\mathbb{D}} \frac{g(u)\, dA(u)}{(1-w\overline{u})^4}.$$

Hence,

$$|S_\alpha f(z)| \le 3\|g\|_\infty (1-|z|^2)^2 \int_{\mathbb{D}} \frac{dA(u)}{|1-z\overline{u}|^4} = 3\|g\|_\infty, \qquad (z \in \mathbb{D}).$$

This gives

$$\|S_\alpha f\|_\infty \le 3\|g\|_\infty \quad \text{for all } g \in L^\infty(\mathbb{D}) \text{ with } R_\alpha g = f.$$

By Proposition 4.2.1 (iii), there exists a constant $C > 0$ such that

$$\|S_\alpha f\|_\infty \le 3C\|f\|_\alpha, \qquad f \in \mathcal{B}_\alpha.$$

This shows that S_α is a bounded linear operator from $\mathcal{B}_\alpha$ into $L^\infty(\mathbb{D})$.

Next, for $f \in \mathcal{B}_\alpha$, again by Proposition 4.1.1 (iii) and Fubini's theorem, we have

$$R_\alpha S_\alpha f(z) = 3\alpha \int_{\mathbb{D}} \frac{dA(w)}{(1-z\overline{w})^{1+\alpha}} \int_{\mathbb{D}} \frac{(1-|w|^2)^2}{(1-w\overline{u})^4}(1-|u|^2)^{\alpha-1}f(u)\, dA(u)$$

$$= \alpha \int_{\mathbb{D}} f(u)(1-|u|^2)^{\alpha-1}\, dA(u) \cdot 3 \int_{\mathbb{D}} \frac{(1-|w|^2)^2\, dA(w)}{(1-w\overline{u})^4(1-z\overline{w})^{\alpha+1}}$$

$$= \alpha \int_{\mathbb{D}} \frac{(1-|u|^2)^{\alpha-1}}{(1-z\overline{u})^{\alpha+1}}f(u)\, dA(u) = f(z).$$

This means that $S_\alpha f \in L^\infty(\mathbb{D})$ implies that $f \in \mathcal{B}_\alpha$ by Proposition 4.2.1 (i). Then there exists a constant $C > 0$ such that

$$\|f\|_\infty = \|R_\alpha S_\alpha f\|_\alpha \le C\|S_\alpha f\|_\infty, \quad f \in \mathcal{B}_\alpha.$$

Also Proposition 4.2.1 (ii) shows that $S_\alpha f \in C_0(\mathbb{D})$ implies that $f \in \mathcal{B}_{\alpha,0}$.

Finally, we show that S_α maps $\mathcal{B}_{\alpha,0}$ into $C_0(\mathbb{D})$. Indeed, by the symmetry of the disc, the operator S_α maps each polynomial to a polynomial times the function $(1 - |z|^2)^2$. In particular, S_α maps each polynomial to a function in $C_0(\mathbb{D})$. Moreover, since $S_\alpha : \mathcal{B}_\alpha \to L^\infty(\mathbb{D})$ is bounded, $\mathcal{B}_{\alpha,0}$ is the closure of the set of polynomials in $\mathcal{B}_\alpha$, and $C_0(\mathbb{D})$ is closed in $L^\infty(\mathbb{D})$, we conclude that S_α maps $\mathcal{B}_{\alpha,0}$ into $C_0(\mathbb{D})$.   □

4.3   The Pre-dual of α-Bloch Space

We have now developed enough tools to completely characterize the pre-dual of α-Bloch spaces.

Theorem 4.3.1   (i) For each $\alpha > 0$, the dual of $L^1(dA)$ can be identified with $\mathcal{B}_\alpha$ (with equivalent norm) under the pairing

$$\langle f, g \rangle_\alpha = \lim_{t \to 1^-} \int_\mathbb{D} f(tz)\overline{g(tz)}(1 - |z|^2)^{\alpha-1}\,dA(z), \quad f \in L^1(dA), g \in \mathcal{B}_\alpha.$$

(ii) For each $\alpha > 0$, the dual of $\mathcal{B}_{\alpha,0}$ can be identified with $L^1(dA)$ (with equivalent norm) under the pairing

$$\langle f, g \rangle = \lim_{t \to 1^-} \int_\mathbb{D} f(tz)\overline{g(tz)}(1 - |z|^2)^{\alpha-1}\,dA(z), \quad f \in \mathcal{B}_{\alpha,0}, g \in L^1(dA).$$

Proof (i) Recall that

$$\int_\mathbb{D} |g(z)|(1 - |z|^2)^{\alpha-1}\,dA(z) < \infty, \quad g \in \mathcal{B}_\alpha.$$

First of all, we note that if $f \in L^1(dA)$ is bounded (i.e., $f \in H^\infty(\mathbb{D})$ and $g \in \mathcal{B}_\alpha$), then

$$\left| \int_\mathbb{D} f(z)\overline{g(z)}(1 - |z|^2)^{\alpha-1}\,dA(z) \right| \le C\|f\|_{L^1}\|g\|_\alpha,$$

where $C > 0$ is some constant independent of both f and g. Indeed, we write $g = R_\alpha \ell$
for some $\ell \in L^\infty(\mathbb{D})$ and apply Fubini's theorem to get

$$\int_{\mathbb{D}} f(z)\overline{g(z)}(1 - |z|^2)^{\alpha-1}\, dA(z)$$

$$\alpha \int_{\mathbb{D}} f(z)(1 - |z|^2)^{\alpha-1}\, dA(z) \int_{\mathbb{D}} \frac{\overline{\ell(w)}\, dA(w)}{(1 - w\bar{z})^{\alpha+1}}$$

$$\alpha \int_{\mathbb{D}} \overline{\ell(w)}\, dA(w) \frac{(1 - |z|^2)^{\alpha-1} f(z)}{(1 - w\bar{z})^{\alpha+1}}\, dA(z).$$

Furthermore, by Proposition 4.1.1 (iii),

$$\int_{\mathbb{D}} f(z)\overline{g(z)}(1 - |z|^2)^{\alpha-1}\, dA(z) = \int_{\mathbb{D}} \overline{\ell(w)}\, dA(w),$$

which gives

$$\left| \int_{\mathbb{D}} f(z)\overline{g(z)}(1 - |z|^2)^{\alpha-1}\, dA(z) \right| \le \|f\|_{L^1} \|\ell\|_\infty.$$

Taking the infimum over ℓ and applying Proposition 4.2.1 (i), we obtain a constant $C > 0$
for which the desired result holds.

Next, we prove that if $f \in L^1(dA)$ and $g \in \mathcal{B}_\alpha$, then there exists

$$\lim_{t\to 1^-} \int_{\mathbb{D}} f(tz)\overline{g(z)}(1 - |z|^2)^{\alpha-1}\, dA(z) = A,$$

with

$$|A| \le C \|f\|_{L^1} \|g\|_\alpha,$$

where $C > 0$ is some constant independent of both f and g. Indeed, for $g \in \mathcal{B}_\alpha$, as shown
above,

$$F_g(f) = \int_{\mathbb{D}} f(tz)\overline{g(z)}(1 - |z|^2)^{\alpha-1}\, dA(z), \quad f \in H^\infty(\mathbb{D}),$$

can be extended to a bounded linear functional on L^1 with $\|F_g\| \le C\|g\|_\alpha$. Therefore, we
may assume that F_g is defined on the whole space L^1 (but not via the above formula). We
fix $f \in L^1(dA)$ and $g \in \mathcal{B}_\alpha$ and let $f_t(z) = f(tz)$ for $0 < t < 1$. Since $f_t \in H^\infty$ and
$\lim_{t\to 1^-} \|f_t - f\|_{L^1} = 0$, we have

$$\lim_{t \to 1^-} \int_{\mathbb{D}} f(tz)\overline{g(z)}(1 - |z|^2)^{\alpha-1} \, dA(z) = \lim_{t \to 1^-} F_g(f_t) = F_g(f)$$

and

$$|F_g(f)| \leq \|F_g\| \|f\|_{L^1} \leq C\|g\|_\alpha \|f\|_{L^1}.$$

Furthermore, we show that if F is bounded linear functional on L^1, then there exists a function $g \in \mathcal{B}_\alpha$ such that

$$F(f) = \lim_{t \to 1^-} \int_{\mathbb{D}} f(tz)\overline{g(z)}(1 - |z|^2)^{\alpha-1} \, dA(z), \quad f \in L^1(dA).$$

Indeed, by the Hahn–Banach extension theorem, F can be extended to a bounded linear functional on $L^1(\mathbb{D}, dA)$ without increasing the norm. Since $(L^1)^* = L^\infty$, there is a function $\ell \in L^\infty(\mathbb{D})$ such that

$$F(f) = \int_{\mathbb{D}} f(tz)\overline{\ell(z)} \, dA(z), \quad f \in L^1.$$

By the first fact above, for each $f \in L^1$, we have

$$F(f) = \lim_{t \to 1^-} \int_{\mathbb{D}} f_t(z)\overline{\ell(z)} \, dA(z)$$

$$= \lim_{t \to 1^-} \int_{\mathbb{D}} f_t(z)\overline{R_\alpha \ell(z)}(1 - |z|^2)^{\alpha-1} \, dA(z).$$

Let $g = R_\alpha \ell$; then by Proposition 4.2.1 (i), $g \in \mathcal{B}_\alpha$ and

$$F(f) = \lim_{t \to 1^-} \int_{\mathbb{D}} f(tz)\overline{g(z)}(1 - |z|^2)^{\alpha-1} \, dA(z), \quad f \in L^1.$$

Finally, by the rotational invariance of the measure $(1 - |z|^2)^{\alpha-1} \, dA(z)$, we have

$$\int_{\mathbb{D}} f(tz)\overline{g(z)}(1 - |z|^2)^{\alpha-1} \, dA(z) = \int_{\mathbb{D}} f(sz)\overline{g(sz)}(1 - |z|^2)^{\alpha-1} \, dA(z)$$

where $s = \sqrt{t}$. From this, it follows that

$$\lim_{t \to 1^-} \int_{\mathbb{D}} f(tz)\overline{g(z)}(1 - |z|^2)^{\alpha-1} \, dA(z) = \lim_{t \to 1^-} \int_{\mathbb{D}} f(tz)\overline{g(tz)}(1 - |z|^2)^{\alpha-1} \, dA(z),$$

which completes the proof of the theorem.

(ii) In view of part (i), we only need to show that each bounded linear functional F on $\mathcal{B}_{\alpha,0}$ arises from a function $g \in L^1$ in the following manner:

$$F(f) = \lim_{t \to 1^-} \int_{\mathbb{D}} f(tz)\overline{g(tz)}(1 - |z|^2)^{\alpha-1} \, dA(z), \quad f \in \mathcal{B}_{\alpha,0}.$$

Indeed, by Proposition 4.2.1 (iii), the operator $S_\alpha : \mathcal{B}_{\alpha,0} \to C_0(\mathbb{D})$ satisfies

$$C^{-1}\|f\|_\alpha \leq \|S_\alpha f\|_\infty \leq C\|f\|_\alpha$$

for some constant $C > 0$ independent of f. We denote by $E = S_\alpha(\mathcal{B}_{\alpha,0})$, then E is a closed subspace of $C_0(\mathbb{D})$ and $F \circ S_\alpha^{-1} : E \to \mathbb{C}$ is a bounded linear functional. We extend it to the whole space $C_0(\mathbb{D})$ and apply the well-known Riesz representation theorem, to get a finite Borel measure μ on $\mathbb{D}$ so that

$$F \circ S_\alpha^{-1}(\ell) = \int_{\mathbb{D}} \ell(z) d\overline{\mu}(z), \quad \ell \in E.$$

From this, it follows that for each $f \in \mathcal{B}_{\alpha,0}$, we have

$$F(f) = \int_{\mathbb{D}} \ell(z) \, d\overline{\mu}(z).$$

Now by Fubini's theorem,

$$F(f) = 3 \int_{\mathbb{D}} (1 - |z|^2)^2 \, d\overline{\mu}(z) \int_{\mathbb{D}} \frac{(1 - |w|^2)^{\alpha-1} f(w)}{(1 - z\overline{w})^4} \, dA(w)$$

$$= 3 \lim_{t \to 1^-} \int_{\mathbb{D}} (1 - |z|^2)^2 \, d\overline{\mu}(z) \int_{\mathbb{D}} \frac{(1 - |w|^2)^{\alpha-1} f(w)}{(1 - tz\overline{w})^4} \, dA(w)$$

$$= \lim_{t \to 1^-} \int_{\mathbb{D}} f(z)\overline{g(tz)}(1 - |z|^2)^{\alpha-1} \, dA(z)$$

$$= \lim_{t \to 1^-} \int_{\mathbb{D}} f(tz)\overline{g(tz)}(1 - |z|^2)^{\alpha-1} \, dA(z).$$

Here, g is a holomorphic function defined by

$$g(w) = 3 \int_{\mathbb{D}} \frac{(1 - |z|^2)^2 \, d\mu(z)}{(1 - w\overline{z})^4},$$

which belongs to L^1, because

$$\int_{\mathbb{D}} |g(w)| \, dA(w)$$

$$\leq 3 \int_{\mathbb{D}} (1 - |z|^2)^2 \, d|\mu|(z) \int_{\mathbb{D}} \frac{d(w)}{|1 - z\overline{w}|^4} = 3 \int_{\mathbb{D}} d|\mu|(z) = 3\|\mu\|.$$

The theorem is proved completely. $\square$

4.4 The Space H_α^∞

Let $\alpha \in (0, \infty)$. The *weighted-type space* H_α^∞ is the family of all $f \in \mathrm{Hol}(\mathbb{D})$ such that

$$\|f\|_{H_\alpha^\infty} = \sup_{z \in \mathbb{D}} (1 - |z|^2)^\alpha |f(z)| < \infty.$$

Note that, based on our previous notation in Sect. 1.1, we have

$$H_\alpha^\infty = A^{-\alpha},$$

the Bergman-type space.

The closed subspace $H_{\alpha,0}^\infty$ consists of elements $f \in H_\alpha^\infty$ which satisfy the extra condition

$$(1 - |z|^2)^\alpha |f(z)| \to 0, \qquad (|z| \to 1).$$

The spaces H_α^∞ and $H_{\alpha,0}^\infty$ are related to α-Bloch spaces as follows:

$$H_\alpha^\infty = \mathcal{B}_{\alpha+1}, \qquad (\alpha > 0), \tag{4.1}$$

and

$$H_{\alpha,0}^\infty = \mathcal{B}_{\alpha+1,0}, \qquad (\alpha > 0). \tag{4.2}$$

We also have the inclusions

$$\mathcal{B} \subsetneq \mathcal{N}_p \subsetneq H_1^\infty, \qquad (0 < p \leq 1),$$

and

$$\mathcal{N}_p = H_1^\infty, \qquad (1 < p < \infty).$$

The above relations are very similar to the corresponding ones between the $\mathcal{Q}_p$-space and $\mathcal{B}$. That is, $\mathcal{Q}_p \subsetneq \mathcal{B}$ if $p \in (0, 1]$ and $\mathcal{Q}_p = \mathcal{B}$ if $p \in (1, \infty)$.

We reformulate Theorem 4.3.1 in the previous section in terms of spaces H_α^∞.

Theorem 4.4.1 *Under the pairing*

$$\langle f, g \rangle_\alpha = \lim_{t \to 1^-} \int_{\mathbb{D}} f(tz)\overline{g(tz)}(1 - |z|^2)^\alpha \, dA(z),$$

we have

$$\left(H_{\alpha,0}^\infty\right)^* \cong L_\alpha^1(dA) \quad and \quad \left(L_\alpha^1(dA)\right)^* \cong H_\alpha^\infty.$$

This result plays a crucial role in the study of weighted composition operators between spaces H_α^∞ and $\mathcal{N}_p$ in the sequel.

4.5   A Test Function for H_α^∞

We introduce a test function for H_α^∞ which will be used later in the proofs of Theorem 5.3.1. Let

$$f_{\theta,r}(z) = \sum_{k=0}^\infty 2^{k\alpha}(re^{i\theta})^{2^k} z^{2^k}, \qquad (z \in \mathbb{D}), \tag{4.3}$$

where $\alpha \in (0, \infty)$, $\theta \in [0, 2\pi)$, and $r \in (0, 1]$. The main properties of this test function are summarized below.

Lemma 4.5.1 *The function $f_{\theta,r}$ belongs to H_α^∞ and, moreover,*

$$\sup_{r,\theta} \| f_{\theta,r} \|_{H_\alpha^\infty} < \infty.$$

If $r \in (0, 1)$, then $f_{\theta,r} \in H_{\alpha,0}^\infty$.

Proof For each $z \in \mathbb{D} \setminus \{0\}$, we have

$$|f_{\theta,r}(z)| \leq \sum_{k=0}^\infty 2^{k\alpha}(\sqrt{r|z|})^{2^{k+1}}$$

$$\leq \sum_{k=0}^\infty \int_k^{k+1} 2^{kx}(\sqrt{r|z|})^{2^x} \, dx$$

$$= \int_0^\infty 2^{kx} (\sqrt{r|z|})^{2^x} dx$$

$$\lesssim \frac{1}{\left(\log \frac{1}{\sqrt{r|z|}}\right)^\alpha} \int_{\log \frac{1}{r|z|}}^\infty s^{\alpha-1} e^{-s} ds$$

$$\lesssim \frac{1}{\left(\log \frac{1}{\sqrt{r|z|}}\right)^\alpha}.$$

Therefore,

$$(1 - |z|^2)^\alpha |f_{\theta,r}(z)| \lesssim \left(\frac{1 - |z|^2}{\log \frac{1}{\sqrt{r|z|}}}\right)^\alpha.$$

Since $\log \frac{1}{x} \geq 1 - x$, this estimation implies

$$(1 - |z|^2)^\alpha |f_{\theta,r}(z)| \lesssim 1, \qquad (z \in \mathbb{D}).$$

Moreover, it shows $f_{\theta,r} \in H_{\alpha,0}^\infty$ provided that $r \neq 1$.   □

For each $n \in \mathbb{N}$, consider the function

$$g_{\theta,r}^n(z) = z^n f_{\theta,r}(z),$$

where $f_{\theta,r}$ is the test function defined in (4.3). Since $z^n H_{\alpha,0}^\infty \subset H_{\alpha,0}^\infty$, it follows from Lemma 4.5.1 that $\{g_{\theta,r}^n\}_{n\in\mathbb{N}} \subset H_{\theta,0}^\infty$ and the norm $\|g_{\theta,r}^n\|_{H_\alpha^\infty}$ is uniformly bounded with respect to θ, r and n. The following lemma says that $\{g_{\theta,r}^n\}_{n\in\mathbb{N}}$ weakly converges to 0 in H_α^∞.

Lemma 4.5.2 *For every* $\Lambda \in (H_\alpha^\infty)^*$,

$$\sup_{\theta,r} |\Lambda(g_{\theta,r}^n)| \to 0, \qquad (n \to \infty).$$

Proof It suffices to prove the claim for every $\Lambda \in (H_{\alpha,0}^\infty)^*$. By Theorem 4.4.1, there exists $h \in L_a^1(dA)$ such that $\Lambda(f) = \langle h, f \rangle_\alpha$ for $f \in H_{\alpha,0}^\infty$. By using the estimate in the proof of Lemma 4.5.1, we have

$$\sup_{\theta,r} |\Lambda(g_{\theta,r}^n)| \leq \sup_{\theta,r} | \lim_{t \to 1} \int_\mathbb{D} |tz|^n |f_{\theta,r}(tz)| |h(tz)| (1 - |z|^2)^\alpha dA(z)$$

$$\lesssim \lim_{t \to 1} \int_{\mathbb{D}} |tz|^n |h(tz)| \, dA(z)$$

$$\leq \lim_{t \to 1} \frac{1}{t^2} \int_{\mathbb{D}} |z|^n |h(z)| \, dA(z)$$

$$= \int_{\mathbb{D}} |z|^n |h(z)| \, dA(z).$$

Since $|z|^n |h(z)| \to 0$ as $n \to \infty$ for each fixed $z \in \mathbb{D}$, and $|z^n h(z)| \leq |h(z)|$ with $h \in L^1(dA)$, the dominated convergence theorem ensures that

$$\lim_{n \to \infty} \int_{\mathbb{D}} |z|^n |h(z)| \, dA(z) = 0.$$

$\square$

Notes on Chapter 4

A paper by J. Anderson [2] is a good survey of the theory of classical Bloch functions. In this chapter, due to our needs, we considered more general spaces. The results provided in this chapter are justified by at least two considerations: firstly it gives a unified approach to some known duality results and secondly we developed several results along the way which are of independent interest. The results of this chapter are mostly from [112]. The last section is taken from [95]. Lemma 4.5.2 is similar to [60, Lemma 5]. Dualities of Bloch spaces' some special cases have been obtained in [85] for $\alpha > 1$ and in [3] for $\alpha = 1$. For the cases $0 < \alpha < 1$, it can actually be deduced from the case $1 < \alpha < 2$ by considering f' instead of f.

Throughout this chapter, φ denotes a non-constant holomorphic self-map of $\mathbb{D}$. Then the *composition operator* C_φ on $\text{Hol}(\mathbb{D})$ is defined by

$$C_\varphi f = f \circ \varphi.$$

Let ψ be a holomorphic function on $\mathbb{D}$, which is not identically equal to zero. Then the *multiplication operator* M_ψ on $\text{Hol}(\mathbb{D})$ is defined by

$$M_\psi f = \psi f.$$

As a generalization of both concepts, the *weighted composition operator* $W_{\psi,\varphi}$ on $\text{Hol}(\mathbb{D})$ is the mapping

$$W_{\psi,\varphi} f = \psi \cdot f \circ \varphi.$$

If $\psi(z) \equiv 1$, then we obtain the standard composition operator C_φ, while the choice $\varphi(z) = z$ gives us the standard multiplication operator M_ψ.

We aim to study weighted composition operators $W_{\psi,\varphi}$ acting between H_α^∞ and $\mathcal{N}_p$ spaces and relate operator theoretic properties, like boundedness and compactness, to function theoretic properties of the inducing functions ψ and φ. As consequences, we obtain characterizations for boundedness and compactness of the composition operators C_φ between these spaces.

5.1   The Multiplication Operator

There is a simple characterization of the multipliers on $\mathcal{N}_p$ spaces.

Theorem 5.1.1 *The multiplication operator M_ψ is well-defined and bounded on $\mathcal{N}_p$ if and only if $\psi \in H^\infty$.*

Proof Suppose $\psi \in H^\infty$. It is clear that

$$\|M_\psi f\|_{\mathcal{N}_p}^2 \leq \|\psi\|_{H^\infty}^2 \|f\|_{\mathcal{N}_p}^2.$$

Conversely, using the test function k_w in Lemma 3.2.1, we have

$$1 \geq \|k_w\|_{\mathcal{N}_p}$$
$$\gtrsim \|M_\psi k_w\|_{\mathcal{N}_p}$$
$$\gtrsim \|M_\psi k_w\|_{A^{-1}}$$
$$\geq \frac{1 - |w|^2}{|1 - \overline{w}w|^2} |\psi(w)|(1 - |w|^2) = |\psi(w)|$$

for all $w \in \mathbb{D}$, i.e., $\psi \in H^\infty$.   □

5.2   Carleson Measures for $\mathcal{N}_p$

For each arc I in the unit circle $\mathbb{T}$, the Carleson box based on I is the set of the form

$$S(I) = \{z \in \mathbb{D} : 1 - |I| \leq |z| < 1, z/|z| \in I\},$$

where $|I|$ denotes the normalized length of I. For each $p \in (0, \infty)$, a bounded positive Borel measure μ on $\mathbb{D}$ is called a p-Carleson measure if

$$C_\mu := \sup_{I \subset \mathbb{T}} \frac{\mu[S(I)]}{|I|^p} < \infty,$$

where the supremum is taken over all arcs $I \subset \mathbb{T}$; μ is called a compact p-Carleson measure if

$$\lim_{|I| \to 0} \frac{\mu[S(I)]}{|I|^p} = 0.$$

Lemma 5.2.1 *Let μ be a positive measure on $\mathbb{D}$, and let $0 < p < \infty$. Then the following assertions hold.*

(i) μ is a p-Carleson measure if and only if

$$\sup_{w \in \mathbb{D}} \iint_{\mathbb{D}} \left(\frac{1 - |w|^2}{|1 - \overline{w}z|^2} \right)^p d\mu(z) < \infty. \tag{5.1}$$

(ii) μ is a compact p-Carleson measure if and only if

$$\lim_{|w| \to 1} \iint_{\mathbb{D}} \left(\frac{1 - |w|^2}{|1 - \overline{w}z|^2} \right)^p d\mu(z) = 0. \tag{5.2}$$

Proof (i) Suppose (5.1) is satisfied, and let M denote the supremum. Then the Carleson box

$$S(I) = \{z \in \mathbb{D} : 1 - h \le |z| \le 1, |\theta - \arg z| \le h\}$$

considers the point $w = (1 - h)e^{i(\theta + h/2)}$. Then we have

$$M \ge \iint_{\mathbb{D}} \left(\frac{1 - |w|^2}{|1 - \overline{w}z|^2} \right)^p d\mu(z)$$

$$\ge \iint_{S(I)} \left(\frac{1 - |w|^2}{|1 - \overline{w}z|^2} \right)^p d\mu(z)$$

$$\ge \left[\inf_{z \in S(I)} \left(\frac{1 - |w|^2}{|1 - \overline{w}z|^2} \right)^p \right] \mu(S(I))$$

$$\gtrsim \frac{\mu(S(I))}{|I|^p}. \tag{5.3}$$

From this, it follows that

$$C_\mu = \sup_{I \subset \mathbb{T}} \frac{\mu(S(I))}{|I|^p} \lesssim M < \infty.$$

Conversely, let μ be a bounded p-Carleson measure on $\mathbb{D}$, that is, $C_\mu < \infty$. We have

$$\mu(\mathbb{D}) \lesssim C_\mu. \tag{5.4}$$

Now, we consider two cases. For $|w| \le \frac{3}{4}$, we have the trivial estimate

$$\iint_{\mathbb{D}} \left(\frac{1 - |w|^2}{|1 - \overline{w}z|^2} \right)^p d\mu(z) \lesssim \mu(\mathbb{D}) \lesssim C_\mu.$$

For $|w| > \frac{3}{4}$, let $w = r_0 e^{i\theta_0}$, and consider the sets

$$E_n = \{z \in \mathbb{D} : |z - e^{i\theta_0}| < 2^n(1 - r_0)\}.$$

Then, since E_n is inside a Carleson box,

$$\mu(E_n) \lesssim C_\mu 2^{np}(1 - |w|)^p, \qquad (n \geq 0).$$

We also have

$$\frac{1 - |w|^2}{|1 - \overline{w}z|^2} \lesssim \frac{1}{1 - |w|}, \qquad z \in E_1,$$

which implies that, with $E_0 = \emptyset$ and for all $n \geq 1$,

$$\frac{1 - |w|^2}{|1 - \overline{w}z|^2} \lesssim \frac{1}{2^{2n}(1 - |w|)}, \qquad z \in E_n \setminus E_{n-1}.$$

Consequently,

$$\iint_{\mathbb{D}} \left(\frac{1 - |w|^2}{|1 - \overline{w}z|^2} \right)^p d\mu(z) \leq \sum_{n=1}^{\infty} \iint_{E_n \setminus E_{n-1}} \left(\frac{1 - |w|^2}{|1 - \overline{w}z|^2} \right)^p d\mu(z)$$

$$\lesssim \sum_{n=1}^{\infty} \frac{\mu(E_n)}{2^{np}(1 - |w|)^p} \lesssim C_\mu \sum_{n=1}^{\infty} \frac{1}{2^{np}}, \qquad (5.5)$$

which means that (5.1) holds.

(ii) Suppose that (5.2) is satisfied. Applying (5.3), we immediately see that μ is a compact p-Carleson measure.

Conversely, suppose that μ is a compact p-Carleson measure. Then μ must be bounded. For $t \in (0, 1)$, let $\chi_{\mathbb{D} \setminus \mathbb{D}_t}$ be the characteristic function of the set $\mathbb{D} \setminus \mathbb{D}_t$, where $\mathbb{D}_t = \{z : |z| < t\}$. Let $d\mu_t(z) = \chi_{\mathbb{D} \setminus \mathbb{D}_t}(z) d\mu(z)$. Then from (5.4) and (5.5), it follows that

$$\iint_{\mathbb{D}} \left(\frac{1 - |w|^2}{|1 - \overline{w}z|^2} \right)^p d\mu(z) = \left(\iint_{\mathbb{D}_t} + \iint_{\mathbb{D} \setminus \mathbb{D}_t} \right) \left(\frac{1 - |w|^2}{|1 - \overline{w}z|^2} \right)^p d\mu(z)$$

$$\leq \left(\frac{1 - |w|^2}{(1 - t)^2} \right)^p \mu(\mathbb{D}) + \iint_{\mathbb{D}} \left(\frac{1 - |w|^2}{|1 - \overline{w}z|^2} \right)^p d\mu_t(z)$$

$$\lesssim \left(\frac{1-|w|^2}{(1-t)^2}\right)^p \mu(\mathbb{D}) + C_{\mu_t}$$

$$\lesssim \left(\frac{1-|w|^2}{(1-t)^2}\right)^p C_\mu + \sup_I \frac{\mu[(\mathbb{D} \setminus \mathbb{D}_t) \cap S(I)]}{|I|^p}.$$

This implies that (5.2) is true.

$\square$

Theorem 5.2.2 *Let $f \in Hol(\mathbb{D})$. Then $f \in \mathcal{N}_p$ if and only if*

$$d\mu_{f,p}(z) = |f(z)|^2 (1-|z|^2)^p \, dA(z)$$

is a p-Carleson measure. Furthermore, in this case, we have

$$\|f\|_{\mathcal{N}_p}^2 \asymp \sup_{I \subset \mathbb{T}} \frac{\mu_{f,p}[S(I)]}{|I|^p}.$$

Proof By the equality

$$1 - |\sigma_w(z)|^2 = \frac{(1-|w|^2)(1-|z|^2)}{|1-\overline{w}z|^2},$$

we have

$$\|f\|_{\mathcal{N}_p}^2 = \sup_{w \in \mathbb{D}} \int_\mathbb{D} \left(\frac{(1-|w|^2)}{|1-\overline{w}z|^2}\right)^p d\mu_{f,p}(z).$$

Combining this representation and Lemma 5.2.1 yields the desired result. $\square$

5.3   The Boundedness of $W_{\psi,\varphi}: H_\alpha^\infty \to \mathcal{N}_p$

We have developed all the required tools to study the boundedness of the weighted composition operator $W_{\psi,\varphi}: H_\alpha^\infty \to \mathcal{N}_p$.

Theorem 5.3.1 *Let $\psi \in Hol(\mathbb{D})$ and φ be an analytic self-map of $\mathbb{D}$. Then the following are equivalent:*

(i) *$W_{\psi,\varphi}: H_\alpha^\infty \to \mathcal{N}_p$ is a well-defined bounded operator.*
(ii) *ψ and φ satisfy*

$$\sup_{w \in \mathbb{D}} \int_{\mathbb{D}} \frac{|\psi(z)|^2}{(1 - |\varphi(z)|^2)^{2\alpha}} (1 - |\sigma_w(z)|^2)^p \, dA(z) < \infty. \tag{5.6}$$

(iii) ψ *and* φ *satisfy*

$$\sup_{I \subset \mathbb{T}} |I|^{-p} \int_{S(I)} \frac{|\psi(z)|^2 (1 - |z|^2)^p}{(1 - |\varphi(z)|^2)^{2\alpha}} \, dA(z) < \infty, \tag{5.7}$$

where the supremum is taken over all arcs $I \subset \mathbb{T}$.

Proof *(i)* $\implies$ *(ii):* For each $\theta \in [0, 2\pi)$, we put $f_\theta = f_{\theta,1}$ which is defined in (4.3). Let $w \in \mathbb{D}$. By Lemma 4.5.1 and Fubini's theorem, we have

$$1 \gtrsim \int_0^{2\pi} \|W_{\psi,\varphi} f_\theta\|_{\mathcal{N}_p}^2 \frac{d\theta}{2\pi}$$

$$\geq \int_{\mathbb{D}} |\psi(z)|^2 (1 - |\sigma_w(z)|^2)^p \left\{ \int_0^{2\pi} |f_\theta(\varphi(z))|^2 \frac{d\theta}{2\pi} \right\} dA(z).$$

Furthermore, by Parseval's formula,

$$\int_0^{2\pi} |f_\theta(\varphi(z))|^2 \frac{d\theta}{2\pi} = \int_0^{2\pi} \left| \sum_{k=0}^{\infty} 2^{k\alpha} \left(e^{i\theta} \varphi(z) \right)^{2^k} \right|^2 \frac{d\theta}{2\pi}$$

$$= \sum_{k=0}^{\infty} (2^k)^{2\alpha} \left(\varphi(z) \right)^{2^k}. \tag{5.8}$$

In case $|\varphi(z)| > \frac{1}{2}$, we have

$$\sum_{k=0}^{\infty} (2^k)^{2\alpha} \left(\varphi(z) \right)^{2^k} = 2^{-2\alpha} \sum_{k=0}^{\infty} (2^{k+1})^{2\alpha} \left(\varphi(z) \right)^{2^k}$$

$$\geq 2^{-2\alpha} \int_0^{\infty} (2^x)^{2\alpha} \left(\varphi(z) \right)^{2^x} dx$$

$$= \frac{2^{-2\alpha}}{\log 2} \left(\log \frac{1}{|\varphi(z)|^2} \right)^{-2\alpha} \int_{\log \frac{1}{|\varphi(z)|^2}}^{\infty} s^{2\alpha-1} e^{-s} \, ds$$

$$\gtrsim \frac{1}{(1 - |\varphi(z)|^2)^{2\alpha}}, \tag{5.9}$$

where the last inequality follows from the simple fact that

$$\log \frac{1}{x} \le (1-x) \log 4, \qquad (1/2 \le x \le 1).$$

Hence, we obtain

$$\int_{|\varphi(z)| > \frac{1}{2}} \frac{|\psi(z)|^2}{(1 - |\varphi(z)|^2)^{2\alpha}} (1 - |\sigma_w(z)|^2)^p \, dA(z) \lesssim 1, \tag{5.10}$$

for all $w \in \mathbb{D}$. Furthermore, since $\psi \in \mathcal{N}_p$, we also have

$$\int_{|\varphi(z)| \le \frac{1}{2}} \frac{|\psi(z)|^2}{(1 - |\varphi(z)|^2)^{2\alpha}} (1 - |\sigma_w(z)|^2)^p \, dA(z) \lesssim \|\psi\|_{\mathcal{N}_p}^2, \tag{5.11}$$

for any $w \in \mathbb{D}$. The inequalities (5.10) and (5.11) together yield (5.6).

$(ii) \implies (i)$: It is rather obvious. Indeed, for each $f \in H_\alpha^\infty$, we have

$$\|W_{\psi,\varphi} f\|_{\mathcal{N}_p}^2 = \sup_{w \in \mathbb{D}} \int_{\mathbb{D}} |\psi(z)|^2 |f \circ \varphi(z)|^2 (1 - \sigma_w(z)|^2)^p \, dA(z)$$

$$= \sup_{w \in \mathbb{D}} \int_{\mathbb{D}} \left[\frac{|\psi(z)|^2}{(1 - |\varphi(z)|^2)^{2\alpha}} (1 - |\sigma_w(z)|^2)^p \right] \cdot \left[(1 - |\varphi(z)|^2)^{2\alpha} |f \circ \varphi(z)|^2 \right] dA(z)$$

$$\lesssim \sup_{w \in \mathbb{D}} \int_{\mathbb{D}} \frac{|\psi(z)|^2}{(1 - |\varphi(z)|^2)^{2\alpha}} (1 - |\sigma_w(z)|^2)^p \, dA(z) \cdot \left[\sup_{z \in \mathbb{D}} (1 - |\varphi(z)|^2)^{2\alpha} |f \circ \varphi(z)|^2 \right]$$

$$\le \sup_{w \in \mathbb{D}} \int_{\mathbb{D}} \frac{|\psi(z)|^2}{(1 - |\varphi(z)|^2)^{2\alpha}} (1 - |\sigma_w(z)|^2)^p \, dA(z) \cdot \sup_{z \in \mathbb{D}} (1 - |z|^2)^{2\alpha} |f(z)|^2$$

$$\lesssim \|f\|_{H_\alpha^\infty}^2,$$

which shows that $W_{\psi,\varphi} : H_\alpha^\infty \to \mathcal{N}_p$ is bounded.

$(i) \implies (iii)$: Suppose $W_{\psi,\varphi} : H_\alpha^\infty \to \mathcal{N}_p$ is bounded. We fix an arc $I \subset \mathbb{T}$ and consider the test function f_θ, which was used in the proof of $(i) \implies (ii)$. By Theorem 5.2.2, Lemma 4.5.1, and Fubini's theorem, we have

$$|I|^{-p} \int_{S(I) \cap |\varphi(z)| > \frac{1}{\sqrt{2}}} \frac{|\psi(z)|^2 (1 - |z|^2)^p}{(1 - |\varphi(z)|^2)^{2\alpha}} \, dA(z) \lesssim 1. \tag{5.12}$$

The boundedness of $W_{\psi,\varphi}$ ensures that $\psi \in \mathcal{N}_p$. Hence, by Theorem 5.2.2, $|\psi(z)|^2 (1 - |z|^2)^p \, dA(z)$ is a p-Carleson measure, and

$$\sup_{I \subset \mathbb{T}} |I|^{-p} \int_{S(I)} |\psi(z)|^2 (1 - |z|^2)^p \, dA(z) \lesssim \|\psi\|_{\mathcal{N}_p}^2.$$

Consequently,

$$|I|^{-p} \int_{S(I) \cap |\varphi(z)| \le \frac{1}{\sqrt{2}}} \frac{|\psi(z)|^2 (1 - |z|^2)^p}{(1 - |\varphi(z)|^2)^{2\alpha}} \, dA(z) \lesssim \|\psi\|_{\mathcal{N}_p}^2. \tag{5.13}$$

Combining (5.12) and (5.13) yields (5.7).

$(iii) \implies (i)$: For every $f \in H_\alpha^\infty$, we have

$$\sup_{I \subset \mathbb{T}} |I|^{-p} \int_{S(I)} |\psi(z)|^2 (1 - |z|^2)^p |f \circ \varphi(z)|^2 \, dA(z)$$

$$\le \|f\|_{H_\alpha^\infty}^2 \cdot \sup_{I \subset \mathbb{T}} |I|^{-p} \int_{S(I)} \frac{|\psi(z)|^2 (1 - |z|^2)^p}{(1 - |\varphi(z)|^2)^{2\alpha}} \, dA(z).$$

From this, together with condition (5.7), it follows that

$$d\mu(z) := |\psi(z)|^2 |f \circ \varphi(z)|^2 (1 - |z|^2)^p \, dA(z)$$

is a p-Carleson measure. Therefore, by Theorem 5.2.2, $W_{\psi,\varphi} f \in \mathcal{N}_p$, and moreover,

$$\|W_{\psi,\varphi} f\|_{\mathcal{N}_p}^2 = \sup_{w \in \mathbb{D}} \int_{\mathbb{D}} \left(\frac{1 - |w|^2}{|1 - \overline{w}z|^2} \right)^p |\psi(z)|^2 |f \circ \varphi(z)|^2 (1 - |z|^2)^p \, dA(z)$$

$$\asymp \sup_{I \subset \mathbb{T}} |I|^{-p} \int_{S(I)} |\psi(z)|^2 (1 - |z|^2)^p |f \circ \varphi(z)|^2 \, dA(z)$$

$$\lesssim \|f\|_{H_\alpha^\infty}^2,$$

which shows that $W_{\psi,\varphi} : H_\alpha^\infty \to \mathcal{N}_p$ is bounded. $\square$

5.4   The Compactness of $W_{\psi,\varphi} : H_\alpha^\infty \to \mathcal{N}_p$

We now consider the compactness of $W_{\psi,\varphi}$.

As is usually the case in operator theory, if a "big-O" condition, like the one in Theorem 5.5.2, describes the bounded operators of a function space, the corresponding "little-o" condition describes the compact operators. Our weighted composition operator is not an exception. To verify this fact, we use a reformulation of weighted compactness for operators between $\mathcal{N}_p$-spaces and the Bergman-type spaces H_α^∞. More explicitly, a weighted composition operator $W_{\psi,\varphi} : H_\alpha^\infty \to \mathcal{N}_p$ is compact if and only if for every bounded sequence (f_n) in H_α^∞ that converges to 0 uniformly on compact subsets of $\mathbb{D}$, we have

$$\lim_{n\to\infty} \|W_{\psi,\varphi} f_n\|_{\mathcal{N}_p} = 0.$$

To do this, we need some preliminary results. Recall from Section 4.5 that

$$g_{\theta,r}^n(z) = z^n f_{\theta,r}(z),$$

where $f_{\theta,r}$ is the test function defined in (4.3). By Lemmas 4.5.1 and 4.5.2, we see that for each fixed $\theta \in [0, 2\pi)$ and $r \in (0, 1)$ and any compact operator $T: H_\alpha^\infty \to \mathcal{N}_p$, we have $T g_{\theta,r}^n \to 0$ in $\mathcal{N}_p$ as $n \to \infty$. However, for proving the main theorem of this section, we need a stronger result.

Before stating and proving such a result, we recall that a bounded linear operator T from a Banach space X to a Banach space Y is called *completely continuous* if, for every weakly convergent sequence (x_n) from X, the sequence $(T x_n)$ is norm-convergent in Y (see, e.g., [15, §VI.3]). Note that compact operators on Banach spaces are always completely continuous.

The following result is a straightforward consequence of a complete continuity of a compact operator and Lemma 4.5.2.

Lemma 5.4.1 *For any compact operator $T: H_\alpha^\infty \to \mathcal{N}_p$, we have*

$$\lim_{n\to\infty} \sup_{\theta,r} \|T g_{\theta,r}^n\|_{\mathcal{N}_p} = 0,$$

where the supremum is taken over all $\theta \in [0, 2\pi)$ and $r \in (0, 1)$.

Proof Since T is a compact operator from a Banach space H_α^∞ to $\mathcal{N}_p$, it is completely continuous. Furthermore, by Lemma 4.5.2, a sequence $(g_{\theta,r}^n)$ is weakly convergent in H_α^∞. Hence, the sequence $(T(g_{\theta,r}^n))$ is norm-convergent in $\mathcal{N}_p$, from which the desired result follows. $\qquad\square$

In the sequel, we study the essential norm of the weighted composition operators $W_{\psi,\varphi}$ between spaces H_α^∞ and $\mathcal{N}_p$.

Recall, in a general setting, that for a bounded linear operator L acting from a Banach space X to a Banach space Y, the essential norm of L on X, denoted by $\|L\|_e$, is the distance from L to the set of all compact operators from X to Y in the operator norm, i.e.,

$$\|L\|_e = \inf\{\|L - K\|_{op}\},$$

where the infimum is taken over all compact operators $K: X \to Y$. Clearly, L is compact if and only if $\|L\|_e = 0$.

Let us denote by $\mathcal{K} = \mathcal{K}(H_\alpha^\infty, \mathcal{N}_p,)$ the set of all compact operators acting from H_α^∞ into $\mathcal{N}_p$. Then the essential norm of $W_{u,\varphi}$ is

$$\|W_{\psi,\varphi}\|_e = \inf_{K \in \mathcal{K}} \left\{ \|W_{\psi,\varphi} - K\| \right\}.$$

We have the following estimates of the essential norm of $W_{\psi,\varphi}$.

Proposition 5.4.2 *Let $\psi \in Hol(\mathbb{D})$, and let φ be a holomorphic self-map of $\mathbb{D}$. Suppose that $W_{\psi,\varphi} : H_\alpha^\infty \to \mathcal{N}_p$ is bounded. Then*

$$\|W_{\psi,\varphi}\|_e^2 \lesssim \limsup_{r \to 1} \sup_{w \in \mathbb{D}} \int_{|\varphi(z)|>r} \frac{|\psi(z)|^2}{(1 - |\varphi(z)|^2)^{2\alpha}} (1 - |\sigma_w(z)|^2)^p \, dA(z).$$

Proof In order to prove upper estimates, consider $C_k f(z) = f\left(\frac{k}{k+1}z\right)$ for each positive integer k and $f \in Hol(\mathbb{D})$. It is clear that every C_k is bounded on H_α^∞. In fact, by the above reformulation of weighted compactness for operators between $\mathcal{N}_p$-spaces and the Bergman-type spaces H_α^∞, applying to the case $\psi \equiv 1$ and $\varphi(z) = \frac{k}{k+1}z$, C_k is also compact on H_α^∞. Consequently,

$$\|W_{\psi,\varphi}\|_e \leq \liminf_{k \to \infty} \|\psi C_\varphi - \psi C_\varphi C_k\|$$

$$= \liminf_{k \to \infty} \sup_{\|f\|_{H_\alpha^\infty} \leq 1} \|\psi C_\varphi (Id - C_k) f\|_{\mathcal{N}_p}. \tag{5.14}$$

Here, I denotes the identity operator on H_α^∞.

Fix a number $k \in \mathbb{N}$ and $f \in H_\alpha^\infty$ with $\|f\|_{H_\alpha^\infty} \leq 1$. For any $r \in (0, 1)$, we have

$$\|\psi C_\varphi (I - C_k) f\|_{\mathcal{N}_p}^2$$

$$\leq \sup_{w \in \mathbb{D}} \int_{|\varphi(z)| \leq r} |\psi(z)|^2 \left| f \circ \varphi(z) - f\left(\frac{k}{k+1}\varphi(z)\right) \right|^2 (1 - |\sigma_w(z)|^2)^p \, dA(z)$$

$$+ \sup_{w \in \mathbb{D}} \int_{|\varphi(z)| > r} |\psi(z)|^2 \left| f \circ \varphi(z) - f\left(\frac{k}{k+1}\varphi(z)\right) \right|^2 (1 - |\sigma_w(z)|^2)^p \, dA(z).$$

Furthermore, by the growth estimate for $f \in H_\alpha^\infty$, we also have

$$\left| f \circ \varphi(z) - f\left(\frac{k}{k+1}\varphi(z)\right) \right| \leq \frac{2\|f\|_{H_\alpha^\infty}}{(1 - |\varphi(z)|^2)^\alpha}. \tag{5.15}$$

Consequently, for any $r \in (0, 1)$ and any $k \in \mathbb{N}$, we get

$$\sup_{w \in \mathbb{D}} \int_{|\varphi(z)| > r} |\psi(z)|^2 \left| f \circ \varphi(z) - f\left(\frac{k}{k+1}\varphi(z)\right) \right|^2 (1 - |\sigma_w(z)|^2)^p \, dA(z)$$

$$\lesssim \sup_{w \in \mathbb{D}} \int_{|\varphi(z)|>r} \frac{|\psi(z)|^2}{(1-|\varphi(z)|^2)^{2\alpha}} (1-|\sigma_w(z)|^2)^p \, dA(z). \tag{5.16}$$

Next we prove that

$$\sup_{\|f\|_{H_\alpha^\infty}} \sup_{w \in \mathbb{D}} \int_{|\varphi(z)| \leq r} |\psi(z)|^2 \left| f \circ \varphi(z) - f\left(\frac{k}{k+1}\varphi(z)\right) \right|^2 (1-|\sigma_w(z)|^2)^p \, dA(z) \to 0 \tag{5.17}$$

as $k \to \infty$. Put $v = \varphi(z)$ and consider the radial segment by $[\frac{k}{k+1}v, v]$. Integrating f' along this radial segment gives

$$\left| f(v) - f\left(\frac{k}{k+1}v\right) \right| \leq \frac{1}{k+1} |v| |f'(\xi(v))|, \tag{5.18}$$

for some $\xi(v) \in [\frac{k}{k+1}v, v]$. By Cauchy's estimate for f' on the circle centered at $\xi(v)$ of radius $R \in (0, 1-r)$, we have

$$|f'(\xi(v))| \leq \frac{1}{R} \max_{|\zeta|=R+r} |f(\zeta)|. \tag{5.19}$$

Combining (5.18) and (5.19) shows that the quantity

$$\sup_{\|f\|_{H_\alpha^\infty}} \sup_{w \in \mathbb{D}} \int_{|\varphi(z)| \leq r} |\psi(z)|^2 \left| f \circ \varphi(z) - f\left(\frac{k}{k+1}\varphi(z)\right) \right|^2 (1-|\sigma_w(z)|^2)^p \, dA(z)$$

is bounded above by

$$\frac{r^2}{R^2(k+1)^2} \cdot \frac{1}{(1-(R+r)^2)^{2\alpha}} \cdot \|\psi\|_{\mathcal{N}_p}^2.$$

Since $\psi \in \mathcal{N}_p$, by the boundedness of $W_{\psi,\varphi}: H_\alpha^\infty \to \mathcal{N}_p$, we obtain (5.17). Now by (5.14), (5.16), (5.18), and (5.19), we deduce

$$\|W_{\psi,\varphi}\|_e^2 \lesssim \sup_{w \in \mathbb{D}} \int_{|\varphi(z)|>r} \frac{|\psi(z)|^2}{(1-|\varphi(z)|^2)^{2\alpha}} (1-|\sigma_w(z)|^2)^p \, dA(z),$$

from which the upper estimate for $\|W_{\psi,\varphi}\|_e^2$ follows.   □

Proposition 5.4.3 *Let $\psi \in Hol(\mathbb{D})$, and let φ be a holomorphic self-map of $\mathbb{D}$. Suppose that $W_{\psi,\varphi}: H_\alpha^\infty \to \mathcal{N}_p$ is bounded. Then*

$$\|W_{\psi,\varphi}\|_e^2 \gtrsim \limsup_{r\to 1} \sup_{w\in\mathbb{D}} \int_{|\varphi(z)|>r} \frac{|\psi(z)|^2}{(1-|\varphi(z)|^2)^{2\alpha}}(1-|\sigma_w(z)|^2)^p\,dA(z).$$

Proof Consider the functions $(g_{\theta,t}^n)_{n\in\mathbb{N}}$ defined in Lemma 4.5.2. By Lemma 4.5.1, the norm $\|g_{\theta,t}^n\|_{H_\alpha^\infty}$ is uniformly bounded with respect to θ, t, and n. So for any compact operator $K\colon H_\alpha^\infty \to \mathcal{N}_p$, we have

$$\|W_{\psi,\varphi}-K\| \gtrsim \|(W_{\psi,\varphi}-K)g_{\theta,t}^n\|_{\mathcal{N}_p} \geq \|W_{\psi,\varphi}g_{\theta,t}^n\|_{\mathcal{N}_p} - \|Kg_{\theta,t}^n\|_{\mathcal{N}_p}. \tag{5.20}$$

By Fatou's lemma, for any $r \in (0,1)$,

$$\sup_{\theta,t} \|W_{\psi,\varphi}g_{\theta,t}^n\|_{\mathcal{N}_p} \geq \liminf_{t\to 1} \int_{\mathbb{D}} |\psi(z)|^2 |g_{\theta,t}^n(\varphi(z))|^2 (1-|\sigma_w(z)|^2)^p\,dA(z)$$

$$\geq \int_{|\varphi(z)|>r} |\psi(z)|^2 |\varphi(z)|^{2n} |f_\theta(\varphi(z))|^2 (1-|\sigma_w(z)|^2)^p\,dA(z).$$

Here, $f_\theta(w)$ denotes the function $f_{\theta,1}(w)$. Integrating these inequalities with respect to θ from 0 to 2π, by Fubini's theorem, we get

$$\sup_{\theta,t} \|W_{\psi,\varphi}g_{\theta,t}^n\|_{\mathcal{N}_p} \tag{5.21}$$

$$\geq \int_{|\varphi(z)|>r} |\psi(z)|^2 |\varphi(z)|^{2n} (1-|\sigma_w(z)|^2)^p \left\{ \int_0^{2\pi} |f_\theta(\varphi(z))|^2 \frac{d\theta}{2\pi} \right\} dA(z).$$

Furthermore, by (5.8) and (5.9), for any $z \in \mathbb{D}$ with $\varphi(z)| > \frac{1}{\sqrt{2}}$, we have

$$\int_0^{2\pi} |f_\theta(\varphi(z))|^2 \frac{d\theta}{2\pi} \gtrsim \frac{1}{(1-|\varphi(z)|^2)^{2\alpha}}. \tag{5.22}$$

Combining (5.21) and (5.22) yields

$$\sup_{\theta,t} \|W_{\psi,\varphi}g_{\theta,t}^n\|_{\mathcal{N}_p} \gtrsim r^{2n} \sup_{w\in\mathbb{D}} \int_{|\varphi(z)|>r} \frac{|\psi(z)|^2}{(1-|\varphi(z)|^2)^{2\alpha}}(1-|\sigma_w(z)|^2)^p\,dA(z),$$

for all $r \in (\frac{1}{\sqrt{2}},1)$. Letting $r \to 1$, we obtain

$$\sup_{\theta,t} \|W_{\psi,\varphi}g_{\theta,t}^n\|_{\mathcal{N}_p}$$

$$\gtrsim \limsup_{r\to 1} \sup_{w\in\mathbb{D}} \int_{|\varphi(z)|>r} \frac{|\psi(z)|^2}{(1-|\varphi(z)|^2)^{2\alpha}}(1-|\sigma_w(z)|^2)^p\,dA(z) \tag{5.23}$$

Note that the last estimate does not depend on n. Since $\sup_{\theta,t}\|Kg_{\theta,t}^n\|_{\mathcal{N}_p} \to 0$ as $n \to \infty$ for any compact operator $K\colon H_\alpha^\infty \to \mathcal{N}_p$, by Lemma 5.4.1, (5.20) (5.23), we get

$$\|W_{\psi,\varphi}\|_e^2 \gtrsim \limsup_{r \to 1}\ \sup_{w \in \mathbb{D}} \int_{|\varphi(z)|>r} \frac{|\psi(z)|^2}{(1 - |\varphi(z)|^2)^{2\alpha}}(1 - |\sigma_w(z)|^2)^p\, dA(z).$$

$\square$

Combining Propositions 5.4.2 and 5.4.3 yields the estimate of the weighted composition operator $W_{\psi,\varphi}$ acting from the space H_α^∞ to the space $\mathcal{N}_p$.

Theorem 5.4.4 *Let $\psi \in Hol(\mathbb{D})$, and let φ be a holomorphic self-map of $\mathbb{D}$. Suppose that $W_{\psi,\varphi}\colon H_\alpha^\infty \to \mathcal{N}_p$ is bounded. Then*

$$\|W_{\psi,\varphi}\|_e^2 \asymp \limsup_{r \to 1}\ \sup_{w \in \mathbb{D}} \int_{|\varphi(z)|>r} \frac{|\psi(z)|^2}{(1 - |\varphi(z)|^2)^{2\alpha}}(1 - |\sigma_w(z)|^2)^p\, dA(z).$$

We provide another estimation which has the same spirit as Theorem 5.4.4.

Theorem 5.4.5 *Let $\psi \in Hol(\mathbb{D})$, and let φ be a holomorphic self-map of $\mathbb{D}$. Suppose that $W_{\psi,\varphi}\colon H_\alpha^\infty \to \mathcal{N}_p$ is bounded. Then*

$$\|W_{\psi,\varphi}\|_e^2 \asymp \limsup_{r \to 1}\ \sup_{I \subset \mathbb{T}} |I|^{-p} \int_{S(I) \cap \{|\varphi(z)|>r\}} \frac{|\psi(z)|^2(1 - |z|^2)^p}{(1 - |\varphi(z)|^2)^{2\alpha}}(1 - |\sigma_w(z)|^2)^p\, dA(z).$$

Here, the supremum is taken over all arcs $I \subset \mathbb{T}$.

Proof By Theorem 5.2.2, for each $r \in (0, 1)$, we have

$$\|W_{\psi,\varphi}(I - C_k)f\|_{\mathcal{N}_p}^2 \asymp \left\{ \sup_{I \subset \mathbb{T}} |I|^{-p} \int_{S(I) \cap \{|\varphi(z)|\leq r\}} + \sup_{I \subset \mathbb{T}} |I|^{-p} \int_{S(I) \cap \{|\varphi(z)|>r\}} \right\}$$

$$\left\{ |\psi(z)|^2 \left| f \circ \varphi(z) - f\left(\frac{k}{k+1}\varphi(z)\right) \right|^2 (1 - |z|^2)^p\, dA(z) \right\}.$$

We write J_1 and J_2 for the last two integrals. On the one hand, from inequality (5.15), it follows that for any $r \in (0, 1)$ and $f \in H_\alpha^\infty$ with $\|f\|_{H_\alpha^\infty} \leq 1$,

$$J_2 \lesssim \sup_{I \subset \mathbb{T}} |I|^{-p} \int_{S(I) \cap \{|\varphi(z)|>r\}} \sup_{w \in \mathbb{D}} \int_{|\varphi(z)|>r} \frac{|\psi(z)|^2}{(1 - |\varphi(z)|^2)^{2\alpha}}(1 - |\sigma_w(z)|^2)^p\, dA(z).$$

On the other hand, inequalities (5.18) and (5.19) and Theorem 5.2.2 imply that for $f \in H_\alpha^\infty$ with $\|f\|_{H_\alpha^\infty} \leq 1$,

$$J_1 \lesssim \frac{r^2}{R^2(k+1)^2} \cdot \frac{1}{(1-(R+r)^2)^{2\alpha}} \cdot \|\psi\|_{\mathcal{N}_p}^2,$$

and the last expression tends to 0 as $k \to \infty$. Note that this estimation holds uniformly on the unit ball of H_α^∞. Hence, by (5.14) and the estimations for J_1 and J_2, and by letting $r \to 1$, we get

$$\|W_{\psi,\varphi}\|_e^2 \lesssim \limsup_{r \to 1} \sup_{I \subset \mathbb{T}} |I|^{-p} \int_{S(I) \cap \{|\varphi(z)| > r\}} \frac{|\psi(z)|^2 (1-|z|^2)^p}{(1-|\varphi(z)|^2)^{2\alpha}} (1-|\sigma_w(z)|^2)^p \, dA(z).$$

The upper estimate is proved.

Also, by Theorem 5.2.2,

$$\sup_{\theta,t} \|W_{\psi,\varphi} g_{\theta,t}^n\|_{\mathcal{N}_p} \gtrsim |I|^{-p} \int_{S(I)} |\psi(z)|^2 |g_{\theta,t}^n(\varphi(z))|^2 (1-|z|^2)^p \, dA(z),$$

for all arcs I. A similar argument in the proof of Theorem 5.4.4 implies

$$\|W_{\psi,\varphi}\|_e^2 \gtrsim \limsup_{r \to 1} \sup_{w \in \mathbb{T}} |I|^{-p} \int_{S(I) \cap |\varphi(z)| > r} \frac{|\psi(z)|^2}{(1-|\varphi(z)|^2)^{2\alpha}} (1-|\sigma_w(z)|^2)^p \, dA(z).$$

The lower estimate is proved. $\square$

Corollary 5.4.6 *Let $\psi \in \mathrm{Hol}(\mathbb{D})$, and let φ be a holomorphic self-map of $\mathbb{D}$.*

(i) $W_{\psi,\varphi} : H_\alpha^\infty \to \mathcal{N}_p$ is a well-defined compact operator.
(ii) ψ and φ satisfy

$$\limsup_{r \to 1} \sup_{w \in \mathbb{D}} \int_{|\varphi(z)| > r} \frac{|\psi(z)|^2}{(1-|\varphi(z)|^2)^{2\alpha}} (1-|\sigma_w(z)|^2)^p \, dA(z) = 0.$$

(iii) ψ and φ satisfy

$$\limsup_{r \to 1} \sup_{I \subset \mathbb{T}} |I|^{-p} \int_{S(I) \cap \{|\varphi(z)| > r\}} \frac{|\psi(z)|^2 (1-|z|^2)^p}{(1-|\varphi(z)|^2)^{2\alpha}} (1-|\sigma_w(z)|^2)^p \, dA(z) = 0,$$

where the supremum is taken over all arcs $I \subset \mathbb{T}$.

The final remark of this section is that taking a special case $\psi(z) \equiv 1$ in theorems above, we can obtain the corresponding results for essential norm and compactness of composition operators C_φ acting from the space H_α^∞ to the space $\mathcal{N}_p$.

5.5   The Boundedness of $W_{\psi,\varphi}: \mathcal{N}_p \to H_\alpha^\infty$

In this section, we consider the weighted composition operator $W_{\psi,\varphi}$ acting from the space $\mathcal{N}_p$ to the space H_α^∞.

As it is proved in Proposition 3.4.2, the $\mathcal{N}_p$-spaces are functional Banach spaces. Thus, the weighted composition operator $W_{\psi,\varphi}: \mathcal{N}_p \to H_\alpha^\infty$ is well-defined and bounded if and only if it maps $\mathcal{N}_p$ into H_α^∞. Therefore, our first concern is boundedness of $W_{\psi,\varphi}: \mathcal{N}_p \to H_\alpha^\infty$, and we have the following result.

Theorem 5.5.1 *Let* $\psi \in \mathrm{Hol}(\mathbb{D})$ *and* φ *be a holomorphic self-map of* $\mathbb{D}$. *Then* $W_{\psi,\varphi}: \mathcal{N}_p \to H_\alpha^\infty$ *is a well-defined bounded operator if and only if* ψ *and* φ *satisfy*

$$\sup_{z \in \mathbb{D}} \frac{|\psi(z)|(1 - |z|^2)^\alpha}{1 - |\varphi(z)|^2} < \infty. \tag{5.24}$$

Proof First suppose $W_{\psi,\varphi}: \mathcal{N}_p \to H_\alpha^\infty$ is bounded. Consider the function

$$k_w(z) = \frac{1 - |w|^2}{(1 - \overline{w}z)^2}, \quad (z \in \mathbb{D}),$$

with $w = \varphi(z_0)$, where $z_0 \in \mathbb{D}$ is fixed. Since $k_w \in \mathcal{N}_p$ and $\|k_w\|_{\mathcal{N}_p} \leq 1$, the boundedness of $W_{\psi,\varphi}$ implies

$$1 \gtrsim \|W_{\psi,\varphi} k_w\|_{H_\alpha^\infty} \geq (1 - |z_0|^2)^\alpha |\psi(z_0)| |k_w(\varphi(z_0))| = \frac{|\psi(z_0)|(1 - |z_0|^2)^\alpha}{1 - |\varphi(z_0)|^2},$$

from which (5.24) follows.

Conversely, suppose that (5.24) is satisfied. Recall that $\mathcal{N}_p \hookrightarrow H_1^\infty$. Then for each $z \in \mathbb{D}$ and $f \in \mathcal{N}_p$, we have

$$(1 - |z|^2)^\alpha |W_{\psi,\varphi} f(z)| \leq \frac{|\psi(z)|(1 - |z|^2)^\alpha}{1 - |\varphi(z)|^2} \|f\|_{H_1^\infty} \lesssim \frac{|\psi(z)|(1 - |z|^2)^\alpha}{1 - |\varphi(z)|^2} \|f\|_{\mathcal{N}_p}.$$

From this, it follows that

$$\|W_{\psi,\varphi} f\|_{H_\alpha^\infty} \lesssim \|f\|_{\mathcal{N}_p},$$

which shows a boundedness of $W_{\psi,\varphi}: \mathcal{N}_p \to H_\alpha^\infty$. $\square$

As an immediate consequence of Theorem 5.5.1, we have the following characterization of bounded composition operator $C_\varphi: \mathcal{N}_p \to H_\alpha^\infty$.

Corollary 5.5.2 *The composition operator* $C_\varphi: \mathcal{N}_p \to H_\alpha^\infty$ *is well-defined and bounded
if and only if*

$$\sup_{z\in\mathbb{D}} \frac{(1-|z|^2)^\alpha}{1-|\varphi(z)|^2} < \infty. \tag{5.25}$$

5.6   The Compactness of $W_{\psi,\varphi}: \mathcal{N}_p \to H_\alpha^\infty$

In this section, we give a complete characterization of the compactness of $W_{\psi,\varphi}: \mathcal{N}_p \to H_\alpha^\infty$. The characterization is purely based on the function theoretic properties of ψ and φ.
Recall that the notation $\|\cdot\|_e$ denotes the essential norm of an operator.

Proposition 5.6.1 *Let* $\psi \in \mathrm{Hol}(\mathbb{D})$ *and* φ *be a holomorphic self-map of* $\mathbb{D}$. *Suppose that*
$W_{\psi,\varphi}: \mathcal{N}_p \to H_\alpha^\infty$ *is bounded. Then*

$$\|W_{\psi,\varphi}\|_e \lesssim \limsup_{|\varphi(z)|\to 1} \frac{|\psi(z)|(1-|z|^2)^\alpha}{1-|\varphi(z)|^2}. \tag{5.26}$$

Proof Consider the operator C_k as in the proof of Theorem 5.4.4. For each positive integer
k, we have

$$\|C_k f\|_{\mathcal{N}_p}^2 = \sup_{w\in\mathbb{D}} \int_\mathbb{D} \left| f\left(\frac{k}{k+1}z\right) \right|^2 (1-|\sigma_w(z)|^2)^p \, dA(z)$$

$$\leq \sup_{w\in\mathbb{D}} \int_\mathbb{D} \frac{\|f\|_{H_1^\infty}^2}{1-\frac{k}{k+1}z)^2} (1-|\sigma_w(z)|^2)^p \, dA(z)$$

$$\lesssim \frac{(k+1)^4}{(2k+1)^2} \|f\|_{\mathcal{N}_p}^2.$$

This shows that C_k is bounded on $\mathcal{N}_p$. Moreover, again by a reformulation of weighted
compactness for operators between $\mathcal{N}_p$-spaces and the Bergman-type spaces H_α^∞, C_k is
compact on $\mathcal{N}_p$, and hence, $W_{\psi,\varphi}C_k$ is compact from $\mathcal{N}_p$ to H_α^∞. Hence, we have

$$\|W_{\psi,\varphi}\|_e \leq \|W_{\psi,\varphi} - W_{\psi,\varphi}C_k\| = \sup_{\|f\|_{\mathcal{N}_p}\leq 1} \|W_{\psi,\varphi}(I-C_k)f\|_{H_\alpha^\infty}, \tag{5.27}$$

where I is the identity operator on $\mathcal{N}_p$. Fix a positive number k and $f \in \mathcal{N}_p$ with $\|f\|_{\mathcal{N}_p} \le 1$. For any $r \in (0, 1)$, we have

$$\|W_{\psi,\varphi}(I - C_k)f\|_{H_\alpha^\infty} \le \sup_{|\varphi(z)| \le r} (1 - |z|^2)|\psi(z)| \left| f \circ \varphi(z) - f\left(\frac{k}{k+1}\varphi(z)\right) \right|$$

$$+ \sup_{|\varphi(z)| > r} (1 - |z|^2)|\psi(z)| \left| f \circ \varphi(z) - f\left(\frac{k}{k+1}\varphi(z)\right) \right|$$

Furthermore,

$$\left| f \circ \varphi(z) - f\left(\frac{k}{k+1}\varphi(z)\right) \right| \lesssim \frac{2\|f\|_{\mathcal{N}_p}}{1 - |\varphi(z)|^2},$$

and thus, for each $r \in (0, 1)$,

$$\sup_{|\varphi(z)| > r} (1 - |z|^2)|\psi(z)| \left| f \circ \varphi(z) - f\left(\frac{k}{k+1}\varphi(z)\right) \right| \lesssim sup_{|\varphi(z)| > r} \frac{|\psi(z)|(1 - |z|^2)^\alpha}{1 - |\varphi(z)|^2}.$$

Moreover, by the same argument in the proof of Theorem 5.4.4, inequalities (5.18) and (5.19) show that for $R \in (0, 1 - r)$, we have

$$\sup_{|\varphi(z)| \le r} (1 - |z|^2)|\psi(z)| \left| f \circ \varphi(z) - f\left(\frac{k}{k+1}\varphi(z)\right) \right|$$

$$\lesssim \frac{r}{R(k+1)} \frac{\|\psi\|_{H_\alpha^\infty}}{1 - (R+r)^2}, \tag{5.28}$$

which tends to 0 as $k \to \infty$. By (5.26), (5.27), and (5.28), we have for any $r \in (0, 1)$

$$\|W_{\psi,\varphi}\|_e \lesssim \sup_{|\varphi(z)| > r} \frac{|\psi(z)|(1 - |z|^2)^\alpha}{1 - |\varphi(z)|^2}.$$

Letting $r \to 1$, we get the desired result. $\qquad\square$

Proposition 5.6.2 *Let $\psi \in Hol(\mathbb{D})$ and φ be a holomorphic self-map of $\mathbb{D}$. Suppose that $W_{\psi,\varphi} : \mathcal{N}_p \to H_\alpha^\infty$ is bounded. Then*

$$\|W_{\psi,\varphi}\|_e \gtrsim \limsup_{|\varphi(z)| \to 1} \frac{|\psi(z)|(1 - |z|^2)^\alpha}{1 - |\varphi(z)|^2}. \tag{5.29}$$

Proof Take any sequence $(z_n) \in \mathbb{D}$ with $|\varphi(z_n)| \to 1$ as $j \to \infty$. Put $w_n = \varphi(z_n)$ and $f_n(z) = k_{w_n}(z)$, where k_w is the function defined in the proof of Theorem 5.5.1. Then (f_n)

forms a bounded sequence in $\mathcal{N}_p$ which converges uniformly to 0 on compact subsets of $\mathbb{D}$. More precisely, inequality

$$\|f_n\|_{H_1^\infty} \lesssim \|f_n\|_{\mathcal{N}_p} \leq 1$$

shows that (f_n) is a bounded sequence in $H_{1,0}^\infty$. Note that $(f_n) \to 0$ weakly in $\mathcal{N}_p$. Then $\|Kf_n\|_{H_\alpha^\infty} \to 0$ as $j \to \infty$ for any compact operator $K : \mathcal{N}_p \to H_\alpha^\infty$.

We also have

$$\|W_{\psi,\varphi}\|_{H_\alpha^\infty} \geq (1 - |z_n|^2)^\alpha \|W_{\psi,\varphi}f(z_n)\| = \frac{|\psi(z_n)|(1 - |z_n|^2)^\alpha(1 - |\varphi(z_n)|^2)}{(1 - |\varphi(z_n)|^2)^2},$$

and hence, for any $j \in \mathbb{N}$,

$$\|W_{\psi,\varphi}\|_{H_\alpha^\infty} \geq \frac{|\psi(z_n)|(1 - |z_n|^2)^\alpha}{1 - |\varphi(z_n)|^2}.$$

Combining the above arguments and the following inequalities

$$\|W_{\psi,\varphi} - K\| \gtrsim \|(W_{\psi,\varphi} - K)f_n\|_{H_\alpha^\infty} \geq \|W_{\psi,\varphi}f_n\|_{H_\alpha^\infty} - \|Kf_n\|_{H_\alpha^\infty},$$

for any compact operator $K : \mathcal{N}_p \to H_\alpha^\infty$, yields

$$\|W_{\psi,\varphi}\|_e \gtrsim \limsup_{j \to \infty} \frac{|\psi(z_n)|(1 - |z_n|^2)^\alpha}{1 - |\varphi(z_n)|^2}.$$

Since $(z_n) \subset \mathbb{D}$ with $|\varphi(z_n)| \to 1$ as $j \to \infty$ is arbitrary, this implies

$$\|W_{\psi,\varphi}\|_e \gtrsim \limsup_{|\varphi(z)| \to 1} \frac{|\psi(z)|(1 - |z|^2)^\alpha}{1 - |\varphi(z)|^2}.$$

$\square$

Combining Propositions 5.6.1 and 5.6.2 yields the following result.

Theorem 5.6.3 *Let* $\psi \in Hol(\mathbb{D})$ *and* φ *be a holomorphic self-map of* $\mathbb{D}$. *Suppose that* $W_{\psi,\varphi} : \mathcal{N}_p \to H_\alpha^\infty$ *is bounded. Then*

$$\|W_{\psi,\varphi}\|_e \asymp \limsup_{|\varphi(z)| \to 1} \frac{|\psi(z)|(1 - |z|^2)^\alpha}{1 - |\varphi(z)|^2}. \tag{5.30}$$

From Theorem 5.6.3, we obtain a complete characterization of boundedness of weighted composition operator $W_{\psi,\varphi}$ acting from the space $\mathcal{N}_p$ to the space H_α^∞.

Theorem 5.6.4 *Let $\psi \in \mathrm{Hol}(\mathbb{D})$ and φ be a holomorphic self-map of $\mathbb{D}$. The weighted composition operator $W_{\psi,\varphi}: \mathcal{N}_p \to H_\alpha^\infty$ is a well-defined compact operator if and only if ψ and φ satisfy*

$$\lim_{|\varphi(z)| \to 1} \frac{|\psi(z)|(1 - |z|^2)^\alpha}{1 - |\varphi(z)|^2} = 0.$$

As an immediate consequence of Theorem 5.5.1, we have the following characterization of bounded composition operator $C_\varphi: \mathcal{N}_p \to H_\alpha^\infty$.

Corollary 5.6.5 *The composition operator $C_\varphi: \mathcal{N}_p \to H_\alpha^\infty$ is compact if and only if*

$$\lim_{r \to 1} \sup_{|\varphi(z)| > r} \frac{(1 - |z|^2)^\alpha}{1 - |\varphi(z)|^2} = 0. \tag{5.31}$$

Corollary 5.6.5 can also be proved directly in another way. Below we provide such a proof which may have its own interest.

Proof Suppose $C_\varphi: \mathcal{N}_p \to H_\alpha^\infty$ is compact. Assume, on the contrary, that there exist $\varepsilon_0 > 0$ and a sequence $(z_n) \subset \mathbb{D}$ such that

$$\frac{(1 - |z_n|^2)^\alpha}{1 - |\varphi(z_n)|^2} \geq \varepsilon_0 \quad \text{whenever } |\varphi(z_n)| > 1 - \frac{1}{n}.$$

By passing to a subsequence if needed, we may assume that $w_n = \varphi(z_n)$ tends to $w_0 \in \mathbb{T}$ as $n \to \infty$.

Let us consider the sequence k_{w_n}, which clearly satisfies $k_{w_n} \to k_{w_0}$ with respect to the compact-open topology. Put $f_n = k_{w_n} - k_{w_0}$. Then, by Lemma 3.2.1, $\|f_n\|_{\mathcal{N}_p} \leq 1$ and $f_n \to 0$ uniformly on compact subsets of $\mathbb{D}$. Hence, $f_n \circ \varphi \to 0$ in the norm of H_α^∞. However, for n large enough, we have

$$\|C_\varphi f_n\|_{H_\alpha^\infty} \geq |k_{w_n}(\varphi(z_n)) - k_{w_0}(\varphi(z_n))|(1 - |z_n|^2)^\alpha$$

$$= \frac{(1 - |z_n|^2)^\alpha}{1 - |\varphi(z_n)|^2} \left| 1 - \frac{(1 - |w_0|^2)(1 - |w_n|^2)}{|1 - \overline{w_0}w_n|^2} \right| \geq \varepsilon_0,$$

which is a contradiction.

Conversely, assume that (5.31) holds. Let (f_n) be a bounded sequence of functions in $\mathcal{N}_p$ which converges to 0 on compact subsets of $\mathbb{D}$. We may surely assume that $\varphi(z)| > \delta$. Then we have

$$\|C_\varphi f_n\|_{H_\alpha^\infty} = \sup_{z\in\mathbb{D}} \frac{(1-|z|^2)^\alpha}{1-|\varphi(z)|^2} |f_n(\varphi(z))|(1-|\varphi(z)|^2).$$

By Proposition 3.4.1, we obtain

$$\|C_\varphi f_n\|_{H_\alpha^\infty} \leq \varepsilon \|f_n\|_{H^\infty} \lesssim \varepsilon \|f_n\|_{\mathcal{N}_p} \leq \varepsilon.$$

Hence, $C_\varphi f_n \to 0$ is the norm topology of H_α^∞, and this fact implies that C_φ is compact.

$\square$

Notes on Chapter 5

The results of this chapter are mostly from [95]. Lemma 5.2.1 is proved in [6, Lemma 2.1]. The counterpart of Theorem 5.1.1 for $\mathcal{Q}_p$-spaces is still unknown. In [101,102], Xiao conjectured that M_ψ is bounded on $\mathcal{Q}_p$ if and only if

$$\psi \in H^\infty \quad \text{and} \quad \sup_{a\in\mathbb{D}} \log^2(1-|a|) \int_{\mathbb{D}} |\psi'(z)|^2 (1-|\sigma_a(z)|^2)^p \, dA(z) < \infty.$$

In [102, p. 22], Xiao also stated as an open problem to characterize the bounded composition operators on $\mathcal{Q}_p$. The problem was completely solved by Pau and Peláez [75]. We state here as an open problem to give a full characterization of when C_φ is bounded on $\mathcal{N}_p$. By doing so, one should be able to combine this with Theorem 5.1.1 and thereby to give a full description of when $W_{\varphi,\psi}$ is bounded on $\mathcal{N}_p$. Having done that, the open problem about bounded composition operators on $\mathcal{Q}_p$ should be solvable. A similar interplay between weighted composition operators (on Bergman-type spaces) and composition operators (on Bloch-type spaces) has been done in [14] and also in [74].

Using the derivative operator $f \mapsto f'$, $\mathcal{Q}_p$-spaces are closely related to $\mathcal{N}_p$-spaces and Bloch-type spaces $\mathcal{B}_\alpha$ related to H_α^∞. Hence, the results of this chapter also cover the corresponding results for C_φ (with $\psi = \varphi'$) acting between $\mathcal{B}_\alpha$ and $\mathcal{Q}_p$-spaces which are presented in [102]. In [100], Xiao characterized the boundedness and compactness of $C_\varphi \colon \mathcal{B} \to \mathcal{Q}_p$ by using a p-Carleson measure. For the case $W_{\psi,\varphi}$, an argument based on a p-Carleson measure is adopted. In [6], the authors characterized the $\mathcal{Q}_p$-space in terms of a p-Carleson measure. For the $\mathcal{N}_p$-space, an analogous characterization holds. The well-known result on the compactness of the composition operator on the Hardy spaces is provided in [17, Proposition 3.11].

In this chapter, the Hadamard gap series and the growth rate of the functions in the space H_μ^∞ in the unit disc are studied.

6.1   Hadamard Gaps

An analytic function $f(z)$ on the open unit disc $\mathbb{D}$ is said to have *Hadamard gaps* if its Taylor series expansion has the form

$$f(z) = \sum_{k=1}^{\infty} a_k z^{n_k},$$

where $n_{k+1}/n_k \geq q > 1$, for all $k \geq 1$. Our goal is to find a necessary and sufficient condition on μ which guarantees that f belongs to the weighted space H_μ^∞ or to the corresponding little weighted space $H_{\mu,0}^\infty$. We recall two classical results about lacunary series.

Theorem 6.1.1 (Hardy-Littlewood)   *Suppose that* $f(z) = \sum_{k=1}^{\infty} c_k z^{n_k}$ *has a Hadamard gap. Assume that the radial limit*

$$\lim_{r \to 1} f(re^{i\theta_0}) = f(e^{i\theta_0})$$

exists. Then the series $\sum_{k=1}^{\infty} c_k e^{in_k\theta_0}$ *is convergent.*

© The Author(s), under exclusive license to Springer Nature Switzerland AG 2023   91
L. H. Khoi, J. Mashreghi, *Theory of N_p Spaces*, Frontiers in Mathematics,
https://doi.org/10.1007/978-3-031-39704-2_6

Theorem 6.1.2 (Zygmund) *Suppose that the trigonometric series*

$$\sum_{k=1}^{\infty} a_k \cos(n_k\theta) + b_k \sin(n_k\theta),$$

in which $n_{k+1}/n_k \geq q > 1$, converges on a set of positive measure. Then the numerical series

$$\sum_{k=1}^{\infty}(a_k^2 + b_k^2)$$

converges.

6.2   α-Bloch Spaces

Recall that for $\alpha \in (0, \infty)$, the α-Bloch space $\mathcal{B}_\alpha$ consists of all functions $f \in \mathrm{Hol}(\mathbb{D})$ which satisfy the condition

$$\|f\|_{\mathcal{B}_\alpha} = \sup_{z \in \mathbb{D}}(1 - |z|)^\alpha |f'(z)| < \infty.$$

The space $\mathcal{B}^1$ is the Bloch space $\mathcal{B}$, while for $\alpha \in (0, 1)$, $\mathcal{B}_\alpha$ is precisely the classical $(1 - \alpha)$-Lipschitz classes. It is also well-known that $\mathcal{B}_\alpha$ is a Banach space with respect to the norm $|f(0)| + \|f\|_{\mathcal{B}_\alpha}$. The little α-Bloch space $\mathcal{B}_{\alpha,0}$ consists of functions $f \in \mathrm{Hol}(\mathbb{D})$ with

$$\lim_{|z| \to 1} (1 - |z|)^\alpha |f'(z)| = 0.$$

For α-Bloch spaces, the following result holds.

Lemma 6.2.1 *Let $g(z) = \sum_{n=0}^{\infty} b_n z^n \in \mathrm{Hol}(\mathbb{D})$. If $g \in \mathcal{B}_\alpha$, then*

$$\limsup_{n \to \infty} |b_n| n^{1-\alpha} < \infty.$$

Similarly, if $g \in \mathcal{B}_{\alpha,0}$, then

$$\lim_{n \to \infty} |b_n| n^{1-\alpha} = 0.$$

Proof Suppose $g \in \mathcal{B}_\alpha$. By the Cauchy formula, we have

$$|b_n| = \left| \frac{1}{2\pi i n} \int_0^{2\pi} g'(re^{i\theta}) r^{1-n} e^{i(1-n)\theta} \, d\theta \right| \lesssim \frac{1}{n} (1-r)^{-\alpha} r^{1-n},$$

for all $0 < r < 1$ and $n \geq 1$. Hence, for $n > 1$ and the optimal choice $r = 1 - \frac{1}{n}$, we get

$$|b_n| \lesssim n^{\alpha-1} \left(1 - \frac{1}{n}\right)^{1-n},$$

which implies that

$$\limsup_{n \to \infty} |b_n| n^{1-\alpha} < \infty.$$

The proof for the case $g \in \mathcal{B}_{\alpha,0}$ is similar.   □

6.3   α-Bloch Spaces and Hadamard Gaps

We continue to complete Lemma 6.2.1 and provide a characterization of analytic functions with Hadamard gaps in α-Bloch spaces.

Theorem 6.3.1 *Let f be a holomorphic function on $\mathbb{D}$ with Hadamard gaps. Then the following assertions hold.*

(i) $f \in \mathcal{B}_\alpha$ if and only if $\limsup_{k \to \infty} |a_k| n_k^{1-\alpha} < \infty$.
(ii) $f \in \mathcal{B}_{\alpha,0}$ if and only if $\lim_{k \to \infty} |a_k| n_k^{1-\alpha} = 0$.

Proof By Lemma 6.2.1, it suffices to prove the sufficiency. First, consider the case

$$\limsup_{k \to \infty} |a_k| n_k^{1-\alpha} < \infty. \tag{6.1}$$

Recall the expansion

$$\frac{1}{(1-|z|)^{1+\alpha}} = \sum_{n=0}^{\infty} A_n |z|^n, \quad A_n \sim \Gamma(1+\alpha)^{-1} n^\alpha.$$

Thus, we have

$$\sum_{n=0}^{\infty} (n+1)^\alpha |z|^n \leq \frac{C_2}{(1-|z|)^{1+\alpha}}, \quad (z \in \mathbb{D}). \tag{6.2}$$

It then follows from (6.1) that

$$|zf'(z)| = \left|\sum_{k=1}^{\infty} a_k n_k z^{n_k}\right| \lesssim \sum_{k=1}^{\infty} n_k^\alpha |z|^{n_k}.$$

Hence, making use of the Cauchy product, we obtain

$$\frac{|zf'(z)|}{1-|z|} \lesssim \sum_{n=1}^{\infty} \left(\sum_{n_k \leq n} n_k^\alpha\right)|z|^n.$$

Let $K = \max\{k : n_k \leq n\}$. We have

$$n^{-\alpha} \sum_{n_k \leq n} n_k^\alpha = \left(\frac{n_K}{n}\right)^\alpha \left[1 + \left(\frac{n_{K-1}}{n_K}\right)^\alpha + \cdots + \left(\frac{n_1}{n_K}\right)^\alpha\right]$$

$$\leq 1 + q^{-\alpha} + q^{-2\alpha} + \cdots = \frac{q^\alpha}{q^\alpha - 1}. \tag{6.3}$$

Therefore, by (6.2), we get

$$\frac{|zf'(z)|}{1-|z|} \lesssim \sum_{n=1}^{\infty} n^\alpha |z|^n$$

$$\lesssim \sum_{n=0}^{\infty} (n+1)^\alpha |z|^n$$

$$\lesssim \frac{|z|}{(1-|z|)^{\alpha+1}}, \qquad (z \in \mathbb{D}),$$

which implies that $f \in \mathcal{B}_\alpha$.

Next, suppose that

$$\lim_{k \to \infty} |a_k| n_k^{1-\alpha} = 0. \tag{6.4}$$

For any $\varepsilon > 0$, there exists $k_0 \geq 2$ such that

$$|a_k| n_k^{1-\alpha} < \varepsilon, \qquad (k \geq k_0).$$

Put

$$P(z) = \frac{1}{|z|} \sum_{k=1}^{k_0-1} |a_k| n_k |z|^{n_k},$$

which is bounded in $\mathbb{D}$. Then there exists $r \in (0, 1)$ such that

$$(1 - |z|)^\alpha P(z) < \varepsilon, \qquad (r < |z| < 1). \tag{6.5}$$

Then we have

$$|zf'(z)| \le \sum_{k=1}^\infty |a_k| n_k |z|^{n_k} \le |z| P(z) + \varepsilon \sum_{k=k_0}^\infty n_k^\alpha |z|^{n_k},$$

which implies that

$$\frac{|zf'(z)|}{1 - |z|} \le \frac{|z| P(z)}{1 - |z|} + \varepsilon \sum_{n=1}^\infty \Big(\sum_{\substack{n_k \le n \\ k \ge k_0}} n_k^\alpha \Big) |z|^n.$$

By (6.2) and (6.3), we have

$$\frac{|zf'(z)|}{1 - |z|} \lesssim \frac{|z| P(z)}{1 - |z|} + \frac{\varepsilon |z|}{(1 - |z|)^{1+\alpha}}. \tag{6.6}$$

Combining (6.5) and (6.6) yields

$$(1 - |z|^\alpha)|f'(z)| \lesssim \varepsilon, \qquad (r < |z| < 1),$$

which shows that $f \in \mathcal{B}_{\alpha,0}$.   □

We recall a celebrated theorem of Fatou, which says that a bounded holomorphic function in the unit disc has radial limits almost everywhere on the unit circle [27]. Moreover, we need the notion of Fatou's set whose elements are called Fatou's points. Namely, let $f(z)$ be a non-constant holomorphic function from the Riemann sphere onto itself. Such functions are precisely the non-constant complex rational functions, that is, $f(z) = p(z)/q(z)$, where $p(z)$ and $q(z)$ are complex polynomials. We may assume that p and q have no common roots and at least one has degree larger than 1. Then there is a finite number of open sets $F_1, \ldots, F_k$, which satisfy the following conditions.

(i) They are invariant under f.
(ii) The union of the sets F_j is dense in the plane.
(iii) $f(z)$ behaves in a regular and equal way on each of the sets F_j.

The last statement means that either the termini of the sequences of iterations generated by the points of F_j are precisely the same set, which is then a finite cycle, or they are finite cycles of circular- or annular-shaped sets that are lying concentrically. In the first case, the cycle is called attracting and in the second, neutral. The sets F_j are the Fatou domains of $f(z)$, and their union is called the Fatou set $F(f)$ of f. Each of Fatou's domains contains at least one critical point of f, that is, a finite point z satisfying $f'(z) = 0$ or $f(z) = \infty$ if the degree of the numerator p is at least twice larger than the degree of the denominator q or if $f = 1/q + c$ for some c and a rational function g.

As a corollary of Theorem 6.3.1, we have the following result for Fatou's points.

Corollary 6.3.2 *Let $f \in Hol(\mathbb{D})$ with the Hadamard gap series expansion. If*

$$\sum_{k=1}^{\infty} |a_k|^2 = \infty \quad and \quad \lim_{k \to \infty} |a_k| = 0,$$

then $f \in \mathcal{B}_{1,0}$, and f has no finite radial limit at almost every point of the unit circle $\mathbb{T}$.

Proof We first note that (6.4) holds whenever $\alpha = 1$, because $|a_k| \to 0$ as $k \to \infty$. Hence, by the Hardy–Littlewood and Zygmund Theorems 6.1.1 and 6.1.2, f has no finite radial limit at almost every point of $\mathbb{T}$.

Let g be meromorphic in $\mathbb{D}$, $F(g)$ the set of all Fatou's points of g, and $F^*(g)$ the set of $\xi \in F(g)$ where g has a finite angular limit. Then $F(g) - F^*(g)$ is of Lebesgue measure zero. Since g is pole-free in a terminal part of each angular domain at $\xi \in F^*(g)$, it follows from the Cauchy formula for g' that $F^*(g) \subset F'(g)$, where $F'(g)$ is the set of $\xi \in \mathbb{T}$ where $(1 - |z|)|f'(z)|$ has zero as the angular limit. Now considering the function f in the corollary, we see that $F'(f) = \mathbb{T}$ and $F^*(f)$ is of measure zero. In other words, the set $F'(f) - F^*(f)$ has full measure 2π.. $\qquad\qquad \square$

Applying Theorem 6.3.1, we now prove the following result.

Corollary 6.3.3 *Let $\alpha \in (0, \infty)$. Then there exist two functions $f_1, f_2 \in \mathcal{B}_\alpha$ and a positive constant C, such that*

$$|f_1'(z)| + |f_2'(z)| \geq \frac{C}{(1 - |z|^2)^\alpha}, \qquad (z \in \mathbb{D}).$$

Equivalently, there exist two functions $g_1, g_2 \in H_\alpha^\infty$, such that

$$|g_1(z)| + |g_2(z)| \geq \frac{C}{(1 - |z|^2)^\alpha}, \qquad (z \in \mathbb{D}).$$

Proof For a large integer q, consider the gap series

$$f_\alpha(z) = \sum_{j=0}^{\infty} q^{j(\alpha-1)} z^{q^j}, \qquad (z \in \mathbb{D}).$$

By Theorem 6.3.1, with $b_j = q^{j(\alpha-1)}$ and $n_j = q^j$, we see that

$$(1 - |z|^2)^\alpha |f'_\alpha(z)| \lesssim 1, \qquad (z \in \mathbb{D}).$$

Now we show that

$$(1 - |z|^2)^\alpha |f'_\alpha(z)| \gtrsim 1, \quad 1 - q^{-k} \le |z| \le 1 - q^{-(k+1/2)}, \quad (k \ge 1). \tag{6.7}$$

Indeed, for each $z \in \mathbb{D}$, we have

$$|f'_\alpha(z)| \ge q^{k\alpha} |z|^{q^\alpha} - \sum_{j=0}^{k-1} q^{j\alpha} |z|^{q^j} - \sum_{j=k+1}^\infty q^{j\alpha} |z|^{q^j} = S_1 - S_2 - S_3.$$

For the part S_1, we fix z with $|z| \in [1 - q^{-k}, 1 - q^{-(k+1/2)}], k \ge 1$, and put $\tau = |z|^{q^k}$. Then

$$(1 - q^{-k})^{q^k} \le \tau \le \left((1 - q^{-(k+1/2)})^{q^{k+1/2}}\right)^{q^{-1/2}}.$$

With q large enough, $k \ge 1$, we have

$$\frac{1}{3} \le \tau \le \left(\frac{1}{2}\right)^{q^{-1/2}}, \tag{6.8}$$

and hence $S_1 \ge \frac{q^{k\alpha}}{3}$. Next, it is easy to check that

$$S_2 \le \sum_{j=0}^{k-1} q^{j\alpha} \le \frac{q^{k\alpha}}{q^\alpha - 1}.$$

Lastly, note that

$$|z|^{q^n(q-1)} \le |z|^{q^{k+1}(q-1)}, \qquad (n \ge k+1).$$

In the sum S_3, the quotient of two successive terms is not bigger than the ratio of the first two terms. Hence, the series of S_3 is controlled by the geometric series which has the same first two terms. Thus, by (6.8), we get

$$S_3 \le q^{(k+1)\alpha} |z|^{q^{k+1}} \sum_{j=0}^\infty \left(q^\alpha |z|^{(q^{k+2} - q^{k+1})}\right)^j$$

$$= \frac{q^{(k+1)\alpha}|z|^{q^{k+1}}}{1 - q^\alpha|z|^{(q^{k+2}-q^{k+1})}}$$

$$= \frac{q^{k\alpha}q^\alpha\tau^q}{1 - q^\alpha\tau^{(q^2-q)}}$$

$$\leq \frac{q^{k\alpha}q^\alpha 2^{-q^{1/2}}}{1 - q^\alpha 2^{-(q^{3/2}-q^{1/2})}}.$$

Therefore, we obtain

$$|f_\alpha'(z)| \geq \frac{q^{k\alpha}}{4} = \frac{(q^\alpha)^{k+1/2}}{4q^{\alpha/2}} \geq \frac{1}{4q^{\alpha/2}(1 - |z|)^\alpha},$$

which gives (6.7).   □

6.4   Normal Weights

A positive continuous function μ on $[0, 1)$ is called *normal* if there exist positive constants α and β with $\alpha < \beta$, and $\delta \in (0, 1)$, such that

$$\frac{\mu(r)}{(1 - r)^\alpha} \text{ is decreasing on } [\delta, 1), \quad \lim_{r \to 1} \frac{\mu(r)}{(1 - r)^\alpha} = 0,$$

and

$$\frac{\mu(r)}{(1 - r)^\beta} \text{ is increasing on } [\delta, 1), \quad \lim_{r \to 1} \frac{\mu(r)}{(1 - r)^\beta} = \infty.$$

The normal function itself is decreasing in a neighborhood of 1 and satisfies $\lim_{r \to 1^-} \mu(r) = 0$. For example,

$$\mu(r) = 1/\log\log e^2(1 - r^2)^{-1}$$

and

$$\mu(r) = (1 - r^2)^a \log^b\left(e^{b/a}(1 - r^2)^{-1}\right), \qquad (a > 0, b \geq 0),$$

are normal functions.

Lemma 6.4.1 *Suppose μ is normal and $0 < \alpha < \min\left\{\frac{1}{e}, 1 - \delta\right\}$. Then there is a positive constant C, independent of α, such that*

$$\int_e^\infty \frac{e^{-\alpha t}}{\mu(1-\frac{1}{t})} \, dt \leq \frac{C}{\alpha\mu(1-\alpha)}.$$

Proof We have

$$\int_e^\infty \frac{e^{-\alpha t}}{\mu(1-\frac{1}{t})} \, dt = \int_e^{1/\alpha} \frac{e^{-\alpha t}}{\mu(1-\frac{1}{t})} \, dt + \int_{1/\alpha}^\infty \frac{e^{-\alpha t}}{\mu(1-\frac{1}{t})} \, dt = A + B.$$

On the one hand, since μ is essentially decreasing on $[0, 1)$,

$$A \leq \int_e^{1/\alpha} \frac{dt}{\mu(1-\frac{1}{t})} \leq \frac{C}{\mu(1-\alpha)} \int_e^{1/\alpha} dt \leq \frac{C}{\alpha\mu(1-\alpha)}.$$

On the other hand, since μ is normal,

$$B \leq \int_{1/\alpha}^\infty \frac{(\alpha t)^b e^{-\alpha t}}{\mu(1-\frac{1}{t})} \, dt = \frac{\int_1^\infty s^b e^{-s} \, ds}{\alpha\mu(1-\alpha)} \leq \frac{C}{\alpha\mu(1-\alpha)}.$$

Combining these inequalities yields the desired result.   □

In the sequel, positive constants are denoted by C, and they may have different values at different places.

6.5 H_μ^∞ and $H_{\mu,0}^\infty$ Spaces

The weighted space H_μ^∞ is defined as

$$H_\mu^\infty = \left\{ f \in \mathrm{Hol}(\mathbb{D}) : \|f\| = \sup_{z \in \mathbb{D}} |f(z)| \mu(|z|) < \infty \right\},$$

where μ is normal on $[0, 1)$. It is straightforward to verify that H_μ^∞ is a Banach space. The little weighted space $H_{\mu,0}^\infty$ is the closed subspace of H_μ^∞ that consists of $f \in H_\mu^\infty$ satisfying

$$\lim_{|z| \to 1^-} |f(z)| \mu(|z|) = 0.$$

In the special but important case $\mu(|z|) = (1 - |z|^2)^\alpha$, $\alpha > 0$, the induced spaces H_μ^∞ and $H_{\mu,0}^\infty$ are denoted by H_α^∞ and $H_{\alpha,0}^\infty$, respectively.

Lemma 6.5.1 *Suppose* $f(z) = \sum\limits_{n=0}^{\infty} a_n z^n \in H_\mu^\infty$. *Then*

$$\limsup_{n\to\infty} |a_n| \mu\left(1 - \frac{1}{n}\right) < \infty. \tag{6.9}$$

Proof By the Cauchy integral formula, we have

$$|a_n| = \left| \frac{1}{2\pi} \int_0^{2\pi} f(re^{i\theta}) r^{-n} e^{-in\theta}\, d\theta \right|$$

$$\leq \frac{1}{2\pi} \int_0^{2\pi} |f(re^{i\theta})| r^{-n}\, d\theta \leq \|f\|_{H_\mu^\infty} \frac{r^{-n}}{\mu(r)}. \tag{6.10}$$

For $n \geq 2$, we choose the optimal value $r = 1 - \frac{1}{n}$ to get the estimation

$$|a_n| \leq \|f\|_{H_\mu^\infty} \left(\mu\left(1 - \frac{1}{n}\right) \right)^{-1} \left(1 - \frac{1}{n}\right)^{-n}.$$

Hence,

$$\limsup_{n\to\infty} |a_n| \left(1 - \frac{1}{n}\right) \leq \|f\|_{H_\mu^\infty} \cdot e < \infty.$$

$\square$

For the space $H_{\mu,0}^\infty$, we have the following corresponding result.

Lemma 6.5.2 *Suppose* $f(z) = \sum\limits_{n=0}^{\infty} a_n z^n \in H_{\mu,0}^\infty$. *Then*

$$\lim_{n\to\infty} |a_n| \mu\left(1 - \frac{1}{n}\right) = 0. \tag{6.11}$$

Proof Since $f \in H_{\mu,0}^\infty$, for every $\varepsilon > 0$, there exists $\delta \in (0, 1)$, such that whenever $\delta < |z| < 1$,

$$\mu(|z|)|f(z)| < \varepsilon. \tag{6.12}$$

Let n_0 be the least integer satisfying $1 - \frac{1}{n_0} > \delta$. By (6.10) and (6.12), for $r \in (\delta, 1)$, we have

$$|a_n| \leq \frac{\varepsilon}{2\pi} \int_0^{2\pi} \frac{r^{-n}}{\mu(r)} \, d\theta = \varepsilon \frac{r^{-n}}{\mu(r)}.$$

Substituting $r = 1 - \frac{1}{n}$ into the last inequality, we obtain

$$\mu\left(1 - \frac{1}{n}\right) |a_n| < \varepsilon e, \qquad (n \geq n_0),$$

from which the result follows. $\qquad\qquad\qquad\qquad\qquad\qquad\qquad\qquad\qquad\qquad\qquad$ $\square$

6.6 H_μ^∞ and $H_{\mu,0}^\infty$ Spaces and Hadamard Gaps

These results of Sect. 6.3 give the motivation to investigate the situation for the more general weighted space H_μ^∞. This is logical since the spaces with normal weights are the most natural generalization of H_α^∞.

Theorem 6.6.1 *Let*

$$f(z) = \sum_{k=0}^\infty a_k z^{n_k} \in Hol(\mathbb{D}),$$

where n_k is a sequence of integers with $n_{k+1}/n_k \geq \lambda > 1$, and let μ be a normal function on the interval $[0, 1)$. Then $f \in H_\mu^\infty$ if and only if

$$\sup_{k\geq 0} \mu\left(1 - \frac{1}{n_k}\right) |a_k| < \infty. \tag{6.13}$$

Proof By Lemma 6.5.1, it suffices to prove the sufficiency. Since μ is normal, without loss of generality, we may assume that the sequence $\left(\mu(1 - \frac{1}{n_k})\right)_{k\geq 0}$ is decreasing. Then we have

$$|f(z)| = \left| \sum_{k=0}^\infty a_k z^{n_k} \right| \leq C \sum_{k=0}^\infty \frac{|z|^{n_k}}{\mu(1 - \frac{1}{n_k})}, \tag{6.14}$$

and hence

$$\frac{|f(z)|}{1 - |z|} \leq C \sum_{n=1}^\infty \left(\sum_{n_k \leq n} \frac{1}{\mu(1 - \frac{1}{n_k})} \right) |z|^n$$

for some positive constant C. Write $K = K(n) = \max\{k : n_k \leq n\}$. Then, by the definition of the normal weight, we have

$$\mu\left(1 - \frac{1}{n}\right) \sum_{n_k \leq n} \frac{1}{\mu\left(1 - \frac{1}{n_k}\right)} = \frac{\mu\left(1 - \frac{1}{n}\right)}{\mu\left(1 - \frac{1}{n_K}\right)} \sum_{n=0}^{K-1} \frac{\mu\left(1 - \frac{1}{n_K}\right)}{\mu\left(1 - \frac{1}{n_K - n}\right)}$$

$$\leq \sum_{n=0}^{K-1} \frac{\mu\left(1 - \frac{1}{n_K}\right)}{\mu\left(1 - \frac{1}{n_K - n}\right)}$$

$$\leq \sum_{n=0}^{K-1} \frac{n_K - n^\alpha}{n_K}$$

$$\leq \frac{\lambda^\alpha}{\lambda^\alpha - 1} < \infty.$$

Consequently,

$$\sum_{n_k \leq n} \frac{1}{\mu\left(1 - \frac{1}{n_k}\right)} \leq \frac{C}{\mu\left(1 - \frac{1}{n}\right)}$$

and, with (6.14),

$$\frac{|f(z)|}{1 - |z|} \leq C \sum_{n=1}^{\infty} \frac{|z|}{\mu\left(1 - \frac{1}{n}\right)}.$$

Note that the function

$$g_x(t) = \frac{x^t}{\mu\left(1 - \frac{1}{t}\right)} = \frac{e^{-t \log \frac{1}{x}}}{\mu\left(1 - \frac{1}{t}\right)}$$

is decreasing in t, for t large enough and each $x \in (0, 1)$, and

$$\sum_{n=1}^{\infty} \frac{|z|}{\mu\left(1 - \frac{1}{n}\right)} \sim \int_e^\infty \frac{e^{-t \log \frac{1}{|z|}}}{\mu\left(1 - \frac{1}{t}\right)} \, dt.$$

By Lemma 6.4.1, we get

$$\int_e^\infty \frac{e^{-t \log \frac{1}{|z|}}}{\mu\left(1 - \frac{1}{t}\right)} \, dt \leq C \frac{1}{\log \frac{1}{|z|} \mu\left(1 - \log \frac{1}{|z|}\right)}.$$

Using the asymptotic estimate

$$\log \frac{1}{|z|} \sim 1 - |z|, \qquad \text{as } |z| \to 1^-,$$

and the normality of μ, we thus obtain

$$\sum_{n=1}^{\infty} \frac{|z|}{\mu(1 - \frac{1}{n})} \leq \frac{C}{(1 - |z|)\mu(|z|)},$$

from which the desired result follows. $\square$

The following result has the same flavor as Theorem 6.6.1. Hence, its proof is considerably shortened.

Theorem 6.6.2 *Let*

$$f(z) = \sum_{k=0}^{\infty} a_k z^{n_k} \in Hol(\mathbb{D}),$$

where n_k is a sequence of integers with $n_{k+1}/n_k \geq \lambda > 1$, and let μ be a normal function on the interval $[0, 1)$. Then $f \in H_{\mu,0}^\infty$ if and only if

$$\lim_{k \to \infty} |a_k| \mu \left(1 - \frac{1}{n_k} \right) = 0. \tag{6.15}$$

Proof By Lemma 6.5.2, it suffices to prove the sufficiency. By (6.15), for every $\varepsilon > 0$, there is an integer $k_0 \geq 1$ such that

$$|a_k| \mu \left(1 - \frac{1}{n_k} \right) < \varepsilon, \qquad (k \geq k_0).$$

Then

$$|f(z)| \leq \sum_{k=0}^{k_0-1} \frac{|z|^{n_k}}{\mu(1 - \frac{1}{n_k})} + \varepsilon \sum_{k=k_0}^{\infty} \frac{|z|^{n_k}}{\mu(1 - \frac{1}{n_k})}.$$

As in the proof of Theorem 6.6.1, we obtain

$$\frac{|f(z)|}{1 - |z|} \leq \frac{P_{k_0}(z)}{1 - |z|} + \frac{\varepsilon C}{\log \frac{1}{|z|} \mu(1 - \log \frac{1}{|z|})},$$

where P_{k_0} is a polynomial of order n_{k_0}, which is bounded on $\mathbb{D}$. From normality of μ, (6.15), and the condition $\lim\limits_{|z| \to 1^-} \mu(|z|) = 0$, the desired result follows.   □

6.7   Lower Estimations

For each $\alpha \in (0, \infty)$, there are two functions $f_1, f_2 \in \mathcal{B}_\alpha$ and a constant $C > 0$, such that

$$|f_1'(z)| + |f_2'(z)| \geq \frac{C}{(1 - |z|^2)^\alpha}, \quad z \in \mathbb{D}. \tag{6.16}$$

In other words, there exist two functions $g_1, g_2 \in H^\infty_\alpha$ such that

$$|g_1(z)| + |g_2(z)| \geq \frac{C}{(1 - |z|^2)^\alpha}, \quad z \in \mathbb{D}. \tag{6.17}$$

These results are still valid for a more general context of weighted space H^∞_μ. It is quite logical, as the spaces with normal weights are the most natural generalization of the space H^∞_α. Applying Theorem 6.6.1, we deduce the generalization below.

Theorem 6.7.1 *Let* $\mu: [0, 1) \to [0, \infty)$ *be nonincreasing radial weight function and normal on the interval* $[0, 1)$. *Then there exist two functions* $f_1, f_2 \in H^\infty_\mu$ *such that*

$$|f_1(z)| + |f_2(z)| \geq \frac{C}{\mu(|z|)}, \quad (z \in \mathbb{D}).$$

Proof Denote

$$f(z) = \sum_{j=1}^{\infty} \frac{z^{q^j}}{\mu(1 - \frac{1}{q^j})}, \quad (z \in \mathbb{D}),$$

where $q \in \mathbb{N}$ large enough. By Theorem 6.6.1, with $a_j = \left(\mu(1 - \frac{1}{q^j})\right)^{-1}$ and $n_j = q^j$, we easily see that $f \in H^\infty_\mu$. We show that

$$|f(z)| \geq \frac{C}{\mu(|z|)}, \quad \text{if } 1 - q^{-k} \leq |z| \leq 1 - q^{-(k+\frac{1}{2})}, \ k \geq 0.$$

For $z \in \mathbb{D}$, we have

$$|f(z)| \geq \frac{|z|^{q^k}}{\mu(1 - \frac{1}{q^k})} - \sum_{j=1}^{k-1} \frac{|z|^{q^j}}{\mu(1 - \frac{1}{q^j})} - \sum_{k+1}^{\infty} \frac{|z|^{q^j}}{\mu(1 - \frac{1}{q^j})}$$

$$= A - B - C.$$

First, since

$$(1 - q^{-k})q^k \le |z|q^k \le \left((1 - q^{-(k+\frac{1}{2})})q^{k+\frac{1}{2}}\right)^{q^{-\frac{1}{2}}},$$

we have

$$\frac{1}{3} \le |z|q^k \le \left(\frac{1}{2}\right)^{q^{-\frac{1}{2}}}.$$

Then for q large enough,

$$A \ge \frac{1}{3\mu(1 - \frac{1}{q^k})}.$$

Next, by the definition of the normal function μ, for q large enough, and for each $j \in \mathbb{N}$, we have

$$\frac{\left(1 - (1 - \frac{1}{q^j})\right)^\alpha}{\left(1 - (1 - \frac{1}{q^{j+1}})\right)^\alpha} \le \frac{\mu(1 - \frac{1}{q^j})}{\mu(1 - \frac{1}{q^{j+1}})} \le \frac{\left(1 - (1 - \frac{1}{q^j})\right)^b}{\left(1 - (1 - \frac{1}{q^{j+1}})\right)^b},$$

and also by the simple computation, we have

$$1 < q^\alpha \le \frac{\mu(1 - \frac{1}{q^j})}{\mu(1 - \frac{1}{q^{j+1}})} \le q^b. \tag{6.18}$$

Then

$$B \le \frac{1}{\mu(1 - \frac{1}{q^k})} \sum_{j=1}^{k-1} \frac{\mu(1 - \frac{1}{q^{j+1}})}{\mu(1 - \frac{1}{q^j})} \cdot \frac{\mu(1 - \frac{1}{q^{j+2}})}{\mu(1 - \frac{1}{q^{j+1}})} \cdots \frac{\mu(1 - \frac{1}{q^k})}{\mu(1 - \frac{1}{q^{k-1}})}$$

$$\le \frac{1}{\mu(1 - \frac{1}{q^k})} \sum_{j=1}^{k-1} \frac{1}{q^{\alpha(k-j)}} \le \frac{1}{\mu(1 - \frac{1}{q^k})} \cdot \frac{1}{q^\alpha - 1}.$$

Last, we have

$$C \leq \frac{|z|^{q^{k+1}}}{\mu(1-\frac{1}{q^k})} \sum_{j=k+1}^{\infty} \frac{\mu(1-\frac{1}{q^k})}{\mu(1-\frac{1}{q^j})} |z|^{(q^j-q^{k+1})}$$

$$\leq \frac{|z|^{q^{k+1}}}{\mu(1-\frac{1}{q^k})} \sum_{j=k+1}^{\infty} \frac{\mu(1-\frac{1}{q^k})}{\mu(1-\frac{1}{q^{k+1}})} \cdot \frac{\mu(1-\frac{1}{q^{k+1}})}{\mu(1-\frac{1}{q^{k+2}})} \cdots \frac{\mu(1-\frac{1}{q^{j-1}})}{\mu(1-\frac{1}{q^j})} \cdot |z|^{(q^j-q^{k+1})}.$$

Then by (6.18), we obtain

$$C \leq \frac{|z|^{q^{k+1}}}{\mu(1-\frac{1}{q^k})} \sum_{j=k+1}^{\infty} q^b \cdot (q^b)^{(j-(k+1))} |z|^{(q^j-q^{k+1})}$$

$$\leq \frac{|z|^{q^{k+1}}}{\mu(1-\frac{1}{q^k})} \sum_{s=0}^{\infty} q^b \cdot \left(q^b |z|^{(q^{k+2}-q^{k+1})}\right)^s = \frac{(|z|^{q^k})^q}{\mu(1-\frac{1}{q^k})} \cdot \frac{q^b}{1-q^b(|z|^{q^k})^{(q^2-q)}}$$

$$= \frac{1}{\mu(1-\frac{1}{q^k})} \cdot \frac{q^b}{1-q^b(|z|^{q^k})^{(q^2-q)}} \leq \frac{1}{\mu(1-\frac{1}{q^k})} \cdot \frac{q^b 2^{-q^{\frac{1}{2}}}}{1-q^b 2^{-(q^{\frac{3}{2}}-q^{\frac{1}{2}})}}.$$

For q large enough and each $k \in \mathbb{N}$, we have

$$|f(z)| \geq \frac{1}{\mu(1-\frac{1}{q^k})} \left(\frac{1}{3} - \frac{1}{q^\alpha - 1} - \frac{q^b 2^{-q^{\frac{1}{2}}}}{1-q^b 2^{-(q^{\frac{3}{2}}-q^{\frac{1}{2}})}} \right) \geq \frac{C}{\mu(1-\frac{1}{q^k})}.$$

By the normality of μ and the inequality

$$1-q^{-k} \leq |z| \leq 1-q^{-(k+\frac{1}{2})},$$

we obtain

$$|f(z)| \geq \frac{C}{\mu(1-\frac{1}{q^k})} \geq \frac{C}{C_q \mu\left(1-\frac{1}{q^{k+\frac{1}{2}}}\right)} \geq \frac{C}{\mu(|z|)}.$$

Similarly, we can easily prove that for

$$g(z) = \sum_{j=1}^{\infty} \frac{z^{n_j}}{\mu(1-\frac{1}{q^{j+\frac{1}{2}}})},$$

where n_j is the largest integer closest to $q^{j+\frac{1}{2}}$, we have

$$|g(z)| \geq \frac{C}{\mu(|z|)}, \quad \text{whenever } 1 - q^{-(k+\frac{1}{2})} \leq |z| \leq 1 - q^{-(k+1)}, k \in \mathbb{N}.$$

Finally, the result of the theorem follows by taking

$$f_1(z) = r + sf(z) \quad \text{and} \quad f_2(z) = tg(z), \ z \in \mathbb{D},$$

for some appropriate positive constants r, s, and t.   □

Notes on Chapter 6

The concept of normal weights is taken from [85]. The Hardy–Littlewood Theorem 6.1.1 is proved in [39] and Zygmund Theorem 6.1.2 is available in [114]. Theorem 6.3.1 and its Corollary 6.3.2 are proved in [103], while the case $\alpha = 1$ is proved in [70]. Basing on Theorem 6.3.1, Corollary 6.3.3 is proved in [102]. However, this latter result had already been proved earlier in [78] for the basic case $\alpha = 1$. Other results are proved in [104]. Fatou's points are introduced in [13]. The lower estimations (6.16) and (6.17) were studied in [102].

The aim of this chapter is to characterize the $\mathcal{N}_p$-spaces in the unit ball as well as the behavior of the weighted composition operators acting on these spaces. We study different properties of the weighted composition operators acting on these spaces.

7.1   On the Unit Ball

Let $\mathbb{B}$ be the open unit ball in the complex vector space $\mathbb{C}^n$; let $\mathrm{Hol}(\mathbb{B})$ denote the space of functions that are holomorphic in $\mathbb{B}$, with the compact-open topology; and let $H^\infty(\mathbb{B})$ denote the Banach space of bounded holomorphic functions on $\mathbb{B}$ with the norm

$$\|f\|_\infty = \sup_{z \in \mathbb{B}} |f(z)|.$$

If $z = (z_1, z_2, \ldots, z_n) \in \mathbb{C}^n$ and $\zeta = (\zeta_1, \zeta_2, \ldots, \zeta_n) \in \mathbb{C}^n$, we define the inner product

$$\langle z, \zeta \rangle = z_1 \bar{\zeta}_1 + \cdots + z_n \bar{\zeta}_n$$

and correspondingly the Euclidean norm

$$|z| = (|z_1|^2 + \cdots + |z_n|^2)^{1/2}.$$

If X and Y are two topological vector spaces, then the symbol $X \hookrightarrow Y$ indicates the continuous embedding of X into Y.

The Beurling-type space, also called the Bergman-type space, $A^{-p}(\mathbb{B})$, $p > 0$, in the unit ball is defined as

$$A^{-p}(\mathbb{B}) = \left\{ f \in \mathrm{Hol}(\mathbb{B}) : |f|_p = \sup_{z \in \mathbb{B}} |f(z)|(1 - |z|^2)^p < \infty \right\}.$$

Weighted composition operators for function spaces defined on the open unit disc $\mathbb{D}$ were defined in Sect. 5. In the present context, the definitions are similar. For the sake of completeness, we mention them here. For φ a holomorphic self-map of $\mathbb{B}$ and a holomorphic function $u : \mathbb{B} \to \mathbb{C}$, the linear operator $W_{u,\varphi} : \mathrm{Hol}(\mathbb{B}) \to \mathrm{Hol}(\mathbb{B})$ defined by

$$W_{u,\varphi}(f)(z) = u(z) \cdot (f \circ \varphi(z)), \qquad f \in \mathrm{Hol}(\mathbb{B}), \quad (z \in \mathbb{B}),$$

is called the *weighted composition operator* with symbols u and φ. We see that $W_{u,\varphi} = M_u C_\varphi$, where $M_u(f) = uf$ is the *multiplication operator* with symbol u and $C_\varphi(f) = f \circ \varphi$ is the *composition operator* with symbol φ. If u is the constant function 1, then $W_{u,\varphi} = C_\varphi$, and if φ is the identity, then $W_{u,\varphi} = M_u$.

Let $a \in \mathbb{B}$. Then the orthogonal projection of $\mathbb{C}^n$ on the (at most) one-dimensional subspace $[a]$ generated by a is given by

$$P_a z = \begin{cases} 0, & \text{if } a = 0, \\ \frac{\langle z, a \rangle}{\langle a, a \rangle} a, & \text{if } a \neq 0. \end{cases}$$

We write $Q_a = I - P_a$ for the projection on the orthogonal complement of $[a]$ and, for the simplicity of notations, let $s_a = (1 - |a|^2)^{1/2}$. Then, as we will see in Sect. 7.2, the mapping

$$\Phi_a(z) = \frac{a - P_a z - s_a Q_a z}{1 - \langle z, a \rangle} \tag{7.1}$$

is an automorphism of the unit ball $\mathbb{B}$, i.e., $\Phi_a \in \mathrm{Aut}(\mathbb{B})$. For $p \in (0, \infty)$ and $f \in \mathrm{Hol}(\mathbb{B})$, we define

$$\|f\|_p := \sup_{a \in \mathbb{B}} \left(\int_{\mathbb{B}} |f(z)|^2 (1 - |\Phi_a(z)|^2)^p \, dV(z) \right)^{1/2}$$

Then $\mathcal{N}_p$ is the space

$$\mathcal{N}_p(\mathbb{B}) = \left\{ f \in \mathrm{Hol}(\mathbb{B}) : \|f\|_p < \infty \right\},$$

where dV is the Lebesgue normalized volume measure on $\mathbb{B}$, i.e., $V(\mathbb{B}) = 1$. When there is no ambiguity, we write $\mathcal{N}_p$ for $\mathcal{N}_p(\mathbb{B})$.

7.2   The Automorphism Φ_a

Recall that Φ_a is defined by (7.1). It is clear that for $n = 1$, we have $P_a = I$ and $Q_a = 0$, and hence $\Phi_a(z)$ becomes the automorphism σ_a of the open unit disc $\mathbb{D}$. We now show that Φ_a satisfies the same property on the ball $\mathbb{B}$. It is easy to directly verify that

$$\Phi_a(0) = a$$

and

$$\Phi_a(a) = 0.$$

In other words, Φ_a exchanges the points 0 and a. Moreover,

$$\Phi_a'(0) = -s_a^2 P_a - s_a Q_a$$

and

$$\Phi_a'(a) = -\frac{P_a}{s_a^2} - \frac{Q_a}{s_a}.$$

Using the definition of inner product, we see that

$$1 - \langle \Phi_a(z), \Phi_a(w) \rangle = \frac{(1 - \langle a, a \rangle)(1 - \langle z, w \rangle)}{(1 - \langle z, a \rangle)(1 - \langle a, w \rangle)} \tag{7.2}$$

holds for all $z \in \overline{\mathbb{B}}$, $w \in \overline{\mathbb{B}}$. In particular, when $w = z$, it leads to the identity

$$1 - |\Phi_a(z)|^2 = \frac{(1 - |a|^2)(1 - |z|^2)}{|1 - \langle z, a \rangle|^2}, \qquad (z \in \overline{\mathbb{B}}). \tag{7.3}$$

Hence, Φ_a maps $\mathbb{B}$ into itself. But, Φ_a is also an involution, i.e.,

$$\Phi_a(\Phi_a(z)) = z, \qquad (z \in \overline{\mathbb{B}}).$$

From this, we deduce that Φ_a is a homeomorphism of the closed unit ball $\overline{\mathbb{B}}$ onto itself and an automorphism of the open unit ball $\mathbb{B}$.

The *pseudo-hyperbolic metric* in the ball is defined by

$$\rho(z, w) = |\Phi_w(z)|, \qquad (z, w \in \mathbb{B}). \tag{7.4}$$

It is easy to verify that $\rho(0, w) = |w|$ and that $\rho(\Phi_w(z), w) = |z|$. Clearly, $\rho(z, w) \geq 0$ and $\rho(z, w) = 0$ if and only if $z = w$, because $\Phi_w(z) = 0$ only for $z = w$. The symmetry property $\rho(z, w) = \rho(w, z)$ follows from (7.3). The triangle inequality is less trivial, and we deduce it below from a stronger result.

Theorem 7.2.1 *The pseudo-hyperbolic metric ρ has the following properties.*

(i) *Rotation invariance:*

$$\rho\big(U(z), U(w)\big) = \rho(z, w)$$

for all $z, w \in \mathbb{B}$ and all unitary matrices U.
(ii) *Möbius invariance:*

$$\rho\big(\Phi_a(z), \Phi_a(w)\big) = \rho(z, w)$$

for all $z, w \in \mathbb{B}$.
(iii) *Strong triangle inequality:*

$$\frac{|\rho(z, a) - \rho(a, w)|}{1 - \rho(z, a)\rho(a, w)} \le \rho(z, w) \le \frac{\rho(z, a) + \rho(a, w)}{1 + \rho(z, a)\rho(a, w)}$$

for all $z, w, a \in \mathbb{B}$.

Proof

(i): Since unitary transformations preserve inner products, the identity

$$1 - \rho\big(U(z), U(w)\big)^2 = 1 - \rho(z, w)^2$$

follows directly from (7.3).
(ii): Due to (7.2) and (7.3), we have

$$
\begin{aligned}
1 - \rho\big(U(z), U(w)\big)^2 &= 1 - \big|\Phi_{\Phi_a(w)}\big(\Phi_a(z)\big)\big|^2 \\
&= \frac{(1 - |\Phi_a(w)|^2)(1 - |\Phi_a(z)|^2)}{|1 - \langle \Phi_a(z), \Phi_a(w)\rangle|^2} \\
&= \frac{(1 - |w|^2)(1 - |z|^2)}{|1 - \langle z, w\rangle|^2} \\
&= 1 - |\Phi_w(z)|^2 = 1 - \rho(z, w)^2.
\end{aligned}
$$

Combining (i) and (ii), we see that ρ is invariant under every automorphism of the ball.

(iii): Since by (ii) ρ is Möbius-invariant, we may assume that $a = 0$. Hence, we need to show that

$$\frac{||z| - |w||}{1 - |z||w|} \le |\Phi_w(z)| \le \frac{|z| + |w|}{1 + |z||w|}. \tag{7.5}$$

Applying the Cauchy–Schwarz inequality, we have

$$1 - |z||w| \le |1 - \langle z, w \rangle| \le 1 + |z||w|.$$

Then (7.5) follows by using the basic identity (7.3).

$\square$

7.3   Surjective Isometries

By using weighted composition operators with particular symbols, we obtain an alternate description of the norm in $\mathcal{N}_p$. Furthermore, for any $\Phi \in \text{Aut}(\mathbb{B})$, as is well-known, there exists a unitary operator U such that $\Phi = U\Phi_a$, where $a = \Phi^{-1}(0)$. This shows that $|\Phi(z)| = |\Phi_a(z)|$, for all $z \in \mathbb{B}$. Consequently, we obtain

$$\|f\|_p = \sup_{a \in \mathbb{B}} \left(\int_{\mathbb{B}} |f(z)|^2 (1 - |\Phi_a(z)|^2)^p \, dV(z) \right)^{1/2}$$

$$= \sup_{\Phi \in \text{Aut}(\mathbb{B})} \left(\int_{\mathbb{B}} |f(z)|^2 (1 - |\Phi(z)|^2)^p \, dV(z) \right)^{1/2}$$

$$= \sup_{\Phi \in \text{Aut}(\mathbb{B})} \left(\int_{\mathbb{B}} |f(z)|^2 (1 - |\Phi^{-1}(z)|^2)^p \, dV(z) \right)^{1/2}.$$

For $\Phi \in \text{Aut}(\mathbb{B})$, let $a = \Phi^{-1}(0)$. For simplicity, write

$$W_\Phi := W_{k_a, \Phi}. \tag{7.6}$$

By the change of variables $z = \Phi(w)$, we obtain

$$\int_{\mathbb{B}} |f(z)|^2 (1 - |\Phi^{-1}(z)|^2)^p \, dV(z)$$

$$= \int_{\mathbb{B}} |f(\Phi(w))|^2 (1 - |w|^2)^p \left(\frac{1 - |a|^2}{|1 - \langle w, a \rangle|^2} \right)^{n+1} dV(w)$$

$$= \int_{\mathbb{B}} |f(\Phi(w))|^2 |k_a(w)|^2 (1 - |w|^2)^p \, dV(w)$$

$$= \|W_\Phi f\|_{A_p^2}^2. \tag{7.7}$$

Here, A_p^2 is the weighted Bergman space over $\mathbb{B}$ defined by

$$A_p^2 = \left\{ f \in \mathrm{Hol}(\mathbb{B}) : \|f\|_{A_p^2} < \infty \right\},$$

where

$$\|f\|_{A_p^2} = \left(\int_{\mathbb{B}} |f(z)|^2 (1 - |z|^2)^p \, dV(z) \right)^{1/2}$$

According to the formulas (7.7), we have

$$\|f\|_p = \sup \left\{ \|W_\Phi f\|_{A_p^2} : \Phi \in \mathrm{Aut}(\mathbb{B}) \right\}. \tag{7.8}$$

Theorem 7.3.1 *For any automorphism Ψ of the unit ball $\mathbb{B}$, the weighted composition operator W_Ψ is a surjective isometry on $\mathcal{N}_p$.*

Proof By a direct calculation that for any two automorphisms Φ and Ψ in $\mathrm{Aut}(\mathbb{B})$, there exists a complex number λ with modulus one such that $W_\Phi W_\Psi = \lambda W_{\Psi \circ \Phi}$. Consequently, by (7.8),

$$\|W_\Psi f\|_p = \sup \left\{ \|W_\Phi W_\Psi f\|_{A_p^2} : \Phi \in \mathrm{Aut}(\mathbb{B}) \right\}$$

$$= \sup \left\{ \|W_{\Psi \circ \Phi} f\|_{A_p^2} : \Phi \in \mathrm{Aut}(\mathbb{B}) \right\} = \|f\|_p.$$

$\square$

7.4 $\mathcal{N}_p(\mathbb{B})$ Is a Chain

As in many function spaces, we show that $\mathcal{N}_p(\mathbb{B})$ is a chain with respect to the parameter p.

Theorem 7.4.1 *For $p > q > 0$, we have*

$$H^\infty(\mathbb{B}) \hookrightarrow \mathcal{N}_q(\mathbb{B}) \hookrightarrow \mathcal{N}_p(\mathbb{B}).$$

Proof Let $f \in H^\infty(\mathbb{B})$. Then we have

$$\|f\|_q^2 = \sup_{a \in \mathbb{B}} \int_{\mathbb{B}} |f(z)|^2 (1 - |\Phi_a(z)|^2)^q \, dV(z)$$

$$\leq \left(\sup_{z \in \mathbb{B}} |f(z)| \right)^2 \sup_{a \in \mathbb{B}} \int_{\mathbb{B}} (1 - |\Phi_a(z)|^2)^q \, dV(z)$$

$$\leq \|f\|_\infty^2,$$

which implies the first embedding. Next, let $f \in \mathcal{N}_q(\mathbb{B})$. Then

$$\|f\|_p^2 = \sup_{a \in \mathbb{B}} \int_{\mathbb{B}} |f(z)|^2 (1 - |\Phi_a(z)|^2)^p \, dV(z)$$

$$= \sup_{a \in \mathbb{B}} \int_{\mathbb{B}} |f(z)|^2 (1 - |\Phi_a(z)|^2)^q \cdot (1 - |\Phi_a(z)|^2)^{p-q} \, dV(z)$$

$$\leq \sup_{a \in \mathbb{B}} \int_{\mathbb{B}} |f(z)|^2 (1 - |\Phi_a(z)|^2)^q \, dV(z) = \|f\|_q^2,$$

which shows the second embedding.   □

7.5   The Embedding $\mathcal{N}_p(\mathbb{B}) \hookrightarrow A^{-\frac{n+1}{2}}(\mathbb{B})$

Recall that n is the dimension of ambient space, i.e., $\mathbb{B} \subset \mathbb{C}^n$.

Theorem 7.5.1 *For $p > 0$, we have*

$$\mathcal{N}_p(\mathbb{B}) \hookrightarrow A^{-\frac{n+1}{2}}(\mathbb{B}).$$

Proof Write $\mathbb{B}_{1/2} = \{z : |z| < 1/2\}$. For each $f \in \mathcal{N}_p(\mathbb{B})$, we have

$$\|f\|_p^2 = \sup_{a \in \mathbb{B}} \int_{\mathbb{B}} |f(z)|^2 (1 - |\Phi_a(z)|^2)^p \, dV(z)$$

$$\geq \int_{\mathbb{B}} |f(z)|^2 (1 - |\Phi_0(z)|^2)^p \, dV(z)$$

$$= \int_{\mathbb{B}} |f(z)|^2 (1 - |z|^2)^p \, dV(z)$$

$$\geq \int_{\mathbb{B}_{1/2}} |f(z)|^2 (1 - |z|^2)^p \, dV(z)$$

$$\geq \left(\frac{3}{4} \right)^p \int_{\mathbb{B}_{1/2}} |f(z)|^2 \, dV(z),$$

that is,

$$\left(\frac{4}{3}\right)^p \|f\|_p^2 \geq \int_{\mathbb{B}_{1/2}} |f(z)|^2 \, dV(z). \tag{7.9}$$

Since $f \in \mathrm{Hol}(\mathbb{B})$, the function $|f|^2$ is subharmonic in $\mathbb{C}^n$ and hence subharmonic in $\mathbb{R}^{2n}$. Then

$$|f(0)|^2 \leq \frac{1}{V(\mathbb{B}_{1/2})} \int_{\mathbb{B}_{1/2}} |f(z)|^2 \, dV(z) = 4^n \int_{\mathbb{B}_{1/2}} |f(z)|^2 \, dV(z). \tag{7.10}$$

Combining (7.9) and (7.10) yields

$$|f(0)|^2 \leq \frac{4^{p+n}}{3^p} \|f\|_p^2, \quad f \in \mathcal{N}_p(\mathbb{B}). \tag{7.11}$$

For every fixed $z \in \mathbb{B}$, we put

$$F_{z,f}(w) = (f \circ \Phi_z(w)) \cdot \frac{(1 - |z|^2)^{\frac{n+1}{2}}}{(1 - \langle w, z \rangle)^{n+1}}, \qquad (w \in \mathbb{B}),$$

which is clearly a holomorphic function in $\mathbb{B}$. We prove that $F_{z,f}(w) \in \mathcal{N}_p(\mathbb{B})$. First, by change of variables $t = \Phi_z(w)$, we have

$$\|F_{z,f}\|_p^2 = \sup_{a \in \mathbb{B}} \int_{\mathbb{B}} |f \circ \Phi_z(w)|^2 \cdot \frac{(1 - |z|^2)^{n+1}}{|1 - \langle w, z \rangle|^{2n+2}} (1 - |\Phi_a(w)|^2)^p \, dV(w)$$

$$= \sup_{a \in \mathbb{B}} \int_{\mathbb{B}} |f(t)|^2 \left(1 - |\Phi_a(\Phi_z(t))|^2\right)^p \, dV(t),$$

due to the fact that

$$\left(\frac{1 - |z|^2}{|1 - \langle w, z \rangle|^2}\right)^{n+1}$$

is the real Jacobian determinant of the automorphism $\Phi_z(w)$. Furthermore, by the transitivity of $\mathrm{Aut}(\mathbb{B})$, $\Phi_a \circ \Phi_z$ belongs to $\mathrm{Aut}(\mathbb{B})$, and so $\Phi_a \circ \Phi_z = U \circ \Phi_{b_a}$, where $b_a = \Phi_a(z)$, U is a unitary transformation of $\mathbb{B}$, and Φ_{b_a} is the involution. Then

$$\sup_{a \in \mathbb{B}} \int_{\mathbb{B}} |f(t)|^2 \left(1 - |\Phi_a(\Phi_z(t))|^2\right)^p \, dV(t)$$

$$= \sup_{a \in \mathbb{B}} \int_{\mathbb{B}} |f(t)|^2 \left(1 - |U(\Phi_{b_a}(t))|^2\right)^p \, dV(t)$$

$$= \sup_{a \in \mathbb{B}} \int_{\mathbb{B}} |f(t)|^2 \left(1 - |\Phi_{b_a}(t)|^2\right)^p dV(t) \qquad (\, |U(z)| = |z|, \; z \in \mathbb{B})$$

$$\leq \sup_{c \in \mathbb{B}} \int_{\mathbb{B}} |f(t)|^2 (1 - |\Phi_c(t)|^2)^p \, dV(t) = \|f\|_p^2.$$

Consequently,

$$\|F_{z,f}\|_p^2 \leq \|f\|_p^2,$$

which shows that $F_{z,f} \in \mathcal{N}_p(\mathbb{B})$. Then by (7.11), we have

$$|f(z)|^2 (1 - |z|^2)^{n+1} = |F_{z,f}(0)|^2 \leq \frac{4^{p+n}}{3^p} \|F_{z,f}\|_p^2 \leq \frac{4^{p+n}}{3^p} \|f\|_p^2, \qquad (z \in \mathbb{B}),$$

which implies that

$$|f|_{\frac{n+1}{2}} = \sup_{z \in \mathbb{B}} |f(z)|(1 - |z|^2)^{\frac{n+1}{2}} \leq \frac{2^{p+n}}{3^{p/2}} \|f\|_p, \qquad (f \in \mathcal{N}_p(\mathbb{B})). \qquad (7.12)$$

That is, $\mathcal{N}_p(\mathbb{B}) \hookrightarrow A^{-\frac{n+1}{2}}(\mathbb{B})$. □

7.6   The Embedding $A^{-k}(\mathbb{B}) \hookrightarrow \mathcal{N}_p(\mathbb{B})$

The following result can be considered as a dual to Theorem 7.5.1.

Theorem 7.6.1 *Let $k \in (0, (n+1)/2]$. Then, for $p > \max\{0, 2k - 1\}$, we have*

$$A^{-k}(\mathbb{B}) \hookrightarrow \mathcal{N}_p(\mathbb{B}).$$

In particular, when $p > n$, $\mathcal{N}_p(\mathbb{B}) = A^{-\frac{n+1}{2}}(\mathbb{B})$.

Proof Let $f \in A^{-k}(\mathbb{B})$. Then

$$\|f\|_p^2 = \sup_{a \in \mathbb{B}} \int_{\mathbb{B}} |f(z)|^2 (1 - |\Phi_a(z)|^2)^p \, dV(z)$$

$$= \sup_{a \in \mathbb{B}} \int_{\mathbb{B}} |f(z)|^2 (1 - |z|^2)^{2k} \cdot \frac{1}{(1 - |z|^2)^{2k}} (1 - |\Phi_a(z)|^2)^p \, dV(z)$$

$$\leq \sup_{z \in \mathbb{B}} |f(z)|^2 (1 - |z|^2)^{2k} \cdot \sup_{a \in \mathbb{B}} \int_{\mathbb{B}} \frac{(1 - |\Phi_a(z)|^2)^p}{(1 - |z|^2)^{2k}} \, dV(z)$$

$$= |f|_k^2 \sup_{a \in \mathbb{B}} \int_{\mathbb{B}} \frac{(1 - |\Phi_a(z)|^2)^p}{(1 - |z|^2)^{2k}} \, dV(z)$$

$$= |f|_k^2 \sup_{a \in \mathbb{B}} \int_{\mathbb{B}} \frac{(1 - |a|^2)^p (1 - |z|^2)^{p-2k}}{|1 - \langle z, a \rangle|^{2p}} \, dV(z) \ \text{(by (7.3))}$$

$$= |f|_k^2 \sup_{a \in \mathbb{B}} (1 - |a|^2)^p \int_{\mathbb{B}} \frac{(1 - |z|^2)^{p-2k}}{|1 - \langle z, a \rangle|^{2p}} \, dV(z).$$

Furthermore, similar to the estimate in Lemma 1.8.1, we have

$$\int_{\mathbb{B}} \frac{(1 - |z|^2)^{p-2k}}{|1 - \langle z, a \rangle|^{2p}} \, dV(z) = \int_{\mathbb{B}} \frac{(1 - |z|^2)^{p-2k}}{|1 - \langle z, a \rangle|^{n+1+(p-2k)+p-n-1+2k}} \, dV(z)$$

$$\asymp \begin{cases} \text{bounded in } \mathbb{B}, & p < n+1-2k, \\ \log \frac{1}{1-|a|^2}, & p = n+1-2k, \quad (7.13) \\ (1 - |a|^2)^{n+1-p-2k}, & p > n+1-2k. \end{cases}$$

In any case, there exists a positive constant C such that

$$(1 - |a|^2)^p \int_{\mathbb{B}} \frac{(1 - |z|^2)^{p-2k}}{|1 - \langle z, a \rangle|^{2p}} \, dV(z) \le C, \qquad (a \in \mathbb{B}),$$

which implies

$$\|f\|_p \le \sqrt{C} |f|_k, \qquad (f \in \mathcal{N}_p(\mathbb{B})).$$

That is, $A^{-k}(\mathbb{B}) \hookrightarrow \mathcal{N}_p(\mathbb{B})$.

In particular, for all $p > n$, by taking $k = \frac{n+1}{2}$, we have $p > 2k - 1$ and hence

$$A^{-\frac{n+1}{2}}(\mathbb{B}) \hookrightarrow \mathcal{N}_p(\mathbb{B}).$$

By Theorem 7.5.1, we always have $\mathcal{N}_p(\mathbb{B}) \hookrightarrow A^{-\frac{n+1}{2}}(\mathbb{B})$. Hence, in fact, $A^{-\frac{n+1}{2}}(\mathbb{B}) = \mathcal{N}_p(\mathbb{B})$. □

7.7 $\mathcal{N}_p(\mathbb{B})$ as a Banach Space

Theorem 7.5.1 and estimations obtained in its proof enable us to show that $\mathcal{N}_p(\mathbb{B})$ is a Banach space in which the point evaluations are continuous linear functionals.

Theorem 7.7.1 $\mathcal{N}_p(\mathbb{B})$ *is a functional Banach space with the norm* $\|\cdot\|_p$, *and moreover, its norm topology is stronger than the compact-open topology.*

Proof We first verify the triangle inequality for $\|\cdot\|_p$. For $f, g \in \mathcal{N}_p(\mathbb{B})$ and $a \in \mathbb{B}$, and by the Cauchy–Schwarz inequality,

$$\left(\int_{\mathbb{B}} |f(z) + g(z)|^2 (1 - |\Phi_a(z)|^2)^p \, dV(z) \right)^{1/2}$$

$$\leq \left(\int_{\mathbb{B}} |f(z)|^2 (1 - |\Phi_a(z)|^2)^p \, dV(z) \right)^{1/2} + \left(\int_{\mathbb{B}} |g(z)|^2 (1 - |\Phi_a(z)|^2)^p \, dV(z) \right)^{1/2}.$$

Taking the supremum of both sides with respect to $a \in \mathbb{B}$ gives the triangle inequality $\|f + g\|_p \leq \|f\|_p + \|g\|_p$. The other properties of a norm are easy to verify.

To establish the completeness of $\mathcal{N}_p(\mathbb{B})$, let (f_m) be a Cauchy sequence in $\mathcal{N}_p(\mathbb{B})$ with respect to the norm $\|\cdot\|_p$. From this, it follows that (f_m) is a Cauchy sequence in the space $\text{Hol}(\mathbb{B})$, and hence it converges to some $f \in \text{Hol}(\mathbb{B})$. It remains to show that $f \in \mathcal{N}_p(\mathbb{B})$. Indeed, there exists a $\ell_0 \in \mathbb{N}$ such that $\|f_m - f_\ell\|_p \leq 1$ for all $m, \ell \geq \ell_0$. Fix an arbitrary $a \in \mathbb{B}$. Then, by Fatou's lemma, we have

$$\int_{\mathbb{B}} |f(z) - f_{\ell_0}(z)|^2 (1 - |\Phi_a(z)|^2)^p \, dV(z)$$

$$\leq \liminf_{\ell \to \infty} \int_{\mathbb{B}} |f_\ell(z) - f_{\ell_0}(z)|^2 (1 - |\Phi_a(z)|^2)^p \, dV(z)$$

$$\leq \liminf_{\ell \to \infty} \|f_\ell - f_{\ell_0}\|_p^2 \leq 1,$$

which implies

$$\|f - f_{\ell_0}\|_p = \sup_{a \in \mathbb{B}} \int_{\mathbb{B}} |f(z) - f_{\ell_0}(z)|^2 (1 - |\Phi_a(z)|^2)^p \, dV(z) \leq 1,$$

and hence $\|f\|_p \leq 1 + \|f_{\ell_0}\|_p < \infty$. In other words, $f \in \mathcal{N}_p(\mathbb{B})$. In the estimations above, if we replace 1 by ε (which of course changes ℓ_0 to another index, say ℓ_ε), we see that the sequence (f_m) actually converges to f in the $\mathcal{N}_p(\mathbb{B})$-norm.

Finally, we prove that $\mathcal{N}_p(\mathbb{B})$ is a functional Banach space, i.e., we show that for each $z \in \mathbb{B}$, the point evaluation $f \longmapsto f(z)$ is continuous on $\mathcal{N}_p(\mathbb{B})$. Indeed, from (7.12), we have

$$|f(z)| \leq \frac{2^{p+n}}{3^{p/2}(1 - |z|^2)^{\frac{n+1}{2}}} \|f\|_p, \qquad (f \in \mathcal{N}_p(\mathbb{B})),$$

which means that the point evaluation is bounded on $\mathcal{N}_p(\mathbb{B})$, and hence it is continuous.

Since the norm topology of $A^{-\frac{n+1}{2}}(\mathbb{B})$ is clearly stronger than the compact-open topology, the second statement follows from (7.12).   □

7.8   The Embedding $\mathcal{B}(\mathbb{B}) \hookrightarrow \mathcal{N}_p(\mathbb{B})$

In the following, $\mathcal{B}(\mathbb{B})$ is the Bloch space in $\mathbb{B}$.

Theorem 7.8.1 *For $0 < p < \infty$, $\mathcal{B}(\mathbb{B}) \hookrightarrow \mathcal{N}_p(\mathbb{B})$.*

Proof Let $f \in \mathcal{B}(\mathbb{B})$. We have

$$|f(z) - f(w)| \le \frac{1}{2}\|f\|_\mathcal{B} \log \frac{1 + |\Phi_z(w)|}{1 - |\Phi_z(w)|}, \qquad (z, w \in \mathbb{B}, z \ne w), \qquad (7.14)$$

where $\|f\|_\mathcal{B} = |f(0)| + \sup_{z \in \mathbb{B}} |\nabla f(z)|(1 - |z|^2)$. From (7.14), it follows that

$$|f(z)| \le |f(0)| + \frac{1}{2}\|f\|_\mathcal{B} \log \frac{4}{1 - |z|^2} \le \frac{3}{2}\|f\|_\mathcal{B} \log \frac{4}{1 - |z|^2}.$$

Consequently, for all $f \in \mathcal{B}(\mathbb{B})$, we have

$$\|f\|_p^2 = \sup_{a \in \mathbb{B}} \left(\int_\mathbb{B} |f(z)|^2 (1 - |\Phi_a(z)|^2)^p \, dV(z) \right)$$

$$\le \frac{3}{2}\|f\|_\mathcal{B}^2 \sup_{a \in \mathbb{B}} \int_\mathbb{B} \left(\log \frac{4}{1 - |z|^2} \right)^2 (1 - |\Phi_a(z)|^2)^p \, dV(z)$$

$$\le \frac{3}{2}\|f\|_\mathcal{B}^2 \cdot \int_\mathbb{B} \left(\log \frac{4}{1 - |z|^2} \right)^2 dV(z) = C\|f\|_\mathcal{B}^2,$$

which shows that $\mathcal{B}(\mathbb{B}) \hookrightarrow \mathcal{N}_p(\mathbb{B})$.   □

Notes on Chapter 7

Composition operators and weighted composition operators acting on spaces of holomorphic functions in the unit disc $\mathbb{D}$ of the complex plane have been extensively studied. We refer the readers to the monographs [17, 83] for detailed information. In the one-dimensional case, composition operators acting on the $\mathcal{N}_p$-space in the unit disc $\mathbb{D}$ were considered in [73]. Some results on boundedness and compactness of these operators were obtained. Moreover, weighted composition operators acting on the $\mathcal{N}_p$-space were considered in [95]. The results of this chapter are mostly the generalization of corresponding results in [73, 95] and are from [42]. The automorphism $\Phi_a \in \text{Aut}(\mathbb{B})$ and its properties are provided in [79, pages 25–27]. Information about pseudo-hyperbolic metric for the ball is given in [22]. For properties of pseudo-hyperbolic metric in Theorem 7.2.1, proofs of parts (i) and (ii) are given in [22], while the proof of part (iii) is implicitly given in [64, Lemma 3]. The integral formula (7.13) is provided in [79, Proposition 1.4.10]. Formula (7.14) is provided in [108, Theorem 3.6]. When $n = 1$, we obtain Proposition 3.1 of [73] as special cases of Theorems 7.5.1, 7.6.1, 7.7.1, and 7.8.1.

In this chapter, we consider the weighted composition operators acting from $\mathcal{N}_p(\mathbb{B})$ to $A^{-q}(\mathbb{B})$.

8.1 A Test Function

The following family of functions in $\mathcal{N}_p$-spaces is important in applications. We usually exploit them as a test function to verify the necessity of conditions.

Lemma 8.1.1 *For each $w \in \mathbb{B}$, put*

$$k_w(z) = \left(\frac{1 - |w|^2}{(1 - \langle z, w \rangle)^2} \right)^{\frac{n+1}{2}}.$$

Then $k_w \in \mathcal{N}_p(\mathbb{B})$ and, moreover,

$$\|k_w\|_p \leq 1, \qquad w \in \mathbb{B}.$$

Proof That $k_w \in \mathrm{Hol}(\mathbb{B})$ is trivial. Then

$$\|k_w\|_p^2 = \sup_{a \in \mathbb{B}} \int_{\mathbb{B}} \left| \left(\frac{1 - |w|^2}{(1 - \langle z, w \rangle)^2} \right)^{n+1} \right| (1 - |\Phi_a(z)|^2)^p \, dV(z)$$

$$\leq \sup_{a \in \mathbb{B}} \int_{\mathbb{B}} \left(\frac{1 - |w|^2}{|1 - \langle z, w \rangle|^2} \right)^{n+1} dV(z) = 1$$

© The Author(s), under exclusive license to Springer Nature Switzerland AG 2023
L. H. Khoi, J. Mashreghi, *Theory of $\mathcal{N}_p$ Spaces*, Frontiers in Mathematics,
https://doi.org/10.1007/978-3-031-39704-2_8

The last equality follows from the fact that $\left(\frac{1-|w|^2}{|1-\langle z,w \rangle|^2} \right)^{n+1}$ is the real Jacobian determinant of $\Phi_w(z)$. $\square$

8.2   Boundedness

First, we study the boundedness of the weighted composition operator $W_{u,\varphi}$. As the first observation, the norm topology in both spaces $\mathcal{N}_p(\mathbb{B})$ and $A^{-q}(\mathbb{B})$ is stronger than the compact-open topology, and hence, it is stronger than the pointwise convergence topology. Hence, if the weighted composition operator $W_{u,\varphi}$ maps $\mathcal{N}_p(\mathbb{B})$ into $A^{-q}(\mathbb{B})$, an application of the closed graph theorem shows that $W_{u,\varphi}$ is automatically bounded from $\mathcal{N}_p(\mathbb{B})$ into $A^{-q}(\mathbb{B})$.

Theorem 8.2.1 *Let $\varphi \colon \mathbb{B} \to \mathbb{B}$ and $u \colon \mathbb{B} \to \mathbb{C}$ be holomorphic mappings, and let $p, q > 0$. Then the weighted composition operator $W_{u,\varphi} \colon \mathcal{N}_p(\mathbb{B}) \to A^{-q}(\mathbb{B})$ is bounded if and only if*

$$\sup_{z \in \mathbb{B}} |u(z)| \frac{(1-|z|^2)^q}{(1-|\varphi(z)|^2)^{\frac{n+1}{2}}} < \infty. \tag{8.1}$$

***Proof* Necessity** Suppose $W_{u,\varphi}$ is a bounded operator acting from $\mathcal{N}_p(\mathbb{B})$ into $A^{-q}(\mathbb{B})$. Then there exists a positive constant M such that

$$\left| W_{u,\varphi}(f) \right|_q \leq M \, \|f\|_p, \qquad (f \in \mathcal{N}_p).$$

Fix $z_0 \in \mathbb{B}$, and consider the test function k_{w_0}, where $w_0 = \varphi(z_0)$. By Lemma 8.1.1, we have

$$\left| W_{u,\varphi}(k_{w_0}) \right|_q \leq M \, \left\| k_{w_0} \right\|_p \leq M.$$

Moreover,

$$\left| W_{u,\varphi}(k_{w_0}) \right|_q = \sup_{z \in \mathbb{B}} \left| u(z) \left(\frac{1-|w_0|^2}{(1-\langle \varphi(z), w_0 \rangle)^2} \right)^{\frac{n+1}{2}} \right| (1-|z|^2)^q$$

$$\geq |u(z_0)| \left(\frac{1-|w_0|^2}{|1-\langle \varphi(z_0), w_0 \rangle|^2} \right)^{\frac{n+1}{2}} (1-|z_0|^2)^q$$

$$= |u(z_0)| \frac{(1-|z_0|^2)^q}{(1-|\varphi(z_0)|^2)^{\frac{n+1}{2}}}.$$

Thus, for every $z_0 \in \mathbb{B}$,

$$|u(z_0)| \frac{(1 - |z_0|^2)^q}{(1 - |\varphi(z_0)|^2)^{\frac{n+1}{2}}} \leq M,$$

from which (8.1) follows.

Sufficiency Suppose (8.1) holds. By Theorem 7.5.1, for some positive constant c_1, we have

$$\sup_{z \in \mathbb{B}} |f(\varphi(z))|(1 - |\varphi(z)|^2)^{\frac{n+1}{2}} \leq |f|_{\frac{n+1}{2}} \leq c_1 \|f\|_p, \qquad (f \in \mathcal{N}_p(\mathbb{B})).$$

Then, for all $f \in \mathcal{N}_p(\mathbb{B})$, we have

$$\begin{aligned}
\left|W_{u,\varphi}(f)\right|_q &= \sup_{z \in \mathbb{B}} |u(z) \cdot f(\varphi(z))| \, (1 - |z|^2)^q \\
&\leq \sup_{z \in \mathbb{B}} |u(z)| \frac{(1 - |z|^2)^q}{(1 - |\varphi(z)|^2)^{\frac{n+1}{2}}} \cdot \sup_{z \in \mathbb{B}} |f(\varphi(z))|(1 - |\varphi(z)|^2)^{\frac{n+1}{2}} \\
&\leq \sup_{z \in \mathbb{B}} |u(z)| \frac{(1 - |z|^2)^q}{(1 - |\varphi(z)|^2)^{\frac{n+1}{2}}} \cdot c_1 \|f\|_p = c_2 \|f\|_p,
\end{aligned}$$

which shows that $W_{u,\varphi}$ is bounded from $\mathcal{N}_p(\mathbb{B})$ into $A^{-q}(\mathbb{B})$. □

8.3 Compactness: Easy Reformulations

To study the compactness of $W_{u,\varphi}$, we start with the following useful test for compactness of weighted composition operators.

Lemma 8.3.1 *Let* $\varphi: \mathbb{B} \to \mathbb{B}$ *and* $u: \mathbb{B} \to \mathbb{C}$ *be holomorphic mappings, and let* $p, q > 0$. *Then the weighted composition operator* $W_{u,\varphi}: \mathcal{N}_p(\mathbb{B}) \to A^{-q}(\mathbb{B})$ *is compact if and only if*

$$|W_{u,\varphi}(f_m)|_q \to 0, \qquad (m \to \infty),$$

for each bounded sequence (f_m) *in* $\mathcal{N}_p(\mathbb{B})$ *such that* (f_m) *uniformly converges to* 0 *on compact subsets of* $\mathbb{B}$.

***Proof* Necessity** Suppose that $W_{u,\varphi}: \mathcal{N}_p(\mathbb{B}) \to A^{-q}(\mathbb{B})$ is compact. Take any bounded sequence (f_m) in $\mathcal{N}_p(\mathbb{B})$ such that (f_m) uniformly converges to 0 on compact subsets of $\mathbb{B}$. We need to prove that $|W_{u,\varphi}(f_m)|_q \to 0$ as $m \to \infty$. Assume that this is not true. Then there exist an $\varepsilon_0 > 0$ and a subsequence (f_{m_k}) of (f_m), which is a fortiori bounded in $\mathcal{N}_p(\mathbb{B})$, such that

$$|W_{u,\varphi}(f_{m_k})|_q \geq \varepsilon_0, \qquad (k \geq 1). \tag{8.2}$$

On the one hand, since $W_{u,\varphi}$ is compact, $\{W_{u,\varphi}(f_{m_k})\}$ has a subsequence, denoted by $\{W_{u,\varphi}(g_j)\}$ that converges to some $g \in A^{-q}(\mathbb{B})$. Since the norm topology of $A^{-q}(\mathbb{B})$ is stronger than the compact-open topology, which in turn is stronger than the pointwise topology, this implies that $\{W_{u,\varphi}(g_j)\}$ converges pointwise to g in $\mathbb{B}$. On the other hand, $f_{m_k} \to 0$ uniformly on every compact subset of $\mathbb{B}$, in particular, on the singleton $\{\varphi(z)\}$, for each fixed $z \in \mathbb{B}$. This means that $\{W_{u,\varphi}(g_j) = u \cdot g_j \circ \varphi\}$ converges pointwise to 0 in $\mathbb{B}$. Thus, $g(z) = 0$, $z \in \mathbb{B}$, and this contradicts (8.2).

Sufficiency We show that for any sequence (g_m) from the unit ball in $\mathcal{N}_p(\mathbb{B})$, the sequence $\{W_{u,\varphi}(g_m)\}$ contains a Cauchy subsequence with respect to the norm $|\cdot|_q$.

First, we prove that (g_m) is a normal sequence in $\text{Hol}(\mathbb{B})$. Indeed, by Theorem 7.5.1,

$$|g_m|_{\frac{n+1}{2}} \leq \frac{2^{p+n}}{3^{p/2}} \|g_m\|_p \leq \frac{2^{p+n}}{3^{p/2}}, \qquad (m \in \mathbb{N}).$$

From this, it follows that for an arbitrary closed ball $\overline{\mathbb{B}_\delta} = \{z \in \mathbb{B} : |z| \leq \delta\}$, $\delta \in (0,1)$, and any $m \in \mathbb{N}$, we have

$$\sup_{z \in \overline{\mathbb{B}_\delta}} |g_m(z)| = \sup_{z \in \overline{\mathbb{B}_\delta}} |g_m(z)|(1 - |z|^2)^{\frac{n+1}{2}} \cdot \frac{1}{(1 - |z|^2)^{\frac{n+1}{2}}}$$

$$\leq \sup_{z \in \mathbb{B}} |g_m(z)|(1 - |z|^2)^{\frac{n+1}{2}} \cdot \sup_{z \in \overline{\mathbb{B}_\delta}} \frac{1}{(1 - |z|^2)^{\frac{n+1}{2}}}$$

$$= |g_m|_{\frac{n+1}{2}} \cdot \sup_{z \in \overline{\mathbb{B}_\delta}} \frac{1}{(1 - |z|^2)^{\frac{n+1}{2}}}$$

$$\leq \frac{2^{p+n}}{3^{p/2}(1 - \delta^2)^{\frac{n+1}{2}}},$$

which means that (g_m) is a locally bounded sequence in $\text{Hol}(\mathbb{B})$. By Montel's theorem, (g_m) is a normal sequence.

Next, we show that $g \in \mathcal{N}_p(\mathbb{B})$. From the normal sequence (g_m), we can extract a subsequence (g_{m_k}) that converges uniformly on compact subsets of $\mathbb{B}$ to some function g. That is, (g_{m_k}) converges to some g in the space $\text{Hol}(\mathbb{B})$. By Fatou's lemma, for each $a \in \mathbb{B}$, we have

$$\int_{\mathbb{B}} |g(z)|^2 (1 - |\Phi_a(z)|^2)^p dV(z) = \int_{\mathbb{B}} \liminf_{k \to \infty} |g_{m_k}(z)|^2 (1 - |\Phi_a(z)|^2)^p dV(z)$$

$$\leq \liminf_{k\to\infty} \int_{\mathbb{B}} |g_{m_k}(z)|^2 (1 - |\Phi_a(z)|^2)^p dV(z)$$

$$\leq \sup_{k\in\mathbb{N}} \int_{\mathbb{B}} |g_{m_k}(z)|^2 (1 - |\Phi_a(z)|^2)^p dV(z)$$

$$= \sup_{k\in\mathbb{N}} \|g_{m_k}\|_p \leq 1.$$

Finally, since $g \in \mathcal{N}_p(\mathbb{B})$, $(g_{m_k} - g)$ is a bounded sequence in $\mathcal{N}_p(\mathbb{B})$ that converges uniformly to zero on every compact subset of $\mathbb{B}$. By the assumption, $|W_{u,\varphi}(g_{m_k} - g)|_q \to 0$, as $k \to \infty$. From this, it follows that $\{W_{u,\varphi}(g_{m_k})\}_{k\in\mathbb{N}}$ is a Cauchy subsequence of $\{W_{u,\varphi}(g_m)\}_{m\in\mathbb{N}}$, and hence $W_{u,\varphi}$ is compact.   □

We also need the following sufficient condition for compactness.

Lemma 8.3.2 *Suppose φ is a self-map of $\mathbb{B}$ such that $\|\varphi\|_\infty < 1$, and let $u \in \mathcal{N}_p$. Then the weighted composition operator $W_{u,\varphi} : \mathcal{N}_p \longrightarrow \mathcal{N}_p$ is compact.*

Proof Let $r = \|\varphi\|_\infty$. Take any $f \in \mathcal{N}_p$. We then have

$$\|W_{u,\varphi}f\|_p = \|u \cdot (f \circ \varphi)\|_p \leq \|f \circ \varphi\|_\infty \|u\|_p \leq \left(\sup_{\{z:|z|\leq r\}} |f(z)| \right) \|u\|_p < \infty.$$

This shows that $W_{u,\varphi}$ maps $\mathcal{N}_p$ into itself.

Now suppose that (f_m) is a bounded sequence in $\mathcal{N}_p$ that converges to zero uniformly on every compact subset of $\mathbb{B}$. Applying the above estimate to $f = f_m$, we obtain

$$\|W_{u,\varphi}f_m\|_p \leq \left(\sup_{\{z:|z|\leq r\}} |f_m(z)| \right) \|u\|_p.$$

Since the set $\{z : |z| \leq r\}$ is compact, the right-hand side of the last quantity converges to 0 as $m \to \infty$. Hence, so does the sequence $\{\|W_{u,\varphi}f_m\|_p\}$. This means that $W_{u,\varphi}$ is compact.   □

8.4   Compactness: Characterization

Using the tools provided in Sect. 8.3, we now formulate the following criteria for compactness of weighted composition operators.

Theorem 8.4.1 *Let $\varphi: \mathbb{B} \to \mathbb{B}$ and $u: \mathbb{B} \to \mathbb{C}$ be holomorphic mappings, and let $p, q > 0$. Then the weighted composition operator $W_{u,\varphi}: \mathcal{N}_p(\mathbb{B}) \to A^{-q}(\mathbb{B})$ is compact if and only if*

$$\lim_{r \to 1^-} \sup_{|\varphi(z)|>r} |u(z)| \frac{(1-|z|^2)^q}{(1-|\varphi(z)|^2)^{\frac{n+1}{2}}} = 0. \tag{8.3}$$

Proof **Necessity** Let $W_{u,\varphi} : \mathcal{N}_p(\mathbb{B}) \to A^{-q}(\mathbb{B})$ be compact. By Theorem 8.2.1,

$$M := \sup_{z \in \mathbb{B}} |u(z)| \frac{(1-|z|^2)^q}{(1-|\varphi(z)|^2)^{\frac{n+1}{2}}} < \infty.$$

Put

$$F(r) := \sup_{|\varphi(z)|>r} |u(z)| \frac{(1-|z|^2)^q}{(1-|\varphi(z)|^2)^{\frac{n+1}{2}}},$$

which is clearly bounded and decreasing on $(0, 1)$. Hence, $\lim_{r \to 1^-} F(r)$ exists. We show that this limit is necessarily zero. Assume that $\lim_{r \to 1^-} F(r) = L > 0$. Then there exists a $r_0 \in (0, 1)$ such that for all $r \in (r_0, 1)$, we have $F(r) > L/2$. We are going to construct a sequence $(z_m) \subset \mathbb{B}$, by the standard diagonal process, to get a contradiction.

Take $r_1 \in (r_0, 1)$. Since

$$F(r_1) = \sup_{|\varphi(z)|>r_1} |u(z)| \frac{(1-|z|^2)^q}{(1-|\varphi(z)|^2)^{\frac{n+1}{2}}} > \frac{L}{2},$$

there exists a point $z_1 \in \{z : |\varphi(z)| > r_1\}$ such that

$$|u(z_1)| \frac{(1-|z_1|^2)^q}{(1-|\varphi(z_1)|^2)^{\frac{n+1}{2}}} > L/4.$$

Next we take $r_2 > \max\left\{|\varphi(z_1)|, \frac{3}{4}\right\}$. Since

$$\max\left\{|\varphi(z_1)|, \frac{3}{4}\right\} \geq |\varphi(z_1)| > r_1 > r_0,$$

we have

$$F(r_2) = \sup_{|\varphi(z)|>r_2} |u(z)| \frac{(1-|z|^2)^q}{(1-|\varphi(z)|^2)^{\frac{n+1}{2}}} > \frac{L}{2},$$

which implies the existence of another point $z_2 \in \{z : |\varphi(z)| > r_2\}$ such that

$$|u(z_2)| \frac{(1-|z_2|^2)^q}{(1-|\varphi(z_2)|^2)^{\frac{n+1}{2}}} > L/4.$$

Note that since $|\varphi(z_2)| > r_2 > |\varphi(z_1)|$, we have $z_2 \neq z_1$. Continuing this process, for any $m \geq 2$, we take $r_m > \max\left\{|\varphi(z_{m-1})|, 1 - \frac{1}{2^m}\right\}$. Since

$$\max\left\{|\varphi(z_{m-1})|, 1 - \frac{1}{2^m}\right\} \geq |\varphi(z_{m-1})| > r_{m-1} > \cdots > r_1 > r_0,$$

we have

$$F(r_m) = \sup_{|\varphi(z)| > r_m} |u(z)| \frac{(1 - |z|^2)^q}{(1 - |\varphi(z)|^2)^{\frac{n+1}{2}}} > \frac{L}{2},$$

and so there exists $z_m \in \{z : |\varphi(z)| > r_m\}$ such that

$$|u(z_m)| \frac{(1 - |z_m|^2)^q}{(1 - |\varphi(z_m)|^2)^{\frac{n+1}{2}}} > L/4$$

and $z_m \neq z_1, \ldots, z_{m-1}$.

Note that for each $m \geq 2$, we have $|\varphi(z_m)| > 1 - \frac{1}{2^m}$, which shows that $|\varphi(z_m)| \to 1$ as $m \to \infty$. Consider the test functions k_{w_m}, where $w_m = \varphi(z_m)$, defined in Lemma 8.1.1. It is easy to see that $k_{w_m} \to 0$ uniformly on every compact subset of $\mathbb{B}$. Moreover, for each $m \geq 1$, $\|k_{w_m}\|_p \leq 1$. Since $W_{u,\varphi}$ is compact, by Lemma 8.3.1, $|W_{u,\varphi}(k_{w_m})|_q \to 0$ as $m \to \infty$. However, for each $m \geq 1$,

$$|W_{u,\varphi}(k_{w_m})|_q = \sup_{z \in \mathbb{B}} |u(z)| \cdot |k_{w_m}(\varphi(z))|(1 - |z|^2)^q$$

$$\geq |u(z_m)| \cdot |k_{w_m}(\varphi(z_m))|(1 - |z_m|^2)^q$$

$$= |u(z_m)| \cdot \frac{(1 - |z_m|^2)^q}{(1 - |\varphi(z_m)|^2)^{\frac{n+1}{2}}} \geq \frac{L}{4},$$

which gives a contradiction.

Sufficiency Suppose that (8.3) holds. Let (f_m) be a bounded sequence in $\mathcal{N}_p(\mathbb{B})$ which converges to zero uniformly on every compact subset of $\mathbb{B}$. We have $\|f_m\|_p \leq M, m \geq 1$, for some $M > 0$. Since $\lim_{r \to 1^-} F(r) = 0$, for each $\varepsilon > 0$, there exists $r_0 \in (0, 1)$ such that for all $r \in (r_0, 1)$,

$$\sup_{|\varphi(z)| > r} |u(z)| \frac{(1 - |z|^2)^q}{(1 - |\varphi(z)|^2)^{\frac{n+1}{2}}} < \frac{\varepsilon}{2c_1 M},$$

where $c_1 = 4^{p+n}/3^p$ is defined in Theorem 7.5.1. Take $r \in (r_0, 1)$. Then we have

$$|W_{u,\varphi}(f_m)|_q = \sup_{z \in \mathbb{B}} |u(z)||f_m(\varphi(z))|(1 - |z|^2)^q = C_{1,m} + C_{2,m},$$

where

$$C_{1,m} = \sup_{|\varphi(z)|>r} |u(z)||f_m(\varphi(z))|(1 - |z|^2)^q$$

and

$$C_{2,m} = \sup_{|\varphi(z)|\leq r} |u(z)||f_m(\varphi(z))|(1 - |z|^2)^q.$$

Note that, for any $m \geq 1$, we have

$$C_{1,m} = \sup_{|\varphi(z)|>r} |u(z)| \frac{(1 - |z|^2)^q}{(1 - |\varphi(z)|^2)^{\frac{n+1}{2}}} \cdot |f_m(\varphi(z))|(1 - |\varphi(z)|^2)^{\frac{n+1}{2}}$$

$$\leq \sup_{|\varphi(z)|>r} |u(z)| \frac{(1 - |z|^2)^q}{(1 - |\varphi(z)|^2)^{\frac{n+1}{2}}} \cdot \sup_{z \in \mathbb{B}} |f_m(\varphi(z))|(1 - |\varphi(z)|^2)^{\frac{n+1}{2}}$$

$$\leq \frac{\varepsilon}{2c_1 M} \sup_{z \in \mathbb{B}} |f_m(\varphi(z))|(1 - |\varphi(z)|^2)^{\frac{n+1}{2}}$$

$$\leq \frac{\varepsilon}{2c_1 M} |f_m|_{\frac{n+1}{2}} \leq \frac{\varepsilon}{2c_1 M} \cdot c_1 \|f_m\|_p$$

$$\leq \frac{\varepsilon}{2c_1 M} \cdot c_1 M = \frac{\varepsilon}{2}.$$

On the other hand, according to the condition, it is easy to verify that

$$\sup_{z \in \mathbb{B}} |u(z)| \cdot \frac{(1 - |z|^2)^q}{(1 - |\varphi(z)|^2)^{\frac{n+1}{2}}} < M_1.$$

Then

$$C_{2,m} = \sup_{|\varphi(z)|\leq r} |u(z)| \frac{(1 - |z|^2)^q}{(1 - |\varphi(z)|^2)^{\frac{n+1}{2}}} \cdot |f_m(\varphi(z))|(1 - |\varphi(z)|^2)^{\frac{n+1}{2}}$$

$$\leq \sup_{|\varphi(z)|\leq r} |u(z)| \frac{(1 - |z|^2)^q}{(1 - |\varphi(z)|^2)^{\frac{n+1}{2}}} \cdot \sup_{|\varphi(z)|\leq r} |f_m(\varphi(z))|(1 - |\varphi(z)|^2)^{\frac{n+1}{2}}$$

$$\leq M_1 \sup_{|\varphi(z)|\leq r} |f_m(\varphi(z))|(1 - |\varphi(z)|^2)^{\frac{n+1}{2}}.$$

Since (f_m) converges to zero on every compact subset of $\mathbb{B}$, it implies that there exists m_ε such that for $m > m_\varepsilon$,

$$\sup_{|\varphi(z)| \leq \delta} |f_m(\varphi(z))|(1 - |\varphi(z)|^2)^{\frac{n+1}{2}} < \frac{\varepsilon}{2M_1}.$$

This shows that for the chosen ε, $C_{2,m} \leq \frac{\varepsilon}{2}$ when $m > m_\varepsilon$.

Combining these two parts, we conclude that for any $\varepsilon > 0$, there exists an m_ε, such that for all $m > m_\varepsilon$, $|W_{u,\varphi}(f_m)|_q \leq \varepsilon$, which means $|W_{u,\varphi}(f_m)|_q \to 0$ as $m \to \infty$. By Lemma 8.3.1, we conclude that $W_{u,\varphi} : \mathcal{N}_p(\mathbb{B}) \to A^{-q}(\mathbb{B})$ is compact. $\qquad\square$

As a special case, we immediately obtain the following useful characterization for the composition operator.

Corollary 8.4.2 *Let* $\varphi : \mathbb{B} \to \mathbb{B}$ *be a holomorphic mapping, and let* $p, q > 0$. *Then the composition operator* C_φ *acting from* $\mathcal{N}_p(\mathbb{B}) \to A^{-q}(\mathbb{B})$ *is:*

(i) Bounded if and only if

$$\sup_{z \in \mathbb{B}} \frac{(1 - |z|^2)^q}{(1 - |\varphi(z)|^2)^{\frac{n+1}{2}}} < \infty$$

(ii) Compact if and only if

$$\lim_{r \to 1^-} \sup_{|\varphi(z)| > r} \frac{(1 - |z|^2)^q}{(1 - |\varphi(z)|^2)^{\frac{n+1}{2}}} = 0$$

8.5 Estimation of $|f(z) - f(w)|$

Recall that the pseudo-hyperbolic metric in the ball was defined by

$$\rho(z, w) = |\Phi_w(z)|, \qquad (z, w \in \mathbb{B}).$$

See (7.4). The following technical lemmas play an important role in the proof of our main result.

Lemma 8.5.1 *For* $z, w \in \mathbb{B}$, *if* $\rho(z, w) \leq \frac{1}{2}$, *then*

$$\frac{1}{6} \leq \frac{1 - |z|^2}{1 - |w|^2} \leq 6.$$

Proof Let $z, w \in \mathbb{B}$. For simplicity, denote $r = \rho(z, w)$. We have

$$\frac{|\rho(\Phi_w(z), 0) - \rho(0, w)|}{1 - \rho(\Phi_w(z), 0)\rho(0, w)} \leq \rho(\Phi_w(z), w) \leq \frac{\rho(\Phi_w(z), 0) + \rho(0, w)}{1 + \rho(\Phi_w(z), 0)\rho(0, w)} \tag{8.4}$$

or, equivalently,

$$\frac{|r - |w||}{1 - r|w|} \leq |z| \leq \frac{r + |w|}{1 + r|w|}.$$

As a matter of fact, for the left inequality, we need only a weaker version. Namely,

$$\frac{|w| - r}{1 - r|w|} \leq |z| \leq \frac{|w| + r}{1 + r|w|}. \tag{8.5}$$

Furthermore, since $r \in (0, \frac{1}{2}]$, from the left inequality of (8.5), it follows that

$$\frac{1 - |z|}{1 - |w|} \leq \frac{1 - \frac{|w| - r}{1 - r|w|}}{1 - |w|} = \frac{1 + r}{1 - r|w|} \leq 3,$$

while the right inequality of (8.5) gives

$$\frac{1 - |z|}{1 - |w|} \geq \frac{1 - \frac{|w| + r}{1 + r|w|}}{1 - |w|} = \frac{1 - |r|}{1 + r|w|} \geq \frac{1}{3}.$$

We also have

$$\frac{1}{2} \leq \frac{1}{1 + |w|} \leq \frac{1 + |z|}{1 + |w|} \leq \frac{2}{1 + |w|} \leq 2.$$

Therefore, we get

$$\frac{1}{6} \leq \frac{1 - |z|^2}{1 - |w|^2} = \frac{1 - |z|}{1 - |w|} \cdot \frac{1 + |z|}{1 + |w|} \leq 6.$$

$\square$

Lemma 8.5.2 *For $f \in \mathcal{N}_p(\mathbb{B})$ and $z, w \in \mathbb{B}$, we have*

$$|f(z) - f(w)| \leq c \, \|f\|_p \, \max \left\{ \frac{1}{(1 - |z|^2)^{\frac{n+1}{2}}}, \frac{1}{(1 - |w|^2)^{\frac{n+1}{2}}} \right\} \rho(z, w),$$

where $A = \dfrac{6^{\frac{n+1}{2}} \cdot 2^{p+n+1}(3 + 2\sqrt{3})\sqrt{n}}{3^{p/2}}.$

Proof We consider two cases.

Cases 1 $\rho(z, w) \geq \frac{1}{4}$. Since $|f(z) - f(w)| \leq |f(z)| + |f(w)|$, by Theorem 7.5.1, we have

$$\min \left\{ (1 - |z|^2)^{\frac{n+1}{2}}, (1 - |w|^2)^{\frac{n+1}{2}} \right\} |f(z) - f(w)|$$

$$\leq (1 - |z|^2)^{\frac{n+1}{2}} |f(z)| + (1 - |w|^2)^{\frac{n+1}{2}} |f(w)|$$

$$\leq 2|f|_{\frac{n+1}{2}} \leq \frac{2^{p+n+1}}{3^{p/2}} \|f\|_p \leq \frac{2^{p+n+3}}{3^{p/2}} \|f\|_p \rho(z, w),$$

which implies

$$|f(z) - f(w)| \leq \frac{2^{p+n+3}}{3^{p/2}} \|f\|_p \max \left\{ \frac{1}{(1 - |z|^2)^{\frac{n+1}{2}}}, \frac{1}{(1 - |w|^2)^{\frac{n+1}{2}}} \right\} \rho(z, w).$$

Cases 2 $\rho(z, w) < \frac{1}{4}$. Fix $w \in \mathbb{B}$. From $\rho(\Phi_w(z), w) = |z|$, it follows that if $z \in \overline{\mathbb{B}_{1/2}}$, then $\rho(\Phi_w(z), w) \leq \frac{1}{2}$. In this case, by Theorem 7.5.1 and Lemma 8.5.1, we have

$$|f(\Phi_w(z))| \leq \frac{|f|_{\frac{n+1}{2}}}{(1 - |\Phi_w(z)|^2)^{\frac{n+1}{2}}}$$

$$\leq \frac{2^{p+n} \|f\|_p}{3^{p/2}(1 - |\Phi_w(z)|^2)^{\frac{n+1}{2}}}$$

$$= \frac{2^{p+n} \|f\|_p}{3^{p/2}(1 - |w|^2)^{\frac{n+1}{2}}} \cdot \left[\frac{1 - |w|^2}{1 - |\Phi_w(z)|^2} \right]^{\frac{n+1}{2}}$$

$$\leq \frac{6^{\frac{n+1}{2}} \cdot 2^{p+n} \|f\|_p}{3^{p/2}(1 - |w|^2)^{\frac{n+1}{2}}}.$$

Now, we follow the standard scheme to estimate a quantity $|f(z) - f(w)|$. Set $g_w = f \circ \Phi_w$; then

$$|f(z) - f(w)| = |f(\Phi_w(\Phi_w(z))) - f(\Phi_w(0))| = |g_w(\Phi_w(z)) - g_w(0)|.$$

For each $z \in \mathbb{B}$ with $\rho(z, w) = |\Phi_w(z)| < \frac{1}{4}$, we have

$$|f(z) - f(w)| = |g_w(\Phi_w(z)) - g_w(0)| \leq |\nabla g_w(t)| \cdot |\Phi_w(z)| = |\nabla g_w(t)| \rho(z, w),$$

where $t = (t_1, t_2, \ldots, t_n)$ is some point in $\mathbb{B}$ with $|t| \leq |\Phi_w(z)| \leq \frac{1}{4}$. Furthermore,

$$|\nabla g_w(t)|\rho(z, w) \le \sqrt{n}\rho(z, w) \max_{1 \le k \le n} \left| \frac{\partial g_w}{\partial z_k}(t) \right|$$

$$\le \sqrt{n}\rho(z, w) \max_{1 \le k \le n} \left| \frac{1}{2\pi i} \int_{|\xi_k|=\frac{\sqrt{3}}{4}} \frac{g_w(t_1, t_2, \ldots, \xi_k, \ldots, t_n)}{(\xi_k - t_k)^2} d\xi_k \right|$$

$$\le \frac{\sqrt{n}\rho(z, w)}{2\pi} \max_{1 \le k \le n} \int_{|\xi_k|=\frac{\sqrt{3}}{4}} \left| \frac{g_w(t_1, t_2, \ldots, \xi_k, \ldots, t_n)}{(\xi_k - t_k)^2} \right| |d\xi_k|.$$

Note that for $(t_1, t_2, \ldots, \xi_k, \ldots, t_n)$ with $|t| \le \frac{1}{4}$ and $|\xi_k| = \frac{\sqrt{3}}{4}$, we have

$$\rho\left(\Phi_w(t_1, t_2, \ldots, \xi_k, \ldots, t_n), w\right)$$

$$= \rho((t_1, t_2, \ldots, \xi_k, \ldots, t_n), 0)$$

$$= |(t_1, t_2, \ldots, \xi_k, \ldots, t_n)|$$

$$\le \sqrt{|t|^2 + |\xi_k|^2} \le \sqrt{\left(\frac{1}{4}\right)^2 + \left(\frac{\sqrt{3}}{4}\right)^2} = \frac{1}{2},$$

and thus

$$|g_w(t_1, t_2, \ldots, \xi_j, \ldots, t_n)| = \left| f\left(\Phi_w(t_1, t_2, \ldots, \xi_k, \ldots, t_n)\right) \right| \le \frac{6^{\frac{n+1}{2}} \cdot 2^{p+n} \|f\|_p}{3^{p/2}(1 - |w|^2)^{\frac{n+1}{2}}}.$$

Also,

$$\max_{1 \le k \le n} \int_{|\xi_k|=\frac{\sqrt{3}}{4}} \frac{|d\xi_k|}{|\xi_k - t_k|^2} \le \max_{1 \le k \le n} \int_{|\xi_k|=\frac{\sqrt{3}}{4}} \frac{|d\xi_k|}{(\frac{\sqrt{3}}{4} - |t_k|)^2}$$

$$\le \max_{1 \le k \le n} \int_{|\xi_k|=\frac{\sqrt{3}}{4}} \frac{|d\xi_k|}{(\frac{\sqrt{3}}{4} - \frac{1}{4})^2} \le 2\pi \frac{\sqrt{3}}{4} \cdot \left(\frac{4}{\sqrt{3} - 1}\right)^2 = 4\pi(3 + 2\sqrt{3}).$$

Consequently,

$$|f(z) - f(w)| \le \frac{\sqrt{n}\rho(z, w)}{2\pi} \cdot \frac{6^{\frac{n+1}{2}} \cdot 2^{p+n} \|f\|_p}{3^{p/2}(1 - |w|^2)^{\frac{n+1}{2}}} \cdot 4\pi(3 + 2\sqrt{3})$$

$$= \frac{6^{\frac{n+1}{2}} \cdot 2^{p+n+1}(3 + 2\sqrt{3})\sqrt{n}}{3^{p/2}} \cdot \frac{1}{(1 - |w|^2)^{\frac{n+1}{2}}} \cdot \|f\|_p \cdot \rho(z, w).$$

Combining the results of the two cases yields

$$|f(z) - f(w)| \leq A\|f\|_p \max \left\{ \frac{1}{(1-|z|^2)^{\frac{n+1}{2}}}, \frac{1}{(1-|w|^2)^{\frac{n+1}{2}}} \right\} \rho(z,w),$$

where $A = \dfrac{6^{\frac{n+1}{2}} \cdot 2^{p+n+1} (3+2\sqrt{3})\sqrt{n}}{3^{p/2}}$.   □

Lemma 8.5.2 gives the estimate for the distance between $f(z)$ and $f(w)$ for $f \in \mathcal{N}_p$ and $z, w \in \mathbb{B}$, in which the constant A is involved. In the special case where z and w are multiple of each other, the above estimate can be simplified. This is explained below.

Lemma 8.5.3 *Let f be in $\mathcal{N}_p$. For any $a \in \mathbb{B}$ and any $\kappa \in (0,1)$, we have*

$$|f(a) - f(\kappa a)| \leq \frac{A(1-\kappa)|a|}{1 - \kappa|a|^2} \cdot \frac{\|f\|_p}{(1-|a|^2)^{\frac{n+1}{2}}} \leq \frac{A\|f\|_p}{(1-|a|^2)^{\frac{n+1}{2}}}. \qquad (8.6)$$

Consequently, for any $0 < r < 1$, we have

$$\sup_{|a| \leq r} |f(a) - f(\kappa a)| \leq \frac{Ar(1-\kappa)\|f\|_p}{(1-r^2)^{\frac{n+3}{2}}}. \qquad (8.7)$$

Here, A is the constant from Lemma 8.5.2.

Proof Lemma 8.5.2 shows that

$$|f(a) - f(\kappa a)| \leq A\|f\|_p \max \left\{ \frac{1}{(1-|a|^2)^{\frac{n+1}{2}}}, \frac{1}{(1-|\kappa a|^2)^{\frac{n+1}{2}}} \right\} \rho(a, \kappa a).$$

By the definition of ρ,

$$|\rho(a, \kappa a)| = |\Phi_a(\kappa a)| = \frac{(1-\kappa)|a|}{1 - \kappa|a|^2} \leq 1.$$

On the other hand, $(1 - |\kappa a|^2)^{-(n+1)/2} \leq (1 - |a|^2)^{-(n+1)/2}$. The inequalities in (8.6) now follow. If $|a| \leq r$, then $1 - \kappa|a|^2 \geq 1 - r^2$. Taking supremum of (8.6) in a yields (8.7). □

8.6 Compactness of Difference

In this section, we study the compactness of the difference of two bounded weighted composition operators acting from $\mathcal{N}_p(\mathbb{B})$ into $A^{-q}(\mathbb{B})$. Let φ_1, φ_2 be two holomorphic self-mappings on $\mathbb{B}$, let $u_1, u_2 : \mathbb{B} \to \mathbb{C}$ be two holomorphic mappings, and let $p, q > 0$.

Consider W_{u_1,φ_1} and W_{u_2,φ_2} the two corresponding weighted composition operators acting from $\mathcal{N}_p(\mathbb{B})$ into $A^{-q}(\mathbb{B})$. We seek conditions under which the difference $W_{u_1,\varphi_1} - W_{u_2,\varphi_2}$ is compact. Inspiring by the pseudo-hyperbolic metric in the unit ball, for two holomorphic mappings $\varphi, \psi : \mathbb{B} \to \mathbb{B}$, we define

$$\rho_{\varphi,\psi}(z) = \left|\Phi_{\varphi(z)}(\psi(z))\right|, \qquad (z \in \mathbb{B}).$$

Evidently, $\rho_{\varphi,\psi} = \rho_{\psi,\varphi}$.

Theorem 8.6.1 *Let φ_1, φ_2 be two self-mapping of $\mathbb{B}$, let $u_1, u_2 : \mathbb{B} \to \mathbb{C}$ be two holomorphic mappings, and let $p, q > 0$. Let, further, W_{u_1,φ_1} and W_{u_2,φ_2} be two weighted composition operators acting from $\mathcal{N}_p(\mathbb{B})$ into $A^{-q}(\mathbb{B})$. Then $W_{u_1,\varphi_1} - W_{u_2,\varphi_2}$ is compact if and only if the following two conditions are satisfied:*

(i)

$$\lim_{r \to 1^-} \sup_{|\varphi_k(z)|>r} \left\{ \frac{|u_k(z)|(1-|z|^2)^q}{(1-|\varphi_k(z)|^2)^{\frac{n+1}{2}}} \rho_{\varphi_1,\varphi_2}(z) \right\} = 0, \qquad (k=1,2).$$

(ii) As $r \to 1^-$,

$$\sup \left[|u_1(z) - u_2(z)| \min \left\{ \frac{(1-|z|^2)^q}{(1-|\varphi_1(z)|^2)^{\frac{n+1}{2}}}, \frac{(1-|z|^2)^q}{(1-|\varphi_2(z)|^2)^{\frac{n+1}{2}}} \right\} \right] \to 0.$$

The supremum is taken over the set $\{\min\{|\varphi_1(z)|, |\varphi_2(z)|\} > r\}$.

Proof It is well-known that the operator $W_{u_1,\varphi_1} - W_{u_2,\varphi_2} : \mathcal{N}_p(\mathbb{B}) \to A^{-q}(\mathbb{B})$ is compact if and only if

$$|(W_{u_1,\varphi_1} - W_{u_2,\varphi_2})(f_m)|_q \to 0, \qquad (m \to \infty), \qquad (8.8)$$

for any bounded sequence (f_m) in $\mathcal{N}_p(\mathbb{B})$ such that (f_m) converges to zero uniformly on every compact subset of $\mathbb{B}$. This is used twice below.

Necessity Suppose $W_{u_1,\varphi_1} - W_{u_2,\varphi_2}$ is a compact operator.

Proof of (i). It suffices to prove for $k = 1$. Since W_{u_1,φ_1} is bounded, by Theorem 8.2.1 and the fact that $\rho_{\varphi_1,\varphi_2}(z) \leq 1$, for all $z \in \mathbb{B}$, we have

$$\sup_{z \in \mathbb{B}} \left\{ \frac{|u_1(z)|(1-|z|^2)^q}{(1-|\varphi_1(z)|^2)^{\frac{n+1}{2}}} \rho_{\varphi_1,\varphi_2}(z) \right\} \leq \sup_{z \in \mathbb{B}} \frac{|u_1(z)|(1-|z|^2)^q}{(1-|\varphi_1(z)|^2)^{\frac{n+1}{2}}} < \infty.$$

Set

$$G(r) = \sup_{|\varphi_1(z))|>r} \left\{ \frac{|u_1(z)|(1-|z|^2)^q}{(1-|\varphi_1(z)|^2)^{\frac{n+1}{2}}} \rho_{\varphi_1,\varphi_2}(z) \right\}, \qquad (0 < r < 1).$$

It is clear that G is bounded and decreasing on $(0,1)$, and hence $\lim_{r \to 1^-} G(r)$ exists.

To get a contradiction, assume that (i) is not true. Then there exists an $L > 0$, such that $\lim_{r \to 1^-} G(r) > L$. By the standard diagonal process, we can choose a sequence $(z_m) \subset \mathbb{B}$, such that $|\varphi_1(z_m)| \to 1$, as $m \to \infty$, and

$$\frac{|u_1(z_m)|(1-|z_m|^2)^q}{(1-|\varphi_1(z_m)|^2)^{\frac{n+1}{2}}} \rho_{\varphi_1,\varphi_2}(z_m) > \frac{L}{4}, \qquad (m \geq 1). \qquad (8.9)$$

Consider the functions

$$g_m(z) = \Phi_{\varphi_2(z_m)}(z) \cdot k_{w_m}(z), \qquad (z \in \mathbb{B}),$$

where $w_m = \varphi_1(z_m)$, $m \geq 1$. Obviously, $g_m \in \mathrm{Hol}(\mathbb{B})$. Moreover, since $|\Phi_{\varphi_2(z_m)}(z)| \leq 1$, $z \in \mathbb{B}$, we have

$$\|g_m(z)\|_p \leq \|k_{w_m}(z)\|_p \leq 1,$$

which shows that $g_m(z) \in \mathcal{N}_p(\mathbb{B})$ for all $m \geq 1$ and that the sequence $\{g_m\}$ is bounded in $\mathcal{N}_p(\mathbb{B})$. Furthermore, by the fact that k_{w_m} converges to zero uniformly on every compact subset of $\mathbb{B}$, and $|g_m(z)| \leq |k_{w_m}(z)|$, $z \in \mathbb{B}$, we see that g_m also converges to zero uniformly on every compact subset of $\mathbb{B}$.

By (8.8),

$$|(W_{u_1,\varphi_1} - W_{u_2,\varphi_2})(g_m)|_q \to 0, \qquad (m \to \infty). \qquad (8.10)$$

Note that for each $m \geq 1$, $\Phi_{\varphi_2(z_m)}(\varphi_2(z_m)) = 0$, which implies $g_m(\varphi_2(z_m)) = 0$. Then,

$$|(W_{u_1,\varphi_1} - W_{u_2,\varphi_2})(g_m)|_q$$

$$= \sup_{z \in \mathbb{B}} \left\{ |u_1(z)g_m(\varphi_1(z)) - u_2(z)g_m(\varphi_2(z))| \cdot (1-|z|^2)^q \right\}$$

$$\geq |u_1(z_m)g_m(\varphi_1(z_m)) - u_2(z_m)g_m(\varphi_2(z_m))| \cdot (1-|z_m|^2)^q$$

$$= |u_1(z_m)g_m(\varphi_1(z_m))| \cdot (1-|z_m|^2)^q$$

$$= \frac{|u_1(z_m)|(1-|z_m|^2)^q}{(1-|\varphi_1(z_m)|^2)^{\frac{n+1}{2}}} \cdot |\Phi_{\varphi_2(z_m)}(\varphi_1(z_m))|$$

$$= \frac{|u_1(z_m)|(1 - |z_m|^2)^q}{(1 - |\varphi_1(z_m)|^2)^{\frac{n+1}{2}}} \rho_{\varphi_1,\varphi_2}(z_m) > \frac{L}{4},$$

by Lemma 8.1.1, which contradicts (8.10). Thus, we must have

$$\lim_{r \to 1^-} \sup_{|\varphi_1(z)| > r} \left\{ \frac{|u_1(z)|(1 - |z|^2)^q}{(1 - |\varphi_1(z)|^2)^{\frac{n+1}{2}}} \rho_{\varphi_1,\varphi_2}(z) \right\} = 0,$$

and (i) is proved.

Proof of (ii). Since both W_{u_1,φ_1} and W_{u_2,φ_2} are bounded, by Theorem 8.2.1, we have

$$\sup_{z \in \mathbb{B}} \left[|u_1(z) - u_2(z)| \min \left\{ \frac{(1 - |z|^2)^q}{(1 - |\varphi_1(z)|^2)^{\frac{n+1}{2}}}, \frac{(1 - |z|^2)^q}{(1 - |\varphi_2(z)|^2)^{\frac{n+1}{2}}} \right\} \right]$$

$$\leq \sup_{z \in \mathbb{B}} \left[|u_1(z)| \min \left\{ \frac{(1 - |z|^2)^q}{(1 - |\varphi_1(z)|^2)^{\frac{n+1}{2}}}, \frac{(1 - |z|^2)^q}{(1 - |\varphi_2(z)|^2)^{\frac{n+1}{2}}} \right\} \right]$$

$$+ \sup_{z \in \mathbb{B}} \left[|u_2(z)| \min \left\{ \frac{(1 - |z|^2)^q}{(1 - |\varphi_1(z)|^2)^{\frac{n+1}{2}}}, \frac{(1 - |z|^2)^q}{(1 - |\varphi_2(z)|^2)^{\frac{n+1}{2}}} \right\} \right]$$

$$\leq \sup_{z \in \mathbb{B}} \frac{|u_1(z)|(1 - |z|^2)^q}{(1 - |\varphi_1(z)|^2)^{\frac{n+1}{2}}} + \sup_{z \in \mathbb{B}} \frac{|u_2(z)|(1 - |z|^2)^q}{(1 - |\varphi_2(z)|^2)^{\frac{n+1}{2}}} < \infty.$$

For each $r \in (0, 1)$, set

$$H(r) = \sup \left[|u_1(z) - u_2(z)| \min \left\{ \frac{(1 - |z|^2)^q}{(1 - |\varphi_1(z)|^2)^{\frac{n+1}{2}}}, \frac{(1 - |z|^2)^q}{(1 - |\varphi_2(z)|^2)^{\frac{n+1}{2}}} \right\} \right],$$

where the supremum is taken over the set $\{\min\{|\varphi_1(z)|, |\varphi_2(z)|\} > r\}$. The function $H(r)$ is bounded and decreasing on $(0, 1)$, and hence, $\lim_{r \to 1^-} H(r)$ exists. We prove that this limit is zero. We follow the same scheme of proving (i), but it requires more delicate arguments. Assume that $\lim_{r \to 1^-} H(r) = L > 0$. Again by the standard diagonal process, we can choose a sequence $(z_m) \subset \mathbb{B}$, such that $\min\{|\varphi_1(z_m)|, |\varphi_2(z_m)|\} \to 1$ as $m \to \infty$ and that, for each $m \geq 1$,

$$|u_1(z_m) - u_2(z_m)| \min \left\{ \frac{(1 - |z_m|^2)^q}{(1 - |\varphi_1(z_m)|^2)^{\frac{n+1}{2}}}, \frac{(1 - |z_m|^2)^q}{(1 - |\varphi_2(z_m)|^2)^{\frac{n+1}{2}}} \right\} > \frac{L}{4}.$$

For each $m \geq 1$, we have either $|\varphi_1(z_m)| \geq |\varphi_2(z_m)|$ or $|\varphi_1(z_m)| \leq |\varphi_2(z_m)|$. Choose a subsequence (z_{m_k}) of (z_m), such that for each $k \geq 1$,

$$|\varphi_1(z_{m_k})| \geq |\varphi_2(z_{m_k})|.$$

Otherwise, there are only finitely many indexes m, such that $|\varphi_1(z_m)| \geq |\varphi_2(z_m)|$, and in this case, we choose a subsequence (z_{m_k}) of (z_m), such that for each $k \geq 1$, $|\varphi_1(z_{m_k})| \leq |\varphi_2(z_{m_k})|$. We only consider the first case, since the second case can be proved by interchanging the role of φ_1 and φ_2, and without loss of generality, we write (z_m) for (z_{m_k}).

Since $|\varphi_1(z_m)| \geq |\varphi_2(z_m)|$ for each $m \geq 1$, we have

$$|u_1(z_m) - u_2(z_m)| \min \left\{ \frac{(1 - |z_m|^2)^q}{(1 - |\varphi_1(z_m)|^2)^{\frac{n+1}{2}}}, \frac{(1 - |z_m|^2)^q}{(1 - |\varphi_2(z_m)|^2)^{\frac{n+1}{2}}} \right\}$$

$$= |u_1(z_m) - u_2(z_m)| \frac{(1 - |z_m|^2)^q}{(1 - |\varphi_2(z_m)|^2)^{\frac{n+1}{2}}} > \frac{L}{4}.$$

Since the sequence $\{\rho_{\varphi_1,\varphi_2}(z_m)\}$ is bounded, it contains a convergent subsequence. Without loss of generality, we can assume that

$$\lim_{m \to \infty} \rho_{\varphi_1,\varphi_2}(z_m) = \ell \geq 0.$$

There are two cases for ℓ to consider.

Case 1: $\ell > 0$. In this case, there exists an $m_0 \geq 1$ such that $\rho_{\varphi_1,\varphi_2}(z_m) > \frac{\ell}{2}$, $m > m_0$. In this case, we have

$$\frac{(1 - |z_m|^2)^q |u_1(z_m) - u_2(z_m)|}{(1 - |\varphi_2(z_m)|^2)^{\frac{n+1}{2}}}$$

$$\leq \frac{(1 - |z_m|^2)^q |u_1(z_m)|}{(1 - |\varphi_2(z_m)|^2)^{\frac{n+1}{2}}} + \frac{(1 - |z_m|^2)^q |u_2(z_m)|}{(1 - |\varphi_2(z_m)|^2)^{\frac{n+1}{2}}}$$

$$\leq \frac{2}{\ell} \rho_{\varphi_1,\varphi_2}(z_m) \left[\frac{(1 - |z_m|^2)^q |u_1(z_m)|}{(1 - |\varphi_2(z_m)|^2)^{\frac{n+1}{2}}} + \frac{(1 - |z_m|^2)^q |u_2(z_m)|}{(1 - |\varphi_2(z_m)|^2)^{\frac{n+1}{2}}} \right]$$

$$\leq \frac{2}{\ell} \rho_{\varphi_1,\varphi_2}(z_m) \left[\frac{(1 - |z_m|^2)^q |u_1(z_m)|}{(1 - |\varphi_1(z_m)|^2)^{\frac{n+1}{2}}} + \frac{(1 - |z_m|^2)^q |u_2(z_m)|}{(1 - |\varphi_2(z_m)|^2)^{\frac{n+1}{2}}} \right],$$

which gives

$$\rho_{\varphi_1,\varphi_2}(z_m) \left[\frac{(1 - |z_m|^2)^q |u_1(z_m)|}{(1 - |\varphi_1(z_m)|^2)^{\frac{n+1}{2}}} + \frac{(1 - |z_m|^2)^q |u_2(z_m)|}{(1 - |\varphi_2(z_m)|^2)^{\frac{n+1}{2}}} \right] \geq \frac{L\ell}{8}, \qquad (m > m_0).$$

However, since $|\varphi_2(z_m)| \leq |\varphi_1(z_m)| \leq 1$, $m \in \mathbb{N}$, from

$$\lim_{m\to\infty} \min\{|\varphi_1(z_m)|, |\varphi_2(z_m)|\} = 1$$

it follows that

$$\lim_{m\to\infty} |\varphi_1(z_m)| = \lim_{m\to\infty} |\varphi_2(z_m)| = 1.$$

Hence, by (i),

$$\lim_{m\to\infty} \rho_{\varphi_1,\varphi_2}(z_m) \frac{(1-|z_m|^2)^q |u_k(z_m)|}{(1-|\varphi_k(z_m)|^2)^{\frac{n+1}{2}}} = 0, \qquad (k=1,2),$$

and thus, as $m \to \infty$,

$$\rho_{\varphi_1,\varphi_2}(z_m) \left(\frac{(1-|z_m|^2)^q |u_1(z_m)|}{(1-|\varphi_1(z_m)|^2)^{\frac{n+1}{2}}} + \frac{(1-|z_m|^2)^q |u_2(z_m)|}{(1-|\varphi_2(z_m)|^2)^{\frac{n+1}{2}}} \right) \to 0,$$

which is impossible.

Cases 2: $\ell = 0$. We claim, for the test functions k_{w_m}, where $w_m = \varphi_1(z_m)$, that

$$\left| 1 - (1-|\varphi_1(z_m)|^2)^{\frac{n+1}{2}} k_{w_m}(\varphi_2(z_m)) \right| \to 0, \qquad \text{(as } m \to \infty\text{)}. \tag{8.11}$$

Indeed, by Lemma 8.5.2, we have

$$\left| 1 - (1-|\varphi_1(z_m)|^2)^{\frac{n+1}{2}} k_{w_m}(\varphi_2(z_m)) \right|$$

$$= (1-|\varphi_1(z_m)|^2)^{\frac{n+1}{2}} \cdot \left| k_{w_m}(\varphi_1(z_m)) - k_{w_m}(\varphi_2(z_m)) \right|$$

$$\leq (1-|\varphi_1(z_m)|^2)^{\frac{n+1}{2}} \cdot c \|k_{w_m}\|_p \cdot \rho_{\varphi_1,\varphi_2}(z_m) \cdot$$

$$\max \left\{ \frac{1}{(1-|\varphi_1(z_m)|^2)^{\frac{n+1}{2}}}, \frac{1}{(1-|\varphi_2(z_m)|^2)^{\frac{n+1}{2}}} \right\}$$

$$\leq \frac{(1-|\varphi_1(z_m)|^2)^{\frac{n+1}{2}} c\rho_{\varphi_1,\varphi_2}(z_m)}{(1-|\varphi_1(z_m)|^2)^{\frac{n+1}{2}}} = c\rho_{\varphi_1,\varphi_2}(z_m).$$

The last expression converges to $\ell = 0$ as $m \to \infty$, and (8.11) follows. Here, c is the constant defined in Lemma 8.5.2. Furthermore, since $\lim_{m\to\infty} \rho_{\varphi_1,\varphi_2}(z_m) = 0$, there exists $m_1 \geq 1$, such that $\rho_{\varphi_1,\varphi_2}(z_m) < \frac{1}{2}$, $m \geq m_1$. Then by Lemma 8.5.1,

$$\frac{1-|\varphi_2(z_m)|^2}{1-|\varphi_1(z_m)|^2} \leq 6.$$

Also, since W_{u_2,φ_2} is bounded from $\mathcal{N}_p(\mathbb{B})$ into $A^{-q}(\mathbb{B})$, by Theorem 8.2.1, there exists some positive number $K > 0$, such that

$$\sup_{z \in \mathbb{B}} \frac{(1-|z|^2)^q |u_2(z)|}{(1-|\varphi_2(z)|^2)^{\frac{n+1}{2}}} < K.$$

Again by $\lim_{m \to \infty} \rho_{\varphi_1,\varphi_2}(z_m) = 0$, there exists $m_2 \geq 1$, such that $\rho_{\varphi_1,\varphi_2}(z_m) < \dfrac{L}{8 \cdot 6^{\frac{n+1}{2}} cK}$, $m > m_2$. Consequently, for $m > \max\{m_1, m_2\}$,

$$|(W_{u_1,\varphi_1} - W_{u_2,\varphi_2})(k_{w_m})|_q$$

$$= \sup_{z \in \mathbb{B}} (1-|z|^2)^q \left| u_1(z)k_{w_m}(\varphi_1(z)) - u_2(z)k_{w_m}(\varphi_2(z)) \right|$$

$$\geq (1-|z_m|^2)^q \left| u_1(z_m)k_{w_m}(\varphi_1(z_m)) - u_2(z_m)k_{w_m}(\varphi_2(z_m)) \right|$$

$$= (1-|z_m|^2)^q \left| \frac{u_1(z_m)}{(1-|\varphi_1(z_m)|^2)^{\frac{n+1}{2}}} - u_2(z_m)k_{w_m}(\varphi_2(z_m)) \right|$$

$$\geq (1-|z_m|^2)^q \left| \frac{u_1(z_m)}{(1-|\varphi_1(z_m)|^2)^{\frac{n+1}{2}}} - \frac{u_2(z_m)}{(1-|\varphi_1(z_m)|^2)^{\frac{n+1}{2}}} \right|$$

$$-(1-|z_m|^2)^q \left| \frac{u_2(z_m)}{(1-|\varphi_1(z_m)|^2)^{\frac{n+1}{2}}} - u_2(z_m)k_{w_m}(\varphi_2(z_m)) \right|$$

$$= (1-|z_m|^2)^q \frac{|u_1(z_m) - u_2(z_m)|}{(1-|\varphi_1(z_m)|^2)^{\frac{n+1}{2}}}$$

$$-(1-|z_m|^2)^q \frac{|u_2(z_m)|}{(1-|\varphi_1(z_m)|^2)^{\frac{n+1}{2}}} \left| 1 - (1-|\varphi_1(z_m)|^2)^{\frac{n+1}{2}} k_{w_m}(\varphi_2(z_m)) \right|.$$

Moreover, we also have

$$(1-|z_m|^2)^q \frac{|u_1(z_m) - u_2(z_m)|}{(1-|\varphi_1(z_m)|^2)^{\frac{n+1}{2}}} \geq \frac{(1-|z_m|^2)^q |u_1(z_m) - u_2(z_m)|}{(1-|\varphi_2(z_m)|^2)^{\frac{n+1}{2}}} > \frac{L}{4}$$

and

$$(1-|z_m|^2)^q \frac{|u_2(z_m)|}{(1-|\varphi_1(z_m)|^2)^{\frac{n+1}{2}}}$$

$$= (1-|z_m|^2)^q \frac{|u_2(z_m)|}{(1-|\varphi_2(z_m)|^2)^{\frac{n+1}{2}}} \frac{(1-|\varphi_2(z_m)|^2)^{\frac{n+1}{2}}}{(1-|\varphi_1(z_m)|^2)^{\frac{n+1}{2}}} \leq K \cdot 6^{\frac{n+1}{2}}.$$

Therefore, we arrive at

$$|(W_{u_1,\varphi_1} - W_{u_2,\varphi_2})(k_{w_m})|_q \geq \frac{L}{4} - 6^{\frac{n+1}{2}}Kc\rho_{\varphi_1,\varphi_2}(z_m) \geq \frac{L}{4} - \frac{L}{8} = \frac{L}{8}. \tag{8.12}$$

However, since $W_{u_1,\varphi_1} - W_{u_2,\varphi_2}$ is compact, and k_{w_m} converges to zero uniformly on every compact subset of $\mathbb{B}$, we must have

$$\lim_{m \to \infty} |(W_{u_1,\varphi_1} - W_{u_2,\varphi_2})(k_{w_m})|_q = 0, \tag{8.13}$$

which contradicts (8.12).

Sufficiency Assume that the conditions (i) and (ii) hold. Take an arbitrary bounded sequence (f_m) in $\mathcal{N}_p(\mathbb{B})$ that converges to zero uniformly on compact subsets of $\mathbb{B}$. By (8.8), it is enough to show that

$$|(W_{u_1,\varphi_1} - W_{u_2,\varphi_2})(f_m)|_q \to 0, \qquad (m \to \infty).$$

To find a contradiction, assume that there are an $\varepsilon_0 > 0$ and a subsequence (f_{m_k}) of (f_m) such that

$$|(W_{u_1,\varphi_1} - W_{u_2,\varphi_2})(f_{m_k})|_q \geq \varepsilon_0, \qquad (k \in \mathbb{N}). \tag{8.14}$$

Indeed, without loss of generality, by relabeling the sequence, we may assume (f_m) itself has the above property. From this, it follows that there exists a sequence $(z_m) \subset \mathbb{B}$, such that

$$H_m = (1 - |z_m|^2)^q |u_1(z_m)f_m(\varphi_1(z_m)) - u_2(z_m)f_m(\varphi_2(z_m))| \geq \frac{\varepsilon_0}{2}, \tag{8.15}$$

for all $m \geq 1$. Here, z_m's are not necessarily distinct. We may also assume that both sequences $\{\varphi_1(z_m)\}$ and $\{\varphi_2(z_m)\}$ converge (as otherwise, we can consider some convergent subsequences). Note that since both W_{u_1,φ_1} and W_{u_2,φ_2} are bounded, by Theorem 8.2.1, there exists $K > 0$, such that

$$\sup_{z \in \mathbb{B}} |u_k(z)|(1 - |z|^2)^q \leq \sup_{z \in \mathbb{B}} \frac{|u_k(z)|(1 - |z|^2)^q}{(1 - |\varphi_k(z)|^2)^{\frac{n+1}{2}}} \leq K, \ k = 1, 2. \tag{8.16}$$

Now we consider the sequence $\max\{|\varphi_1(z_m)|, |\varphi_2(z_m)|\}$. It is clear that

$$\lim_{m \to \infty} \max\{|\varphi_1(z_m)|, |\varphi_2(z_m)|\} = q \leq 1.$$

As a matter of fact, $q = 1$. To verify this, assume that $q < 1$. Then, by (8.15),

$$\frac{\varepsilon_0}{2} \leq H_m \leq (1 - |z_m|^2)^q \left(|u_1(z_m)f_m(\varphi_1(z_m))| + |u_2(z_m)f_m(\varphi_2(z_m))| \right)$$

$$\leq K\Big(|f_m(\varphi_1(z_m))| + |f_m(\varphi_2(z_m))|\Big)$$

for all $m \geq 1$. Furthermore, there exists $m_3 \geq 1$ such that

$$\max\{|\varphi_1(z_m)|, |\varphi_2(z_m)|\} \leq \frac{1+q}{2}, \qquad (m > m_3).$$

In particular, for all $m > m_3$, $\varphi_k(z_m) \in \left\{z : |z| \leq \frac{1+q}{2}\right\}$, $k = 1, 2$. Since $\left\{z : |z| \leq \frac{1+q}{2}\right\}$ is a compact set of $\mathbb{B}$, the sequence $\{f_m(z)\}$ converges uniformly to zero on this set, and hence, both sequences $\{f_m(\varphi_k(z_m))\}$, $k = 1, 2$, converge to zero, as $m \to \infty$, which shows that

$$H_m \leq K\left(|f_m(\varphi_1(z_m))| + |f_m(\varphi_2(z_m))|\right) \to 0, \qquad (m \to \infty).$$

But, this contradicts the fact that $H_m \geq \frac{\varepsilon_0}{2}$, $m \geq 1$.

Thus, we have

$$\max\{|\varphi_1(z_m)|, |\varphi_2(z_m)|\} \to 1, \qquad (m \to \infty). \tag{8.17}$$

Then at least one of the limits $\lim_{m\to\infty} |\varphi_k(z_m)|$ $(k = 1, 2)$ must be 1. So we may assume that

$$\begin{cases} \lim_{m\to\infty} \varphi_1(z_m) = P, \text{ with } |P| = 1, \\ \lim_{m\to\infty} \varphi_2(z_m) = Q, \text{ with } |Q| \leq 1. \end{cases} \tag{8.18}$$

Furthermore, we may also assume that there exists the limit

$$\lim_{m\to\infty} \rho_{\varphi_1,\varphi_2}(z_m) = \ell \geq 0,$$

since otherwise, we consider a convergent subsequence. As a matter of fact, $\ell = 0$. To verify this, assume in contrary that $\ell > 0$. Consider two cases of $|Q| \leq 1$.

Case 1: $|Q| = 1$. In this case, from (i) and (8.18), it follows that

$$\lim_{m\to\infty} \frac{|u_k(z_m)|(1 - |z_m|^2)^q}{(1 - |\varphi_k(z_m)|^2)^{\frac{n+1}{2}}} = 0 \quad (k = 1, 2). \tag{8.19}$$

Then, by Theorem 7.5.1 and (8.15), we have

$$H_m \leq \frac{(1 - |z_m|^2)^q |u_1(z_m)|}{(1 - |\varphi_1(z_m)|^2)^{\frac{n+1}{2}}} (1 - |\varphi_1(z_m)|^2)^{\frac{n+1}{2}} |f_m(\varphi_1(z_m))|$$

$$+\frac{(1-|z_m|^2)^q|u_2(z_m)|}{(1-|\varphi_2(z_m)|^2)^{\frac{n+1}{2}}}(1-|\varphi_2(z_m)|^2)^{\frac{n+1}{2}}|f_m(\varphi_2(z_m))|$$

$$\leq\left[\frac{(1-|z_m|^2)^q|u_1(z_m)|}{(1-|\varphi_1(z_m)|^2)^{\frac{n+1}{2}}}+\frac{(1-|z_m|^2)^q|u_2(z_m)|}{(1-|\varphi_2(z_m)|^2)^{\frac{n+1}{2}}}\right]|f_m|_{\frac{n+1}{2}}$$

$$\leq\frac{2^{p+n}}{3^{p/2}}\left[\frac{(1-|z_m|^2)^q|u_1(z_m)|}{(1-|\varphi_1(z_m)|^2)^{\frac{n+1}{2}}}+\frac{(1-|z_m|^2)^q|u_2(z_m)|}{(1-|\varphi_2(z_m)|^2)^{\frac{n+1}{2}}}\right]\|f_m\|_p.$$

In the last inequality, letting $m\to\infty$, by (8.19) as well as boundedness of (f_m) in $\mathcal{N}_p(\mathbb{B})$, we get

$$\lim_{m\to\infty}H_m=0,$$

which is impossible, because it contradicts (8.15).

Case 2: $|Q|<1$. In this case, the second limit in (8.18) gives $\varphi_2(z_m)\in\left\{z:|z|\leq\frac{1+|Q|}{2}\right\}$, for all m large enough, say $m>m_4$. Then by (8.16) and Theorem 7.5.1, for all $m>m_4$, we have

$$H_m\leq(1-|z_m|^2)^q|u_1(z_m)||f_m(\varphi_1(z_m))|$$

$$+(1-|z_m|^2)^q|u_2(z_m)||f_m(\varphi_2(z_m))|$$

$$=\frac{(1-|z_m|^2)^q|u_1(z_m)|}{(1-|\varphi_1(z_m)|^2)^{\frac{n+1}{2}}}(1-|\varphi_1(z_m)|^2)^{\frac{n+1}{2}}|f_m(\varphi_1(z_m))|$$

$$+(1-|z_m|^2)^q|u_2(z_m)||f_m(\varphi_2(z_m))|$$

$$\leq\frac{(1-|z_m|^2)^q|u_1(z_m)|}{(1-|\varphi_1(z_m)|^2)^{\frac{n+1}{2}}}|f_m|_{\frac{n+1}{2}}+K|f_m(\varphi_2(z_m))|$$

$$\leq\frac{2^{p+n}}{3^{p/2}}\cdot\frac{(1-|z_m|^2)^q|u_1(z_m)|}{(1-|\varphi_1(z_m)|^2)^{\frac{n+1}{2}}}\|f_m\|_p+K|f_m(\varphi_2(z_m))|.$$

Letting $m\to\infty$, from (8.19) and the fact that $f_m(\varphi_2(z_m))$ converges to zero uniformly on the compact set $\left\{z:|z|\leq\frac{1+|Q|}{2}\right\}$, it follows that right-hand side of the last inequality tends to 0 as $m\to\infty$, which again contradicts (8.15). Thus, we have

$$\lim_{m\to\infty}\rho_{\varphi_1,\varphi_2}(z_m)=0.\tag{8.20}$$

We also have $|Q|=1$, that is, $\lim_{m\to\infty}|\varphi_2(z_m)|=1$. Indeed, for each $m\geq1$, we have

$$1-\rho^2_{\varphi_1,\varphi_2}(z_m)=1-|\Phi_{\varphi_2(z_m)}(\varphi_1(z_m)|^2$$

$$= 1 - \frac{(1 - |\varphi_1(z_m)|^2)(1 - |\varphi_2(z_m)|^2)}{|1 - \langle \varphi_1(z_m), \varphi_2(z_m) \rangle|^2}.$$

Since $\lim\limits_{m \to \infty} |\varphi_1(z_m)| = 1$, if $\lim\limits_{m \to \infty} |\varphi_2(z_m)| = |Q| < 1$, then we would have for all m large enough

$$|1 - \langle \varphi_1(z_m), \varphi_2(z_m) \rangle| \geq 1 - |\langle \varphi_1(z_m), \varphi_2(z_m) \rangle|$$

$$\geq 1 - |\varphi_1(z_m)||\varphi_2(z_m)| \geq 1 - |Q| > 0.$$

But, this implies that $\lim\limits_{m \to \infty} \rho_{\varphi_1, \varphi_2}(z_m) = 1$, which contradicts (8.20).

Now, by the same reasoning as in the necessity part, we may assume that

$$|\varphi_1(z_m)| \geq |\varphi_2(z_m)|, \qquad (m \geq 1). \tag{8.21}$$

Then, from (8.15) and (8.16) and Lemma 8.5.2, it follows that for each $m \geq 1$,

$$H_m = (1 - |z_m|^2)^q \, |u_1(z_m) f_m(\varphi_1(z_m)) - u_2(z_m) f_m(\varphi_2(z_m))|$$

$$\leq (1 - |z_m|^2)^q \, |u_1(z_m) f_m(\varphi_1(z_m)) - u_1(z_m) f_m(\varphi_2(z_m))|$$

$$+ (1 - |z_m|^2)^q \, |u_1(z_m) f_m(\varphi_2(z_m)) - u_2(z_m) f_m(\varphi_2(z_m))|$$

$$= \frac{(1 - |z_m|^2)^q |u_1(z_m)|}{(1 - |\varphi_1(z_m)|^2)^{\frac{n+1}{2}}} \cdot (1 - |\varphi_1(z_m)|^2)^{\frac{n+1}{2}} |f_m(\varphi_1(z_m)) - f_m(\varphi_2(z_m))|$$

$$+ \frac{(1 - |z_m|^2)^q |f_m(\varphi_2(z_m))|}{(1 - |\varphi_2(z_m)|^2)^{\frac{n+1}{2}}} \cdot (1 - |\varphi_2(z_m)|^2)^{\frac{n+1}{2}} |u_1(z_m) - u_2(z_m)|$$

$$\leq cK \cdot \|f_m\|_p \max \left\{ \frac{(1 - |\varphi_1(z_m)|^2)^{\frac{n+1}{2}}}{(1 - |\varphi_1(z_m)|^2)^{\frac{n+1}{2}}}, \frac{(1 - |\varphi_1(z_m)|^2)^{\frac{n+1}{2}}}{(1 - |\varphi_2(z_m)|^2)^{\frac{n+1}{2}}} \right\} \rho_{\varphi_1, \varphi_2}(z_m)$$

$$+ \frac{2^{p+n}}{3^{p/2}} \|f_m\|_p \cdot \frac{(1 - |z_m|^2)^q |u_1(z_m) - u_2(z_m)|}{(1 - |\varphi_2(z_m)|^2)^{\frac{n+1}{2}}}.$$

However, by (8.21),

$$\max \left\{ \frac{(1 - |\varphi_1(z_m)|^2)^{\frac{n+1}{2}}}{(1 - |\varphi_1(z_m)|^2)^{\frac{n+1}{2}}}, \frac{(1 - |\varphi_1(z_m)|^2)^{\frac{n+1}{2}}}{(1 - |\varphi_2(z_m)|^2)^{\frac{n+1}{2}}} \right\}$$

$$= \max \left\{ 1, \frac{(1 - |\varphi_1(z_m)|^2)^{\frac{n+1}{2}}}{(1 - |\varphi_2(z_m)|^2)^{\frac{n+1}{2}}} \right\} = 1,$$

and hence,

$$H_m \le cK \|f_m\|_p \cdot \rho_{\varphi_1,\varphi_2}(z_m)$$

$$+ \frac{2^{p+n}}{3^{p/2}} \|f_m\|_p \cdot \frac{(1-|z_m|^2)^q |u_1(z_m) - u_2(z_m)|}{(1-|\varphi_2(z_m)|^2)^{\frac{n+1}{2}}}$$

$$= cK \|f_m\|_p \cdot \rho_{\varphi_1,\varphi_2}(z_m)$$

$$+ \|f_m\|_p \cdot \frac{2^{p+n}}{3^{p/2}} |u_1(z_m) - u_2(z_m)| \cdot$$

$$\min \left\{ \frac{(1-|z_m|^2)^q}{(1-|\varphi_2(z_m)|^2)^{\frac{n+1}{2}}}, \frac{(1-|z_m|^2)^q}{(1-|\varphi_1(z_m)|^2)^{\frac{n+1}{2}}} \right\}.$$

In the inequality above, since $(\|f_m\|_p)$ is bounded, by (8.20),

$$\lim_{m\to\infty} \|f_m\|_p \cdot \rho_{\varphi_1,\varphi_2}(z_m) = 0.$$

Furthermore, by (ii) and $\lim_{m\to\infty} |\varphi_1(z_m)| = \lim_{m\to\infty} |\varphi_2(z_m)| = 1$, we get

$$\lim_{m\to\infty} \|f_m\|_p |u_1(z_m) - u_2(z_m)| \cdot \min \left\{ \frac{(1-|z_m|^2)^q}{(1-|\varphi_2(z_m)|^2)^{\frac{n+1}{2}}}, \frac{(1-|z_m|^2)^q}{(1-|\varphi_1(z_m)|^2)^{\frac{n+1}{2}}} \right\} = 0.$$

These equalities imply that $\lim_{m\to\infty} H_m = 0$, but this contradicts (8.15).

$\square$

8.7   Essential Norm: Upper Bound

In this section and the next, we study the essential norm of the weighted composition operator $W_{u,\varphi} \colon \mathcal{N}_p \to A^{-q}$. Let us denote by $\mathcal{K} = \mathcal{K}(\mathcal{N}_p, A^{-q})$ the set of all compact operators acting from $\mathcal{N}_p$ into A^{-q}. Recall that the essential norm of $W_{u,\varphi}$ is defined by

$$\|W_{u,\varphi}\|_e = \inf_{K\in\mathcal{K}} \|W_{u,\varphi} - K\|.$$

Trivially, the essential norm of a compact operator is zero.

Theorem 8.7.1 *Let p and q be two positive numbers. Let $\varphi \colon \mathbb{B} \to \mathbb{B}$ be a holomorphic self-map of $\mathbb{B}$, and let $u \colon \mathbb{B} \to \mathbb{C}$ be a holomorphic function. Suppose that $W_{u,\varphi}$ is a bounded operator acting from $\mathcal{N}_p$ to A^{-q}. Then*

$$\|W_{u,\varphi}\|_e \le A \lim_{r \to 1^-} \sup_{|\varphi(z)| > r} \frac{|u(z)|(1 - |z|^2)^q}{(1 - |\varphi(z)|^2)^{\frac{n+1}{2}}},$$

where A is the constant from Lemma 8.5.2.

Proof Since $W_{u,\varphi}$ is bounded, we see that u belongs to A^{-q} and Theorem 8.2.1 shows that

$$\lim_{r \to 1^-} \sup_{|\varphi(z)| > r} \frac{|u(z)|(1 - |z|^2)^q}{(1 - |\varphi(z)|^2)^{\frac{n+1}{2}}}$$

exists and is a finite real number. First, we prove that, for any $r \in [0, 1)$,

$$\|W_{u,\varphi}\|_e \le A \sup_{|\varphi(z)| > r} \frac{|u(z)|(1 - |z|^2)^q}{(1 - |\varphi(z)|^2)^{\frac{n+1}{2}}}. \tag{8.22}$$

For each $k \ge 1$, set $\varphi_k(z) = kz/(k+1)$, $z \in \mathbb{B}$. By Lemma 8.3.2, C_{φ_k} is compact on $\mathcal{N}_p$, and hence, $W_{u,\varphi} \circ C_{\varphi_k}$ is compact acting from $\mathcal{N}_p$ into A^{-q}. We then have, for each $k \ge 1$,

$$\|W_{u,\varphi}\|_e \le \|W_{u,\varphi} - W_{u,\varphi} \circ C_{\varphi_k}\| = \sup_{\|f\|_p \le 1} |(W_{u,\varphi} - W_{u,\varphi} \circ C_{\varphi_k})(f)|_q,$$

which implies

$$\|W_{u,\varphi}\|_e \le \inf_{k \in \mathbb{N}} \left\{ \sup_{\|f\|_p \le 1} |(W_{u,\varphi} - W_{u,\varphi} \circ C_{\varphi_k})(f)|_q \right\}. \tag{8.23}$$

For $f \in \mathcal{N}_p$, we estimate

$$|(W_{u,\varphi} - W_{u,\varphi} \circ C_{\varphi_k})(f)|_q$$

$$= \sup_{z \in \mathbb{B}} \left\{ |u(z)| \left| f(\varphi(z)) - f\left(\frac{k}{k+1}\varphi(z)\right) \right| (1 - |z|^2)^q \right\}$$

$$\le \sup_{|\varphi(z)| > r} \left\{ |u(z)| \left| f(\varphi(z)) - f\left(\frac{k}{k+1}\varphi(z)\right) \right| (1 - |z|^2)^q \right\}$$

$$+ \sup_{|\varphi(z)| \le r} \left\{ |u(z)| \left| f(\varphi(z)) - f\left(\frac{k}{k+1}\varphi(z)\right) \right| (1 - |z|^2)^q \right\}.$$

On the one hand, by (8.6),

$$\sup_{|\varphi(z)|>r}\left\{|u(z)|\left|f(\varphi(z))-f\Big(\frac{k}{k+1}\varphi(z)\Big)\right|(1-|z|^2)^q\right\}$$

$$\leq\left(\sup_{|\varphi(z)|>r}\frac{A|u(z)|(1-|z|^2)^q}{(1-|\varphi(z)|^2)^{\frac{n+1}{2}}}\right)\|f\|_p.$$

On the other hand, by (8.7),

$$\sup_{|\varphi(z)|\leq r}\left\{|u(z)|\left|f(\varphi(z))-f\Big(\frac{k}{k+1}\varphi(z)\Big)\right|(1-|z|^2)^q\right\}$$

$$\leq\sup_{|\varphi(z)|\leq r}\left|f(\varphi(z))-f\Big(\frac{k}{k+1}\varphi(z)\Big)\right|\cdot\sup_{z\in\mathbb{B}}\left\{|u(z)|(1-|z|^2)^q\right\}$$

$$\leq\left(\frac{Ar|u|_q}{(k+1)(1-r^2)^{\frac{n+3}{2}}}\right)\|f\|_p.$$

Therefore, if $\|f\|_p\leq 1$, then

$$\left|\left(W_{u,\varphi}-W_{u,\varphi}\circ C_{\varphi_k}\right)(f)\right|_q$$

$$\leq A\left(\sup_{|\varphi(z)|>r}\frac{|u(z)|(1-|z|^2)^q}{(1-|\varphi(z)|^2)^{\frac{n+1}{2}}}\right)+\frac{Ar|u|_q}{(k+1)(1-r^2)^{\frac{n+3}{2}}}.$$

It then follows that

$$\inf_{k\in\mathbb{N}}\left\{\sup_{\|f\|_p\leq 1}\left|\left(W_{u,\varphi}-W_{u,\varphi}\circ C_{\varphi_k}\right)(f)\right|_q\right\}$$

$$\leq\inf_{k\in\mathbb{N}}\left\{A\left(\sup_{|\varphi(z)|>r}\frac{|u(z)|(1-|z|^2)^q}{(1-|\varphi(z)|^2)^{\frac{n+1}{2}}}\right)+\frac{Ar|u|_q}{(k+1)(1-r^2)^{\frac{n+3}{2}}}\right\}$$

$$=A\sup_{|\varphi(z)|>r}\frac{|u(z)|(1-|z|^2)^q}{(1-|\varphi(z)|^2)^{\frac{n+1}{2}}}.$$

Combining this and (8.23), we obtain (8.22). Now, letting $r\to 1^-$ in (8.22), we arrive at the desired inequality

$$\|W_{u,\varphi}\|_e\leq A\lim_{r\to 1^-}\sup_{|\varphi(z)|>r}\frac{|u(z)|(1-|z|^2)^q}{(1-|\varphi(z)|^2)^{\frac{n+1}{2}}}.\qquad\qquad\square$$

8.8 Essential Norm: Lower Bound

We now discuss the estimation for the lower bound of $\|W_{u,\varphi}\|_e$. We will make use of weakly convergent sequences in the Bergman space A^2. The following lemma plays an important role.

Lemma 8.8.1 *Suppose* $(f_m)_{m\geq 1} \subset A^2$ *is a sequence that converges weakly to zero in* A^2. *Then* $(f_m)_{m\geq 1}$ *converges weakly to zero in* $\mathcal{N}_p$ *as well.*

Proof Let Γ be a bounded linear functional on $\mathcal{N}_p$. Then

$$\|\Gamma\|_{(A^2)'} = \sup_{f\in A^2} \frac{|\Gamma(f)|}{\|f\|_{A^2}} \leq \sup_{f\in A^2} \frac{|\Gamma(f)|}{\|f\|_p} \leq \sup_{f\in\mathcal{N}_p} \frac{|\Gamma(f)|}{\|f\|_p} = \|\Gamma\|_{(\mathcal{N}_p)'},$$

which implies Γ is also a bounded linear functional on A^2. The second and third inequalities follow from the fact that $\|f\|_p \leq \|f\|_{A^2}$ for any $f \in A^2$. This means that, as sets, $(\mathcal{N}_p)'$ is a subset of $(A^2)'$. Thus, if $f_m \to 0$ weakly in A^2, then we necessarily have $f_m \to 0$ weakly in $\mathcal{N}_p$ as well. $\square$

Corollary 8.8.2 *Let* $\{w_m\}_{m\in\mathbb{N}} \subset \mathbb{B}$, *and assume that* $|w_m| \to 1$ *as* $m \to \infty$. *Then* $k_{w_m} \to 0$ *weakly in* $\mathcal{N}_p$.

Proof It is well-known that $k_{w_m} \to 0$ weakly in A^2 as $m \to \infty$. Indeed, for any $f \in A^2$, using the reproducing property, we have

$$\langle f, k_{w_m} \rangle = (1 - |w_m|^2)^{(n+1)/2} f(w_m),$$

which converges to zero as $m \to \infty$. The result now follows immediately from Lemma 8.8.1. $\square$

Theorem 8.8.3 *Let* p *and* q *be two positive numbers, let* $\varphi: \mathbb{B} \to \mathbb{B}$ *be a holomorphic self-map of* $\mathbb{B}$, *and let* $u : \mathbb{B} \to \mathbb{C}$ *be a holomorphic function. Suppose that* $W_{u,\varphi}$ *is a bounded operator acting from* $\mathcal{N}_p$ *to* A^{-q}. *Then*

$$\|W_{u,\varphi}\|_e \geq \lim_{r\to 1^-} \sup_{|\varphi(z)|>r} \frac{|u(z)|(1 - |z|^2)^q}{(1 - |\varphi(z)|^2)^{\frac{n+1}{2}}}.$$

Proof The case $\|\varphi\|_\infty < 1$ is obvious since the right-hand side is zero. Now assume that $\|\varphi\|_\infty = 1$. For any $r \in (0, 1)$, the set $S_r = \{z \in \mathbb{B} : |\varphi(z)| > r\}$ is not empty. For each $z \in \mathbb{B}$, consider the test function $k_{\varphi(z)}$ in Lemma 8.1.1. Then, for any compact operator $Q \in \mathcal{K}$, we have

$$\|W_{u,\varphi} - Q\| = \sup_{\|f\|_p \le 1} |(W_{u,\varphi} - Q)(f)|_q$$

$$\ge |(W_{u,\varphi} - Q)(k_{\varphi(z)})|_q$$

$$\ge |W_{u,\varphi}(k_{\varphi(z)})|_q - |Q(k_{\varphi(z)})|_q$$

$$\ge \frac{|u(z)|(1 - |z|^2)^q}{(1 - |\varphi(z)|^2)^{\frac{n+1}{2}}} - |Q(k_{\varphi(z)})|_q,$$

which is equivalent to

$$\|W_{u,\varphi} - Q\| + |Q(k_{\varphi(z)})|_q \ge \frac{|u(z)|(1 - |z|^2)^q}{(1 - |\varphi(z)|^2)^{\frac{n+1}{2}}}. \tag{8.24}$$

Taking the supremum on z over the set S_r on both sides of (8.24) yields

$$\|W_{u,\varphi} - Q\| + \sup_{z \in S_r} |Q(k_{\varphi(z)})|_q \ge \sup_{z \in S_r} \frac{|u(z)|(1 - |z|^2)^q}{(1 - |\varphi(z)|^2)^{\frac{n+1}{2}}},$$

which is

$$\|W_{u,\varphi} - Q\| + \sup_{|\varphi(z)| > r} |Q(k_{\varphi(z)})|_q \ge \sup_{|\varphi(z)| > r} \frac{|u(z)|(1 - |z|^2)^q}{(1 - |\varphi(z)|^2)^{\frac{n+1}{2}}}. \tag{8.25}$$

Denote $H(r) = \sup_{|\varphi(z)| > r} |Q(k_{\varphi(z)})|_q$. Since $H(r)$ decreases as r increases, $\lim_{r \to 1^-} H(r)$ exists. We claim that this limit is necessarily zero. For the purpose of obtaining a contradiction, assume that $\lim_{r \to 1^-} H(r) = L > 0$. Then there is a sequence $\{z_m\} \subset \mathbb{B}$ satisfying $|\varphi(z_m)| \to 1$ as $m \to \infty$, and for each $m \ge 1$,

$$|Q(k_{\varphi(z_m)})|_q > \frac{L}{2}. \tag{8.26}$$

By Corollary 8.8.2, $\{k_{\varphi(z_m)}\}$ converges weakly to zero in $\mathcal{N}_p$. Since Q is compact, we have $\{|Q(k_{\varphi(z_m)})|_q\}$ converges to zero as $m \to \infty$, which contradicts (8.26). Therefore, $\lim_{r \to 1^-} \sup_{|\varphi(z)| > r} |Q(k_{\varphi(z)})|_q = 0$.

Letting $r \to 1^-$ on both sides of (8.25), we conclude that for any compact operator $Q \in \mathcal{K}$,

$$\|W_{u,\varphi} - Q\| \ge \lim_{r \to 1^-} \sup_{|\varphi(z)| > r} \frac{|u(z)|(1 - |z|^2)^q}{(1 - |\varphi(z)|^2)^{\frac{n+1}{2}}}.$$

From this, it follows that

$$\|W_{u,\varphi}\|_e = \inf_{Q\in\mathcal{K}}\{\|W_{u,\varphi} - Q\|\} \geq \lim_{r\to 1^-}\ \sup_{|\varphi(z)|>r}\frac{|u(z)|(1 - |z|^2)^q}{(1 - |\varphi(z)|^2)^{\frac{n+1}{2}}}. \qquad \square$$

In conclusion, combining Theorems 8.7.1 and 8.8.3, we obtain a full description of the essential norm of $W_{u,\varphi}$.

Theorem 8.8.4 *Let p and q be two positive numbers, let $\varphi: \mathbb{B} \to \mathbb{B}$ be a holomorphic self-map of $\mathbb{B}$, and let $u : \mathbb{B} \to \mathbb{C}$ be a holomorphic function. Suppose that $W_{u,\varphi}$ is a bounded operator acting from $\mathcal{N}_p$ to A^{-q}. Then*

$$\|W_{u,\varphi}\|_e \asymp \lim_{r\to 1^-}\ \sup_{|\varphi(z)|>r}\frac{|u(z)|(1 - |z|^2)^q}{(1 - |\varphi(z)|^2)^{\frac{n+1}{2}}}.$$

Theorem 8.8.4 provides us a characterization of compact weighted composition operators from $\mathcal{N}_p$ to A^{-q}, as does Theorem 8.2.1 for the boundedness.

Corollary 8.8.5 *Suppose that $W_{u,\varphi}$ is a bounded operator acting from $\mathcal{N}_p$ to A^{-q} as in Theorem 8.8.4. Then $W_{u,\varphi}$ is compact if and only if*

$$\lim_{r\to 1^-}\ \sup_{|\varphi(z)|>r}\frac{|u(z)|(1 - |z|^2)^q}{(1 - |\varphi(z)|^2)^{\frac{n+1}{2}}} = 0.$$

Notes on Chapter 8

The main results in this chapter are from papers [43, 45]. When $n = 1$, Theorem 8.2.1 reduces to Theorem 3 of [95] as a special case. Moreover, in this special case, Theorem 8.4.1 contains [95, Corollary 2], and Corollary 8.4.2 contains Theorems 4.1 and 4.3 of [73]. The fact that formula (7.4) is a true metric is proved in [22]. Formula (8.4) is provided in [22, Theorem 1(c)]. Lemma 8.5.3 is taken from [43].

9.1 Multipliers and $\mathcal{M}$ Invariance

Recall that a function $u : \mathbb{B} \to \mathbb{C}$ is a *multiplier* of $\mathcal{N}_p$ if uf belongs to $\mathcal{N}_p$ for all $f \in \mathcal{N}_p$. An application of the closed graph theorem shows that for any $u \in \text{Mult}(\mathcal{N}_p)$, the multiplication operator M_u is bounded on $\mathcal{N}_p$. We first describe the space $\text{Mult}(\mathcal{N}_p)$ of multipliers of $\mathcal{N}_p$. This result shows that the situation is similar to Hardy spaces H^p.

Theorem 9.1.1 *For any $p > 0$, we have $\text{Mult}(\mathcal{N}_p) = H^\infty$. Moreover, for any $u \in H^\infty$,*

$$\|M_u\|_{\mathcal{N}_p \to \mathcal{N}_p} = \|u\|_\infty.$$

Proof For $u \in H^\infty$ and $f \in \mathcal{N}_p$, the function uf belongs to $\text{Hol}(\mathbb{B})$, and, from the definition of the norm in $\mathcal{N}_p$, it follows that

$$\|uf\|_p \leq \|u\|_\infty \|f\|_p.$$

This shows that $H^\infty \subset \text{Mult}(\mathcal{N}_p)$ and $\|M_u\| \leq \|u\|_\infty$. Now suppose that u is an element in $\text{Mult}(\mathcal{N}_p)$. For any integer $m \geq 1$, we have

$$\|u^m\|_p = \|M_u^m 1\|_p \leq \|M_u\|^m \|1\|_p.$$

Combining with Theorem 7.5.1, we obtain a positive constant $C > 0$ independent of u, m, and z such that, for all $z \in \mathbb{B}$,

$$|u^m(z)| \leq C(1 - |z|^2)^{-(n+1)/2} \|u^m\|_p \leq C(1 - |z|^2)^{-(n+1)/2} \|1\|_p \|M_u\|^m.$$

Consequently,

$$|u(z)| \leq \left(C(1 - |z|^2)^{-(n+1)/2} \|1\|_p \right)^{1/m} \|M_u\|.$$

Letting $m \to \infty$, we conclude that $|u(z)| \leq \|M_u\|$ for all $z \in \mathbb{B}$. Therefore, u belongs to H^∞ and $\|u\|_\infty \leq \|M_u\|$. This completes the proof of the theorem.   □

A space $\mathcal{X}$ of functions defined on $\mathbb{B}$ is said to be *Möbius-invariant*, or simply $\mathcal{M}$-*invariant*, if $f \circ \Phi \in \mathcal{X}$ for every $f \in \mathcal{X}$ and every $\Phi \in \mathrm{Aut}(\mathbb{B})$. Recall that k_w is the test function introduced in Lemma 8.1.1.

Theorem 9.1.2 *The space $\mathcal{N}_p$ is $\mathcal{M}$-invariant. Moreover, for any $\Phi \in \mathrm{Aut}(\mathbb{B})$, we have*

$$\|C_\Phi\| = \|M_{1/k_a}\| = \left(\frac{1 + |a|}{1 - |a|} \right)^{\frac{n+1}{2}},$$

where $a = \Phi^{-1}(0)$.

Proof Note that for any automorphism Φ on $\mathbb{B}$, we have

$$C_\Phi = M_{1/k_a} \circ W_\Phi.$$

Recall that W_Φ is defined by (7.6). According to Theorem 7.3.1, the weighted composition operator W_Φ is a surjective isometry on $\mathcal{N}_p$. Since $1/k_a$ is a bounded function, it is a multiplier of $\mathcal{N}_p$, and since W_Φ is a surjective isometry, it follows that $\|C_\Phi\| = \|M_{1/k_a}\|$. But, by Theorem 9.1.1, we know that

$$\|M_{1/k_a}\| = \|1/k_a\|_\infty = \left(\frac{1 + |a|}{1 - |a|} \right)^{\frac{n+1}{2}}.$$

                                                                                             □

9.2   An Upper Estimate for $\| \cdot \|_p$

It is immediate from the definition of the norm in $\mathcal{N}_p$ that

$$\|f\|_p \geq \|f\|_{A_p^2}, \qquad (p > 0).$$

We now provide an upper estimate for $\|f\|_p$, whenever $p \leq n$.

Theorem 9.2.1 *Let $0 < p \le n$. Then there exists a positive constant $C = C(n, p)$ such that, for any $f \in \mathcal{N}_p$,*

$$\| f \|_p \le C \left(\int_{\mathbb{B}} \left(\sup_{|w|=|z|} |f(w)|^2 \right) (1 - |z|^2)^p \, dV(z) \right)^{1/2}.$$

Proof For $a \in \mathbb{B}$, integration in polar coordinates gives

$$\int_{\mathbb{B}} |f(z)|^2 (1 - |\Phi_a(z)|^2)^p \, dV(z)$$

$$= \int_{\mathbb{B}} |f(z)|^2 \frac{(1 - |z|^2)^p (1 - |a|^2)^p}{|1 - \langle z, a \rangle|^{2p}} \, dV(z)$$

$$= 2n \int_0^1 r^{2n-1} (1 - |r|^2)^p \left(\int_{\mathbb{S}} |f(r\varsigma)|^2 \frac{(1 - |a|^2)^p}{|1 - \langle \varsigma, ra \rangle|^{2p}} \, d\sigma(\varsigma) \right) dr$$

$$\le 2n \int_0^1 r^{2n-1} (1 - |r|^2)^p \left(\sup_{\varsigma \in \mathbb{S}} |f(r\varsigma)|^2 \right) \left(\int_{\mathbb{S}} \frac{(1 - |a|^2)^p}{|1 - \langle \varsigma, ra \rangle|^{2p}} \, d\sigma(\varsigma) \right) dr.$$

Now, recall that with $a \in \mathbb{B}$ and $0 < r < 1$, we have

$$\int_{\mathbb{S}} \frac{d\sigma(\varsigma)}{|1 - \langle r\varsigma, a \rangle|^{2p}} = \int_{\mathbb{S}} \frac{d\sigma(\varsigma)}{|1 - \langle \varsigma, ra \rangle|^{2p}}$$

$$= \int_{\mathbb{S}} \frac{d\sigma(\varsigma)}{|1 - \langle \varsigma, ra \rangle|^{n+(2p-n)}}$$

$$\asymp \begin{cases} \text{bounded in } \mathbb{B} & \text{for } 0 < p < \frac{n}{2}, \\ \log \frac{1}{1-r^2|a|^2} \le \log \frac{1}{1-|a|^2} & \text{for } p = \frac{n}{2}, \\ (1 - r^2|a|^2)^{n-2p} \le (1 - |a|^2)^{n-2p} & \text{for } \frac{n}{2} < p \le n. \end{cases}$$

Thus, for $0 < p \le n$, there exists a positive constant C independent of a and r, such that

$$\int_{\mathbb{S}} \frac{(1 - |a|^2)^p}{|1 - \langle r\varsigma, a \rangle|^{2p}} \, d\sigma(\varsigma) \le C.$$

It then follows that

$$\int_{\mathbb{B}} |f(z)|^2 (1 - |\Phi_a(z)|^2)^p \, dV(z) \le C \left(2n \int_0^1 r^{2n-1} (1 - |r|^2)^p \sup_{|w|=r} |f(w)|^2 \, dr \right)$$

$$= C \int_{\mathbb{B}} \left(\sup_{|w|=|z|} |f(w)|^2 \right) (1 - |z|^2)^p \, dV(z).$$

Taking supremum over $a \in \mathbb{B}$ gives the required inequality.   □

9.3   The Space $\mathcal{N}_p^0$

The little $\mathcal{N}_p^0$ space is defined as

$$\mathcal{N}_p^0 = \mathcal{N}_p^0(\mathbb{B}) = \left\{ f \in \mathcal{N}_p : \lim_{|a| \to 1^-} \int_{\mathbb{B}} |f(z)|^2 (1 - |\Phi_a(z)|^2)^p dV(z) = 0 \right\}.$$

In this section, we show that polynomials are dense in $\mathcal{N}_p^0$ and that this latter space is closed in $\mathcal{N}_p$.

Theorem 9.3.1 $\mathcal{N}_p^0$ is a closed subspace of $\mathcal{N}_p$, and hence, it is a Banach space by itself.

Proof It is elementary that $\mathcal{N}_p^0$ is a subspace of $\mathcal{N}_p$, and hence, it remains to show that $\mathcal{N}_p^0$ is closed. Consider a sequence $(f_n) \subset \mathcal{N}_p^0$ that converges to some $f \in \mathcal{N}_p$. We need to show that $f \in \mathcal{N}_p^0$.

By the convergence assumption, for any $\varepsilon > 0$, there exists an $N \geq 1$ such that

$$\|f - f_n\|_p < \sqrt{\frac{\varepsilon}{4}}, \qquad (n \geq N).$$

Let $n_0 > N$ be fixed. Since $f_{n_0} \in \mathcal{N}_p^0$, there exists a $\delta \in (0, 1)$ such that

$$\sup_{\delta < |a| < 1} \int_{\mathbb{B}} |f_{n_0}(z)|^2 (1 - |\Phi_a(z)|^2)^p dV(z) < \frac{\varepsilon}{4}.$$

As a consequence,

$$\sup_{\delta < |a| < 1} \int_{\mathbb{B}} |f(z)|^2 (1 - |\Phi_a(z)|^2)^p dV(z)$$

$$\leq \sup_{\delta < |a| < 1} \int_{\mathbb{B}} 2 \left(|f(z) - f_{n_0}(z)|^2 + |f_{n_0}(z)|^2 \right) (1 - |\Phi_a(z)|^2)^p dV(z)$$

$$\leq 2\|f - f_{n_0}\|_p^2 + 2 \sup_{\delta < |a| < 1} \int_{\mathbb{B}} |f_{n_0}|^2 (1 - |\Phi_a(z)|^2)^p dV(z) < \varepsilon,$$

which implies

$$\lim_{|a| \to 1^-} \int_{\mathbb{B}} |f(z)|^2 (1 - |\Phi_a(z)|^2)^p dV(z) = 0.$$

From this, we conclude that $\mathcal{N}_p^0$ is closed. $\square$

Lemma 9.3.2 *For any $p > 0$, we have $A^2 \subset \mathcal{N}_p^0$.*

Proof Let f be an element in A^2. For $a \in \mathbb{B}$, define

$$g_a(z) = |f(z)|^2 (1 - |\Phi_a(z)|^2)^p = |f(z)|^2 \frac{(1 - |a|^2)^p (1 - |z|^2)^p}{|1 - \langle z, a \rangle|^{2p}}, \qquad (z \in \mathbb{B}).$$

We have $0 \leq g_a(z) \leq |f(z)|^2$ and $\lim\limits_{|a| \to 1^-} |g_a(z)| = 0$ for all $z \in \mathbb{B}$. The dominated convergence theorem then implies

$$\lim_{|a| \to 1^-} \int_\mathbb{B} |f(z)|^2 (1 - |\Phi_a(z)|^2)^p \, dV(z) = \lim_{|a| \to 1^-} \int_\mathbb{B} g_a(z) \, dV(z) = 0.$$

This shows that f belongs to $\mathcal{N}_p^0$.   □

In Sect. 10.3, we will see that the range of any compact composition operator C_φ : $\mathcal{N}_p \longrightarrow \mathcal{N}_q$ must be contained in the little space $\mathcal{N}_q^0$. To deduce that result, we need the following important property of elements in $\mathcal{N}_q^0$.

Lemma 9.3.3 *Let h be an element in the space $\mathcal{N}_q^0$. Suppose $\{A_k\}_{k \geq 1}$ is a decreasing sequence of measurable subsets of $\mathbb{B}$ whose intersection is empty. Then*

$$\lim_{k \to \infty} \left[\sup_{a \in \mathbb{B}} \int_{A_k} |h(z)|^2 (1 - |\Phi_a(z)|^2)^q \, dV(z) \right] = 0. \tag{9.1}$$

Proof Let $\varepsilon > 0$ be given. Since $h \in \mathcal{N}_q^0$, there exists a positive number $0 < \delta < 1$ such that

$$\sup_{\delta < |a| < 1} \int_\mathbb{B} |h(z)|^2 (1 - |\Phi_a(z)|^2)^q \, dV(z) < \varepsilon. \tag{9.2}$$

On the other hand, if $a \in \mathbb{B}$ with $|a| \leq \delta$, then for $z \in \mathbb{B}$,

$$1 - |\Phi_a(z)|^2 = \frac{(1 - |a|^2)(1 - |z|^2)}{|1 - \langle z, a \rangle|^2} \leq \frac{1 + |a|}{1 - |a|}(1 - |z|^2) \leq \frac{1 + \delta}{1 - \delta}(1 - |z|^2).$$

Consequently, for any integer $k \geq 1$, we have

$$\sup_{|a| \leq \delta} \int_{A_k} |h(z)|^2 (1 - |\Phi_a(z)|^2)^q \, dV(z) \leq \left(\frac{1 + \delta}{1 - \delta} \right)^q \int_{A_k} |h(z)|^2 (1 - |z|^2)^q \, dV(z).$$

Since $h \in \mathcal{N}_q^0 \subset A_q^2$, the function $|h(z)|^2(1 - |z|^2)^q$ belongs to $L^1(\mathbb{B}, dV)$. Because $\{A_{k \geq 1}\}$ is a decreasing sequence of subsets whose intersection is empty, there exists a positive integer k_ε such that for all $k \geq k_\varepsilon$,

$$\left(\frac{1+\delta}{1-\delta}\right)^q \int_{A_k} |h(z)|^2(1 - |z|^2)^q \, dV(z) < \varepsilon.$$

This implies that for such k,

$$\sup_{|a| \leq \delta} \int_{A_k} |h(z)|^2(1 - |\Phi_a(z)|^2)^q \, dV(z) < \varepsilon. \tag{9.3}$$

Combining (9.2) and (9.3) yields

$$\sup_{a \in \mathbb{B}} \int_{A_k} |h(z)|^2(1 - |\Phi_a(z)|^2)^q \, dV(z) < \varepsilon$$

for all $k \geq k_\varepsilon$. Since ε was arbitrary, (9.1) follows.   □

9.4   The Closure of Polynomials in $\mathcal{N}_p^0$

In this section, we show that polynomials are dense in $\mathcal{N}_p^0$. Our strategy is standard: we first show that the dilation is a well-defined bounded operator on $\mathcal{N}_p^0$. In fact, this result is disguised as Theorem 9.4.1. Then we use that result to show that polynomials are dense.

Theorem 9.4.1 *Suppose $f \in \mathcal{N}_p$. Then $f \in \mathcal{N}_p^0$ if and only if*

$$\|f_r - f\|_p \to 0, \qquad (r \to 1^-),$$

where $f_r(z) = f(rz)$, $z \in \mathbb{B}$.

Proof **Necessity** Suppose $f \in \mathcal{N}_p^0$. This implies that for any $\varepsilon > 0$, there exists $\delta > 0$ such that, for all $\delta < |a| < 1$,

$$\int_{\mathbb{B}} |f(z)|^2(1 - |\Phi_a(z)|^2)^p \, dV(z) < \frac{\varepsilon}{6 \cdot 4^n}. \tag{9.4}$$

Furthermore, by the Schwarz–Pick lemma, we have

$$|\Phi_{ra}(rz)| \leq |\Phi_a(z)| \tag{9.5}$$

for all $r \in (0, 1)$ and $a, z \in \mathbb{B}$. Now, fix $\delta_0 \in (\delta, 1)$, and consider r satisfying $\max\left\{\frac{1}{2}, \frac{\delta}{\delta_0}\right\} < r < 1$. In this case, for all $a \in \mathbb{B}$ with $|a| \in (\delta_0, 1)$, by (9.4) and (9.5), we have

$$\int_{\mathbb{B}} |f(rz)|^2 (1 - |\Phi_a(z)|^2)^p \, dV(z) \leq \int_{\mathbb{B}} |f(rz)|^2 (1 - |\Phi_{ra}(rz)|^2)^p \, dV(z)$$

$$= \left(\frac{1}{r}\right)^{2n} \int_{r\mathbb{B}} |f(w)|^2 (1 - |\Phi_{ra}(w)|^2)^p \, dV(w)$$

$$\leq 4^n \int_{\mathbb{B}} |f(w)|^2 (1 - |\Phi_{ra}(w)|^2)^p \, dV(w) < \frac{\varepsilon}{6}.$$

Since $f \in A_p^2$, f_r converges to f as $r \to 1^-$, in the norm topology of the Bergman space $A_p^2(\mathbb{B})$. This implies that there exists a $r_1 \in (0, 1)$ such that for $r_1 < r < 1$, we have

$$\int_{\mathbb{B}} |f(rz) - f(z)|^2 (1 - |z|^2)^p \, dV(z) < \frac{(1 - \delta_0)^{2p} \varepsilon}{3}.$$

Consequently, for $|a| \leq \delta_0$ and $r_1 < r < 1$, we have

$$\sup_{|a| \leq \delta_0} \int_{\mathbb{B}} |f(rz) - f(z)|^2 (1 - |\Phi_a(z)|^2)^p \, dV(z)$$

$$= \sup_{|a| \leq \delta_0} \left\{ (1 - |a|^2)^p \int_{\mathbb{B}} |f(rz) - f(z)|^2 \frac{(1 - |z|^2)^p}{|1 - \langle z, a \rangle|^{2p}} \, dV(z) \right\}$$

$$\leq \sup_{|a| \leq \delta_0} \int_{\mathbb{B}} |f(rz) - f(z)|^2 \frac{(1 - |z|^2)^p}{|1 - \langle z, a \rangle|^{2p}} \, dV(z)$$

$$\leq \frac{1}{(1 - \delta_0)^{2p}} \int_{\mathbb{B}} |f(rz) - f(z)|^2 (1 - |z|^2)^p \, dV(z) < \frac{\varepsilon}{3}.$$

For all r with $\max\left\{\frac{1}{2}, \frac{\delta}{\delta_0}, r_1\right\} < r < 1$, combining the above estimates yields

$$\|f_r - f\|_p^2 = \sup_{a \in \mathbb{B}} \int_{\mathbb{B}} |f(rz) - f(z)|^2 (1 - |\Phi_a(z)|^2)^p \, dV(z)$$

$$\leq \left(\sup_{|a| \leq \delta_0} + \sup_{\delta_0 < |a| < 1} \right) \int_{\mathbb{B}} |f(rz) - f(z)|^2 (1 - |\Phi_a(z)|^2)^p \, dV(z)$$

$$\leq \frac{\varepsilon}{3} + 2 \sup_{\delta_0 < |a| < 1} \int_{\mathbb{B}} \left(|f(rz)|^2 + |f(z)|^2 \right) (1 - |\Phi_a(z)|^2)^p \, dV(z)$$

$$< \frac{\varepsilon}{3} + 2 \left(\frac{\varepsilon}{6} + \frac{\varepsilon}{6 \cdot 4^n} \right) < \frac{\varepsilon}{3} + 2 \left(\frac{\varepsilon}{6} + \frac{\varepsilon}{6} \right) = \varepsilon,$$

which show that $\|f_r - f\|_p \to 0$ as $r \to 1^-$.

Sufficiency Suppose that $\|f_r - f\|_p \to 0$, as $r \to 1^-$. For each $0 < r < 1$, the holomorphic function f_r is bounded. Hence, by Lemma 9.3.2, it belongs to $\mathcal{N}_p^0$. Since, by Theorem 9.3.1, $\mathcal{N}_p^0$ is a closed subspace of $\mathcal{N}_p$, it follows that f also belongs to $\mathcal{N}_p^0$. $\quad\square$

As a corollary to Theorem 9.4.1, we can easily deduce that polynomials are dense in $\mathcal{N}_p^0$.

Corollary 9.4.2 *Polynomials are dense in $\mathcal{N}_p^0$.*

Proof By Theorem 9.4.1, for each $f \in \mathcal{N}_p^0$, we have

$$\lim_{r \to 1^-} \|f_r - f\|_p = 0.$$

Since each f_r can be uniformly approximated by polynomials, and moreover, by Theorem 7.5.1, the sup-norm dominates the $\mathcal{N}_p$-norm, we conclude that every $f \in \mathcal{N}_p^0$ can be approximated in the $\mathcal{N}_p$-norm by polynomials. $\quad\square$

9.5   Carleson Measures

Let $\xi \in \mathbb{S} := \partial\mathbb{B}$. A Carleson tube at ξ is defined as

$$Q_r(\xi) = \{z \in \mathbb{B} : |1 - \langle z, w\rangle| < r\}, \qquad (r > 0).$$

A positive Borel measure μ on $\mathbb{B}$ is called a *p-Carleson measure* if there exists a constant $C > 0$ such that

$$\mu(Q_r(\xi)) \leq Cr^p \tag{9.6}$$

for all $\xi \in \mathbb{S}$ and $r > 0$. Moreover, if

$$\lim_{r \to 0} \frac{\mu(Q_r(\xi))}{r^p} = 0,$$

uniformly with respect to $\xi \in \mathbb{S}$, then μ is called a *vanishing p-Carleson measure.*

Lemma 9.5.1 *Let $p = n + 1 + \alpha > 0$ and μ be a finite positive Borel measure on $\mathbb{B}$. Then the following assertions are equivalent.*

(i) *μ is a vanishing p-Carleson measure.*
(ii) *For each $s > 0$,*

$$\lim_{|z|\to 1^-} \int_{\mathbb{B}} \frac{(1-|z|^2)^s d\mu(w)}{|1-\langle z, w\rangle|^{p+s}} = 0. \tag{9.7}$$

(iii) For some s > 0, (9.7) holds.

Proof *(i)* $\implies$ *(ii)*: The assumption *(i)* means that for any $\varepsilon > 0$, there exists $\delta > 0$ such that for $0 < r < \delta$, we have

$$\frac{\mu(Q_r(\xi))}{r^p} < \varepsilon \qquad (\xi \in \mathbb{S}). \tag{9.8}$$

Also, since μ is a Carleson measure, there is a positive constant C such that (9.6) holds. Let $s > 0$. For the same ε chosen above, take $N_0 \geq 1$ such that

$$\sum_{k=N_0+1}^{\infty} \frac{1}{2^{sk}} < \varepsilon. \tag{9.9}$$

Fix some $z \in \mathbb{B}$ with

$$\max\left\{\frac{3}{4}, 1 - \frac{\delta}{2^{N_0+1}}\right\} < |z| < 1$$

and set $\xi = z/|z|$. For any nonnegative integer k, let $r_k = 2^{k+1}(1 - |z|)$. We decompose the unit ball $\mathbb{B}$ into the disjoint union of the following sets:

$$E_0 = Q_{r_0}(\xi), \quad E_k = Q_{r_k}(\xi) \setminus Q_{r_{k-1}}(\xi), \qquad (1 \leq k < \infty).$$

For $k \geq 2$ and $w \in E_k$, we have

$$\begin{aligned}
|1 - \langle z, w\rangle| &= \left||z|(1 - \langle \xi, w\rangle) + (1 - |z|)\right| \\
&\geq |z||1 - \langle \xi, w\rangle| - (1 - |z|) \\
&\geq (3/4)2^k(1 - |z|) - (1 - |z|) \\
&\geq 2^{k-1}(1 - |z|).
\end{aligned} \tag{9.10}$$

This estimation also holds for $k = 1$ and $k = 0$, since

$$|1 - \langle z, w\rangle| \geq 1 - |z| \geq \frac{1}{2}(1 - |z|).$$

Now we consider two cases for k.

Case I: $0 \leq k \leq N_0$. In this case, we have

$$0 < r_k = 2^{k+1}(1 - |z|) \leq 2^{N_0+1}(1 - |z|) < \delta.$$

This implies, by (9.8), that

$$\mu(E_k) \leq \mu(Q_{r_k}(\xi)) \leq r_k^p \varepsilon = 2^{p(k+1)}(1 - |z|)^p \varepsilon. \tag{9.11}$$

Case II: $k > N_0$. Using (9.6), we have

$$\mu(E_k) \leq \mu(Q_{r_k}(\xi)) \leq 2^{p(k+1)}(1 - |z|)^p C. \tag{9.12}$$

For

$$\max\left\{\frac{3}{4}, 1 - \frac{\delta}{2^{N_0+1}}\right\} < |z| < 1,$$

using (9.9), (9.10), (9.11), and (9.12), we see that

$$\int_{\mathbb{B}} \frac{(1 - |z|^2)^s d\mu(w)}{|1 - \langle z, w \rangle|^{p+s}}$$

$$= \sum_{k=0}^{\infty} \int_{E_k} \frac{(1 - |z|^2)^s d\mu(w)}{|1 - \langle z, w \rangle|^{p+s}}$$

$$\leq \sum_{k=0}^{\infty} \frac{(1 - |z|^2)^s \mu(E_k)}{(2^{k-1}(1 - |z|))^{p+s}}$$

$$= \left(\sum_{k=0}^{N_0} + \sum_{k=N_0+1}^{\infty}\right) \frac{(1 - |z|^2)^s \mu(E_k)}{(2^{k-1}(1 - |z|))^{p+s}}$$

$$\leq \sum_{k=0}^{N_0} \frac{2^{s+p(k+1)}(1 - |z|)^{p+s} \varepsilon}{2^{(k-1)(p+s)}(1 - |z|)^{p+s}} + \sum_{k=N_0}^{\infty} \frac{2^{s+p(k+1)}(1 - |z|)^{p+s} C}{2^{(k-1)(p+s)}(1 - |z|)^{p+s}}$$

$$= \varepsilon \sum_{k=0}^{N_0} \frac{2^{s+p(k+1)}}{2^{(k-1)(p+s)}} + C \sum_{k=N_0+1}^{\infty} \frac{2^{s+p(k+1)}}{2^{(k-1)(p+s)}}$$

$$= \varepsilon \cdot 4^{s+p} \sum_{k=0}^{N_0} \frac{1}{2^{ks}} + C \cdot 4^{p+s} \sum_{k=N_0+1}^{\infty} \frac{1}{2^{ks}}$$

$$\leq \frac{4^{s+p}}{1 - 2^{-s}} \varepsilon + 4^{p+s} C \varepsilon = M\varepsilon,$$

where M is a constant depending only on s and p. This shows that statement (ii) holds.

The implication $(ii) \implies (iii)$ is obvious.

$(iii) \implies (i)$: Suppose the condition (iii) holds. This means that there exists $s > 0$, such that

$$\lim_{|z| \to 1^-} \int_{\mathbb{B}} \frac{(1 - |z|^2)^s d\mu(w)}{|1 - \langle z, w \rangle|^{p+s}} = 0.$$

Then for any $\varepsilon > 0$, there exists $\delta > 0$, such that when $\delta < |z| < 1$,

$$\int_{\mathbb{B}} \frac{(1 - |z|^2)^s d\mu(w)}{|1 - \langle z, w \rangle|^{p+s}} < \varepsilon. \tag{9.13}$$

We first show that μ is a p-Carleson measure. Indeed, for $|z| \le \delta$, we have

$$\int_{\mathbb{B}} \frac{(1 - |z|^2)^s d\mu(w)}{|1 - \langle z, w \rangle|^{p+s}} \le \int_{\mathbb{B}} \frac{d\mu(w)}{|1 - \langle z, w \rangle|^{p+s}} \le \frac{\mu(\mathbb{B})}{(1 - \delta)^{p+s}} < \infty.$$

This fact and (9.13) show that μ is a p-Carleson measure.

Next, we prove that μ is a vanishing p-Carleson measure. Let ξ be in $\mathbb{S}$. For $r \in (0, 1 - \delta)$, put $z = (1 - r)\xi$. Then $\delta < |z| < 1$, and for any $w \in Q_r(\xi)$,

$$|1 - \langle z, w \rangle| = \left| (1 - r)(1 - \langle \xi, w \rangle) + r \right| \le (1 - r)r + r < 2r.$$

Consequently,

$$\frac{(1 - |z|^2)^s}{|1 - \langle z, w \rangle|^{p+s}} \ge \frac{(1 - |z|)^s}{|1 - \langle z, w \rangle|^{p+s}} \ge \frac{r^s}{(2r)^{p+s}} = \frac{2^{-(p+s)}}{r^p}.$$

Using (9.13), we obtain

$$\frac{\mu(Q_r(\xi))}{r^p} = \frac{1}{r^p} \int_{Q_r(\xi)} d\mu(w) \le 2^{p+s} \int_{Q_r(\xi)} \frac{(1 - |z|^2)^s d\mu(w)}{|1 - \langle z, w \rangle|^{p+s}}$$

$$\le 2^{p+s} \int_{\mathbb{B}} \frac{(1 - |z|^2)^s d\mu(w)}{|1 - \langle z, w \rangle|^{p+s}} < 2^{p+s} \varepsilon,$$

which implies (i). $\qquad\qquad\square$

The following result describes a relationship between functions in $\mathcal{N}_p$ and $\mathcal{N}_p^0$ and Carleson measures.

Theorem 9.5.2 *Let $p > 0$ and $f \in Hol(\mathbb{B})$. Define*

$$d\mu_{f,p}(z) = |f(z)|^2 (1 - |z|^2)^p \, dV(z).$$

Then the following assertions hold.

(i) $f \in \mathcal{N}_p$ if and only if $\mu_{f,p}$ is a p-Carleson measure.
(ii) $f \in \mathcal{N}_p^0$ if and only if $\mu_{f,p}$ is a vanishing p-Carleson measure.

Moreover,

$$\|f\|_p^2 \asymp \sup_{r \in (0,1), \xi \in \mathbb{S}} \frac{\mu_{f,p}(Q_r(\xi))}{r^p} = \sup_{r \in (0,1), \xi \in \mathbb{S}} \frac{1}{r^p} \int_{Q_r(\xi)} |f(z)|^2 (1 - |z|^2)^p \, dV(z).$$

Proof For any for $f \in \mathrm{Hol}(\mathbb{B})$, we can write

$$\|f\|_p^2 = \sup_{a \in \mathbb{B}} \int_{\mathbb{B}} |f(z)|^2 \frac{(1 - |a|^2)^p (1 - |z|^2)^p}{|1 - \langle a, z \rangle|^{2p}} \, dV(z)$$

$$= \sup_{a \in \mathbb{B}} \int_{\mathbb{B}} \left(\frac{1 - |a|^2}{|1 - \langle a, z \rangle|^2} \right)^p \, d\mu_{f,p}(z).$$

From this, statement (i) and the asymptotic formula for the norm follow. The statement (ii) is a consequence of Lemma 9.5.1.   □

9.6 $\mathcal{N}_p$ Spaces Are Not Homeomorphic

In this section, we show that, for $0 < p \le n$, all $\mathcal{N}_p$-spaces are different topological vector spaces. This result together with Theorem 7.5.1 gives a complete relationship between $\mathcal{N}_p$-spaces for all $p > 0$.

We prove this fact by a construction. To do so, we need some additional tools. Let us denote the point $(1, 0, \ldots, 0) \in \mathbb{S}$ by $\mathbf{1}$. On $\mathbb{S}$, we have the natural rotation-invariant probability measure σ. Denote by $W_k(\mathbb{S})$ the space of all k-homogeneous polynomials on $\mathbb{C}^n$ restricted to the unit sphere $\mathbb{S}$. On $W_k(\mathbb{S})$, we can consider a chain of norms: for $f \in W_k(\mathbb{S})$, put

$$\|f\|_p = \left(\int_{\mathbb{S}} |f(\xi)|^p \, d\sigma(\xi) \right)^{1/p}$$

if $1 \le p < \infty$ and

$$\|f\|_\infty = \sup_{\xi \in \mathbb{S}} |f(\xi)|.$$

The Banach space $\left(W_k(\mathbb{S}), \|\cdot\|_p\right)$ is denoted by $W_k^p(\mathbb{S})$.
  Write

$$\alpha(n, p) = \int_{\mathbb{S}_n} |\langle \xi, \mathbf{1}\rangle|^p \, d\sigma(\xi),$$

where $\mathbb{S}_n$ is the unit sphere in $\mathbb{C}^n$. In particular, it can be computed that

$$\alpha(n, p) = \Gamma\left(\frac{1+p}{2}\right)\Gamma(n)\Gamma\left(n + \frac{p}{2}\right).$$

  For two finite-dimensional Banach spaces X and Y, the projection constant $\lambda(X)$ of X is a number defined by

$$\lambda(X) = \inf\{\|T\| \cdot \|S\|,\ T : X \to L^{\infty},\ S : L^{\infty} \to X,\ ST = \mathrm{Id}_X\}$$

and the so-called Banach–Mazur distance $d(X, Y)$ between X and Y is defined by

$$d(X, Y) = \inf\{\|T\| \cdot \|T^{-1}\|, T : X \to Y \text{ is surjective}\}.$$

It is well-known that

$$\lambda(X) \le d(X, Y)\lambda(Y). \tag{9.14}$$

Also,

$$\lambda(\ell_k^2) = \frac{\alpha(k, 1)}{\alpha(k, 2)} \ge \frac{1}{2}\sqrt{k\pi}. \tag{9.15}$$

  Now we give a brief description of the construction of some important polynomials. To do so, we consider the operator

$$(T_k f)(\xi) = \frac{1}{\alpha(n, 2k)} \int_{\mathbb{S}} f(\eta)\langle \xi, \eta \rangle^k \, d\sigma(\eta) \tag{9.16}$$

is an orthogonal projection from $L_2(\mathbb{S})$ onto $W_k^2(\mathbb{S})$. The norm of this projection, considered as an operator from $L^{\infty}(\mathbb{S})$ onto $W_k^{\infty}(\mathbb{S})$, is smaller than 2^{n-1}. We also have

$$\inf\{\|p\|_2 : p \in W_k(\mathbb{S}) \text{ and } \|p\|_{\infty} = 1\} = \sqrt{\alpha(n, 2k)}. \tag{9.17}$$

Moreover,

$$\dim\left(W_k(\mathbb{S})\right) = \frac{1}{\alpha(n, 2k)}. \qquad (9.18)$$

Lemma 9.6.1 *For every k, there exists $P_k \in W_k(\mathbb{S})$ with $\|P_k\|_\infty = 1$ and $\|P_k\|_2 \geq \sqrt{\pi}/2^n$.*

Proof We apply (9.14) for $X = W_k^2(\mathbb{S})$ and $Y = W_k^\infty(\mathbb{S})$. By (9.15), (9.16), and (9.18), we obtain

$$d\left(W_k^2(\mathbb{S}), W_k^\infty(\mathbb{S})\right) \geq \frac{\lambda(W_k^2(\mathbb{S}))}{\lambda(W_k^\infty(\mathbb{S}))}$$

$$\geq 2^{1-n} \frac{\sqrt{\pi}}{2} \sqrt{\frac{1}{\alpha(n, 2k)}}. \qquad (9.19)$$

Let I denote the identity map from $W_k^\infty(\mathbb{S})$ to $W_k^2(\mathbb{S})$. By the definition of the Banach–Mazur distance, we have

$$\|I\| \cdot \|I^{-1}\| \geq d\left(W_k^2(\mathbb{S}), W_k^\infty(\mathbb{S})\right).$$

Then (9.17) gives $\|I^{-1}\| = \sqrt{\dfrac{1}{\alpha(n, 2k)}}$. So finally, by (9.19), we conclude that

$$\|I\| \geq 2^{1-n} \frac{\sqrt{\pi}}{2}.$$

The proof is completed. □

Now we return to the main problem. By Lemma 9.6.1, there exists a sequence of homogeneous polynomials $(P_k)_{k\geq 1}$ satisfying $\deg(P_k) = k$,

$$\|P_k\|_\infty = \sup_{\xi \in \mathbb{S}} |P_k(\xi)| = 1, \qquad (9.20)$$

and

$$\left(\int_{\mathbb{S}} |P_k(\xi)|^2 d\sigma(\xi)\right)^{1/2} \geq \frac{\sqrt{\pi}}{2^n}. \qquad (9.21)$$

Note that the homogeneity of P_k implies that $|P_k(z)| \leq |z|^k$ for all $z \in \mathbb{B}$.

Let $(m_k)_{k=0}^\infty$ be a sequence of positive integers such that $m_{k+1}/m_k \geq c > 1$ for all $k \geq 0$. Let

$$f(z) = \sum_{k=0}^{\infty} b_k P_{m_k}(z), \qquad (z \in \mathbb{B}). \tag{9.22}$$

Such a function is said to belong to the *Hadamard gap class*. In the following result, we obtain an estimate for the $\mathcal{N}_p$-norm and A^{-q}-norm of f.

Theorem 9.6.2 *Let f be defined as in (9.22), and let $p > 0$. Then the following statements hold.*

(i) For $0 < p \le n$, we have

$$\|f\|_p^2 \asymp \sum_{k=0}^{\infty} \frac{|b_k|^2}{m_k^{p+1}}.$$

(ii) For any $q > 0$, we have

$$|f|_q \asymp \sup_k \frac{|b_k|}{m_k^q}.$$

Here, $\|f\|_p$ and $|f|_q$ denote the norm of f in the spaces $\mathcal{N}_p$ and A^{-q}, respectively.

Proof

(i): Since $|P_{m_k}(w)| \le |w|^{m_k}$ for all $k \ge 0$ and $w \in \mathbb{B}$, we have

$$\sup_{|w|=|z|} |f(w)| \le \sum_{k=0}^{\infty} |b_k||z|^{m_k}$$

for all $z \in \mathbb{B}$. Then we do integration in polar coordinates. Hence, by Theorem 9.2.1,

$$\|f\|_p^2 \lesssim \int_0^1 \left(\sum_{k=0}^{\infty} |b_k|r^{m_k}\right)^2 (1-r^2)^p \, dr.$$

On the other hand, we have

$$\int_0^1 \left(\sum_{k=0}^{\infty} |b_k|r^{m_k}\right)^2 (1-r^2)^p \, dr \asymp \sum_{k=0}^{\infty} 2^{-k(p+1)} \left(\sum_{2^k \le m_j < 2^{k+1}} |b_j|\right)^2. \tag{9.23}$$

Since $m_{j+1} \geq c m_j$ for all j, the cardinality of $\{j : 2^k \leq m_j < 2^{k+1}\}$ is at most $1 + \log_c 2$. It then follows that

$$\sum_{k=0}^{\infty} 2^{-k(p+1)} \bigg(\sum_{2^k \leq m_j < 2^{k+1}} |b_j| \bigg)^2 \lesssim \sum_{k=0}^{\infty} 2^{-k(p+1)} \bigg(\sum_{2^k \leq m_j < 2^{k+1}} |b_j|^2 \bigg)$$

$$\lesssim \sum_{k=0}^{\infty} \bigg(\sum_{2^k \leq m_j < 2^{k+1}} m_j^{-(p+1)} |b_j|^2 \bigg)$$

$$= \sum_{k=0}^{\infty} \frac{|b_k|^2}{m_k^{p+1}}.$$

Combining the above estimates, we obtain $\| f \|_p^2 \lesssim \sum_{j=0}^{\infty} \frac{|b_k|^2}{m_k^{p+1}}.$

To prove the reverse inequality, we use the orthogonality of homogeneous polynomials of different degrees in A_p^2 to obtain

$$\| f \|_p^2 \geq \| f \|_{A_p^2}^2 = \int_{\mathbb{B}} \bigg| \sum_{k=0}^{\infty} b_k P_{m_k}(z) \bigg|^2 (1 - |z|^2)^p \, dV(z)$$

$$= \sum_{k=0}^{\infty} |b_k|^2 \int_{\mathbb{B}} |P_{m_k}(z)|^2 (1 - |z|^2)^p \, dv(z)$$

$$= \sum_{k=0}^{\infty} |b_k|^2 \int_0^1 2n r^{2n+2m_k-1}(1 - r^2)^p \, dr \int_{\mathbb{S}} |P_{m_k}(\xi)|^2 \, d\sigma(\xi)$$

$$\gtrsim \sum_{k=0}^{\infty} |b_k|^2 \int_0^1 t^{n+m_k-1}(1 - t)^p \, dt$$

$$= \sum_{k=0}^{\infty} |b_k|^2 \frac{\Gamma(n + m_k)\Gamma(p + 1)}{\Gamma(n + m_k + p + 1)} \gtrsim \sum_{k=0}^{\infty} \frac{|b_k|^2}{m_k^{p+1}}.$$

The last inequality follows from Stirling's formula. In the line before, we applied the change of variables $t = r^2$ and also appealed to (9.21). We have thus completed the proof of (i).

(ii): Assume $f \in A^{-q}$ for $q > 0$. Fix a positive integer k. For $r > 0$ and $\xi \in \mathbb{S}$, we have $|f|_q^2 (1 - r)^{-2q} \geq |f(r\xi)|^2$. Integrating with respect to $\xi \in \mathbb{S}$ and using (9.21) yield

$$\frac{|f|_q^2}{(1 - r)^{2q}} \geq \int_{\mathbb{S}} |f(r\xi)|^2 \, d\sigma(\xi) = \int_{\mathbb{S}} \bigg| \sum_{j=0}^{\infty} b_j r^{m_j} P_{m_j}(\xi) \bigg|^2 \, d\sigma(\xi)$$

$$= \sum_{j=0}^{\infty} |b_j|^2 r^{2m_j} \int_{\mathbb{S}} |P_{m_j}(\xi)|^2 \, d\sigma(\xi) \gtrsim |b_k|^2 r^{2m_k}.$$

Setting $r = m_k/(q + m_k)$, we obtain

$$|b_k| \lesssim \frac{|f|_q}{r^{m_k}(1-r)^q} = |f|_q \left(1 + \frac{q}{m_k}\right)^{m_k} \left(1 + \frac{m_k}{q}\right)^q \lesssim |f|_q m_k^q,$$

which implies $\sup_k \dfrac{|b_k|}{m_k^q} \lesssim |f|_q$.

Put $L = \sup_k \{|b_k| m_k^{-q}\}$ so $|b_k| \le L m_k^q$ for all $k \ge 0$. Recall that $|P_{m_k}(z)| \le |z|^{m_k}$ for all $k \ge 0$. For each $z \in \mathbb{B}$, we have

$$\frac{|f(z)|}{1 - |z|} \le \left(\sum_{k=0}^{\infty} |b_k| |P_{m_k}(z)|\right)\left(\sum_{s=0}^{\infty} |z|^s\right)$$

$$\le L\left(\sum_{k=0}^{\infty} m_k^q |z|^{m^k}\right)\left(\sum_{s=0}^{\infty} |z|^s\right) \tag{9.24}$$

$$= L \sum_{\ell=1}^{\infty} \left(\sum_{m_k \le \ell} m_k^q\right)|z|^\ell.$$

Since $m_{k+1}/m_k \ge c > 1$ for all k, we have

$$\sum_{m_k \le \ell} m_k^q = \ell^q \sum_{m_k \le \ell} \left(\frac{m_k}{\ell}\right)^q \le \ell^q \sum_{s=0}^{\infty} (c^{-q})^s = \frac{\ell^q}{1 - c^{-q}}.$$

By Stirling's formula, it follows that

$$\sum_{m_k \le \ell} m_k^q \le \frac{\ell^q}{1 - c^{-q}} \lesssim \frac{1}{1 - c^{-q}} \frac{\Gamma(\ell + q + 1)}{\Gamma(\ell + 1)\Gamma(q)}. \tag{9.25}$$

Combining (9.24) and (9.25) yields

$$\frac{|f(z)|}{1 - |z|} \lesssim \frac{L}{1 - c^{-q}} \sum_{\ell=0}^{\infty} \frac{\Gamma(\ell + q + 1)}{\Gamma(\ell + 1)\Gamma(q)} |z|^\ell = \frac{L}{1 - c^{-q}} \frac{1}{(1 - |z|)^{q+1}}.$$

Consequently,

$$|f|_q = \sup_{z \in \mathbb{B}} |f(z)|(1 - |z|)^q \lesssim L = \sup_k \frac{|b_k|}{m_k^q}.$$

This completes the proof of (ii).

$\square$

Corollary 9.6.3 *If $0 < p_1 < p_2 \le n$, then we have*

$$\mathcal{N}_{p_1} \subsetneqq \mathcal{N}_{p_2} \subsetneqq A^{-\frac{n+1}{2}}.$$

Proof Define

$$f_1(z) = \sum_{k=0}^{\infty} 2^{\frac{k(n+1)}{2}} P_{2^k}(z) \tag{9.26}$$

and

$$f_2(z) = \sum_{k=0}^{\infty} 2^{\frac{k(1+p_1)}{2}} P_{2^k}(z) \tag{9.27}$$

for $z \in \mathbb{B}$. Using Theorem 9.6.2, it can be checked with a direct computation that $f_1 \in A^{-\frac{n+1}{2}} \setminus \mathcal{N}_{p_2}$ and $f_2 \in \mathcal{N}_{p_2} \setminus \mathcal{N}_{p_1}$.
Indeed, for $f_1(z)$, we have

$$b_k = 2^{\frac{k(n+1)}{2}} \quad \text{and} \quad m_k = 2^k.$$

Thus,

$$|f|_{\frac{n+1}{2}} \asymp \sup_k \frac{|b_k|}{m_k^{\frac{n+1}{2}}} = \sup_k \frac{2^{\frac{k(n+1)}{2}}}{(2^k)^{\frac{n+1}{2}}} = 1 < \infty,$$

while

$$\|f\|_{p_2}^2 \asymp \sum_{k=0}^{\infty} \frac{|b_k|^2}{m_k^{p_2+1}} = \sum_{k=0}^{\infty} \frac{(2^{\frac{k(n+1)}{2}})^2}{(2^k)^{p_2+1}} = \sum_{k=0}^{\infty} \frac{2^{k(n+1)}}{2^{k(p_2+1)}} = \infty.$$

The verification for f_2 is similar.

$\square$

Notes on Chapter 9

The results of this chapter are mainly from [46]. More information on the projection constant and Banach–Mazur distance is given in [80] and also in [77, Items 22.1, 28.1]. The construction of homogeneous polynomials (P_k) is given in [81]. The content of Sect. 9.5 is taken from [107]. A characterization for a Hadamard gap class function to be in a weighted Bergman space was given in [89]. Theorem 9.6.2 is a higher-dimensional version of [73]. Note that Theorem 9.6.2, for $n = 1$, contains the corresponding results in [73] as particular cases. Estimate (9.23) is provided in [69, Theorem 1].

One of the most interesting topics in the study of $\mathcal{N}_p$-spaces is the composition operators and weighted composition operators between them. In this chapter, we investigate composition operators C_φ acting from $\mathcal{N}_p$ into $\mathcal{N}_q$.

10.1   Boundedness of $C_\varphi : \mathcal{N}_p \longrightarrow \mathcal{N}_p$, the Univalent Symbol

We begin with the case where φ is a univalent holomorphic self-map of the unit ball $\mathbb{B}$.

Theorem 10.1.1 *Let φ be a univalent holomorphic self-map of $\mathbb{B}$. Suppose that*

$$\delta = \inf\{|J\varphi(z)| : z \in \mathbb{B}\} > 0,$$

where $J\varphi$ is the complex Jacobian of φ. Then $C_\varphi : \mathcal{N}_p \longrightarrow \mathcal{N}_p$, $p > 0$, is a bounded operator with $\|C_\varphi\| \leq \delta^{-1}$.

Proof Let $f \in \mathcal{N}_p$. For $a, z \in \mathbb{B}$, the Schwarz–Pick lemma gives $|\Phi_{\varphi(a)}(\varphi(z))| \leq |\Phi_a(z)|$. Since φ is univalent, it is biholomorphic from $\mathbb{B}$ onto $\varphi(\mathbb{B})$. This makes the change of variables $w = \varphi(z)$ possible in the following estimates. More explicitly, we have

$$\delta^2 \int_{\mathbb{B}} |f(\varphi(z))|^2 (1 - |\Phi_a(z)|^2)^p \, dV(z)$$

$$\leq \int_{\mathbb{B}} |f(\varphi(z))|^2 (1 - |\Phi_{\varphi(a)}(\varphi(z))|^2)^p |J\varphi(z)|^2 \, dV(z)$$

$$= \int_{\varphi(\mathbb{B})} |f(w)|^2 (1 - |\Phi_{\varphi(a)}(w)|^2)^p \, dV(w)$$

$$\leq \int_{\mathbb{B}} |f(w)|^2 (1 - |\Phi_{\varphi(a)}(w)|^2)^p \, dV(w) \leq \|f\|_p^2.$$

Then taking supremum over $a \in \mathbb{B}$ gives

$$\delta^2 \|C_\varphi f\|_p^2 \leq \|f\|_p^2,$$

which immediately implies $\|C_\varphi f\|_p \leq \delta^{-1} \|f\|_p$. Since f was an arbitrary element in $\mathcal{N}_p$, we conclude that C_φ is a bounded operator on $\mathcal{N}_p$ with $\|C_\varphi\| \leq \delta^{-1}$ as desired.   □

Corollary 10.1.2 *Suppose $A : \mathbb{C}^n \longrightarrow \mathbb{C}^n$ is an invertible linear operator and b is a vector in $\mathbb{C}^n$ such that $\varphi(z) = Az + b$ is a self-mapping of $\mathbb{B}$. Then $C_\varphi : \mathcal{N}_p \longrightarrow \mathcal{N}_p$ is bounded for all $p > 0$.*

Proof Since $|J\varphi(z)| = |\det(A)| > 0$, Theorem 10.1.1 provides the desired conclusion.   □

10.2   Boundedness of $C_\varphi : \mathcal{N}_p \longrightarrow \mathcal{N}_q$, the General Symbol

For a general self-mapping φ of the unit ball, we give a necessary condition and then a sufficient condition for C_φ to be bounded from $\mathcal{N}_p$ into $\mathcal{N}_q$. We make use of a sequence of homogeneous polynomials $\{P_m\}$ (see (9.21) in Chap. 9), which satisfy $\deg(P_m) = m$, and

$$\|P_m\|_\infty = \sup_{\zeta \in \mathbb{S}} |P_m(\zeta)| = 1, \text{ and } \int_{\mathbb{S}} |P_m(\zeta)|^2 \, d\sigma(\zeta) \geq \frac{\pi}{4^n}. \tag{10.1}$$

We also need the inequality

$$1 + \sum_{k=0}^{\infty} 2^{k\gamma} x^{2^k} \gtrsim (1 - x)^{-\gamma} \quad \text{for } 0 \leq x < 1, \tag{10.2}$$

where γ is a positive number. Indeed, this is already proved in (5.9) Chap. 5 for $x > 1/2$. On the other hand, it is clear that $2^\gamma \geq (1 - x)^{-\gamma}$ for $0 \leq x \leq 1/2$. Therefore, (10.2) holds.

The following estimation result will be needed in obtaining the necessary condition for the boundedness of C_φ.

Lemma 10.2.1 *Let p and q be positive numbers, and let φ be a holomorphic self-map of $\mathbb{B}$. Let $\mathbf{B}_{\mathcal{N}_p}$ denote the closed unit ball of $\mathcal{N}_p$. Suppose μ is a positive number and E is a measurable subset of $\mathbb{B}$ such that*

$$\sup_{a \in \mathbb{B}, f \in \mathbf{B}_{\mathcal{N}_p}} \int_E |f(\varphi(z))|^2 (1 - |\Phi_a(z)|^2)^q \, dV(z) \leq \mu.$$

Then, for any $0 < \varepsilon \leq p + 1$, we have

$$\sup_{a \in \mathbb{B}} \int_E \frac{(1 - |\Phi_a(z)|^2)^q}{(1 - |\varphi(z)|^2)^{p+1-\varepsilon}} \, dV(z) \lesssim \mu.$$

Proof The hypothesis implies that for any $a \in \mathbb{B}$ and any $f \in \mathcal{N}_p$, we have

$$\int_E |f(\varphi(z))|^2 (1 - |\Phi_a(z)|^2)^q \, dV(z) \leq \mu \|f\|_p^2. \tag{10.3}$$

Put

$$F(z) = 1 + \sum_{k=0}^{\infty} 2^{k(p+1-\varepsilon)/2} P_{2^k}(z), \qquad (z \in \mathbb{B}).$$

Then, by Theorem 9.6.2,

$$\|F\|_p^2 \asymp \sum_{k=0}^{\infty} \frac{2^{k(p+1-\varepsilon)}}{2^{k(p+1)}} = \sum_{k=0}^{\infty} 2^{-k\varepsilon} < \infty,$$

which shows that F belongs to $\mathcal{N}_p$. Let $\mathfrak{U}_n$ denote the group of all unitary operators on the Hilbert space $\mathbb{C}^n$. For each $U \in \mathfrak{U}_n$, the unitary invariant property of the volume measure shows that $F \circ U$ also belongs to $\mathcal{N}_p$ and $\|F \circ U\|_p = \|F\|_p$. Setting $f = F \circ U$ in (10.3) gives

$$\int_E |F(U\varphi(z))|^2 (1 - |\Phi_a(z)|^2)^q \, dV(z) \leq \mu \|F \circ U\|_p^2 = \mu \|F\|_p^2$$

for all $a \in \mathbb{B}$ and $U \in \mathfrak{U}_n$. Now, fix $a \in \mathbb{B}$. Integration with respect to the Haar measure dU on $\mathfrak{U}_n$ and using Fubini's theorem yield

$$\int_E \left(\int_{\mathfrak{U}_n} |F(U\varphi(z))|^2 \, dU \right) (1 - |\Phi_a(z)|^2)^q \, dV(z) \leq \mu \|F\|_p^2.$$

Then, for any $z \in \mathbb{B}$,

$$\int_{\mathfrak{U}_n} |F(U\varphi(z))|^2 \, dU = \int_{\mathbb{S}} |F(|\varphi(z)|\varsigma)|^2 \, d\sigma(\varsigma)$$

$$= \int_{\mathbb{S}} \left| 1 + \sum_{k=0}^{\infty} 2^{k(p+1-\varepsilon)/2} P_{2^k}(|\varphi(z)|\zeta) \right|^2 d\sigma(\zeta)$$

$$= 1 + \sum_{k=0}^{\infty} 2^{k(p+1-\varepsilon)} (|\varphi(z)|^2)^{2^k} \int_{\mathbb{S}} |P_{2^k}(\zeta)|^2 d\sigma(\zeta)$$

$$\geq \frac{\pi}{4^n} \left(1 + \sum_{k=0}^{\infty} 2^{k(p+1-\varepsilon)} (|\varphi(z)|^2)^{2^k} \right) \quad \text{(by (10.1))}$$

$$\gtrsim \left(1 - |\varphi(z)|^2 \right)^{-(p+1-\varepsilon)} \quad \text{(by (10.2))}.$$

Here, we note that the first equality is a basic identity for unitary operators, while for the third equality, the orthogonality and the homogeneity of $\{P_{2^k}\}$ are used. Consequently,

$$\int_E \frac{(1 - |\Phi_a(z)|^2)^q}{(1 - |\varphi(z)|^2)^{p+1-\varepsilon}} \, dV(z)$$

$$\lesssim \int_E \left(\int_{\mathfrak{U}} |F(U\varphi(z))|^2 \, dU \right) (1 - |\Phi_a(z)|^2)^q \, dV(z) \leq \mu \|F\|_p^2,$$

as desired.   □

We are now ready to prove a necessary condition and a sufficient condition for the boundedness of $C_\varphi : \mathcal{N}_p \longrightarrow \mathcal{N}_q$.

Theorem 10.2.2 *Let p and q be two positive numbers, and let φ be a holomorphic self-map of $\mathbb{B}$. If*

$$\sup_{a \in \mathbb{B}} \int_{\mathbb{B}} \frac{(1 - |\Phi_a(z)|^2)^q}{(1 - |\varphi(z)|^2)^{n+1}} \, dV(z) < \infty, \tag{10.4}$$

then $C_\varphi : \mathcal{N}_p \longrightarrow \mathcal{N}_q$ is bounded. Conversely, if $C_\varphi : \mathcal{N}_p \longrightarrow \mathcal{N}_q$ is bounded, then for any $0 < \varepsilon \leq p + 1$,

$$\sup_{a \in \mathbb{B}} \int_{\mathbb{B}} \frac{(1 - |\Phi_a(z)|^2)^q}{(1 - |\varphi(z)|^2)^{p+1-\varepsilon}} \, dV(z) < \infty. \tag{10.5}$$

Proof Suppose (10.4) holds. By Theorem 7.5.1, there is a constant $C > 0$ such that for each $f \in \mathcal{N}_p$, we have

$$|f(z)|(1 - |z|^2)^{\frac{n+1}{2}} \leq C \|f\|_p, \qquad (z \in \mathbb{B}).$$

Hence,

$$\sup_{a\in\mathbb{B}} \int_\mathbb{B} |f(\varphi(z))|^2 (1 - |\Phi_a(z)|^2)^q \, dV(z) \le (C\|f\|_p)^2 \sup_{a\in\mathbb{B}} \int_\mathbb{B} \frac{(1 - |\Phi_a(z)|^2)^q}{(1 - |\varphi(z)|^2)^{n+1}} \, dV(z),$$

which shows that C_φ is bounded from $\mathcal{N}_p$ into $\mathcal{N}_q$.

Conversely, suppose $C_\varphi : \mathcal{N}_p \longrightarrow \mathcal{N}_q$ is bounded. Then

$$\sup_{a\in\mathbb{B}, f\in\mathbf{B}_{\mathcal{N}_p}} \int_\mathbb{B} |f(\varphi(z))|^2 (1 - |\Phi_a(z)|^2)^q \, dV(z) = \sup_{f\in\mathbf{B}_{\mathcal{N}_p}} \|C_\varphi f\|_q^2 \le \|C_\varphi\|^2.$$

The desired inequality now follows from Lemma 10.2.1. $\square$

An application of Theorem 10.2.2 immediately gives the following result. In fact, the operator C_φ in the corollary is compact, as we see in the next section.

Corollary 10.2.3 *Let φ be a holomorphic self-mapping of $\mathbb{B}$ such that $\|\varphi\|_\infty < 1$. Then $C_\varphi : \mathcal{N}_p \longrightarrow \mathcal{N}_q$ is bounded for all $p, q > 0$.*

10.3 Characterizations of Compactness of $C_\varphi : \mathcal{N}_p \longrightarrow \mathcal{N}_q$

In this section, we study the compactness of composition operators between $\mathcal{N}_p$-spaces. We use the following result which is a reformulation of the well-known criterion for compactness, i.e., weak convergent sequences are transformed to strong convergent sequences: "A bounded composition operator $C_\varphi : \mathcal{N}_p \to \mathcal{N}_q$ is compact if and only if for any bounded sequence $(f_m) \subset \mathcal{N}_p$ converging to zero uniformly on compact subsets of $\mathbb{B}$, the sequence $(\|f_m \circ \varphi\|_q)$ converges to zero as $m \to \infty$." We start with the following result, which provides a necessary condition for C_φ to be a compact operator.

Lemma 10.3.1 *Suppose $C_\varphi : \mathcal{N}_p \longrightarrow \mathcal{N}_q$ is a compact composition operator. Let $\mathbf{B}_{\mathcal{N}_p} = \{f \in \mathcal{N}_p : \|f\|_p \le 1\}$. Then the following statements are true.*

(i) $C_\varphi(\mathcal{N}_p) \subset \mathcal{N}_q^0$ and

$$\lim_{|a|\to 1^-} \sup_{f\in\mathbf{B}_{\mathcal{N}_p}} \int_\mathbb{B} |f(\varphi(z))|^2 (1 - |\Phi_a(z)|^2)^q \, dV(z) = 0. \tag{10.6}$$

(ii) For any decreasing sequence $\{A_k\}_{k\ge 1}$ of measurable subsets of the unit ball whose intersection is empty, we have

$$\lim_{k\to\infty} \sup_{f\in\mathbf{B}_{\mathcal{N}_p}} \left[\sup_{a\in\mathbb{B}} \int_{A_k} |f(\varphi(z))|^2 (1 - |\Phi_a(z)|^2)^q \, dV(z) \right] = 0. \tag{10.7}$$

Proof We first prove that $C_\varphi(\mathcal{N}_p)$ is a subset of $\mathcal{N}_q^0$. Let f be in $\mathcal{N}_p$. For any integer $m \geq 1$, put $f_m(z) = f(mz/(m+1))$. Then each f_m belongs to H^∞, and the sequence (f_m) is bounded on $\mathcal{N}_p$ and converges to f uniformly on compact subsets of $\mathbb{B}$. By the compactness criteria mentioned above, the sequence $(C_\varphi f_m)$ converges to $C_\varphi f$ in $\mathcal{N}_q$ as $m \to \infty$. Since each function $C_\varphi f_m$ belongs to $H^\infty \subset \mathcal{N}_q^0$ and $\mathcal{N}_q^0$ is a closed subspace of $\mathcal{N}_p$, we conclude that $C_\varphi f$ is an element in $\mathcal{N}_q^0$. Since f was arbitrary, it follows that the image of $\mathcal{N}_p$ under C_φ is contained in $\mathcal{N}_q^0$.

Let $\varepsilon > 0$ be given. Since C_φ is compact and its range is contained in $\mathcal{N}_q^0$, the image $C_\varphi(\mathbf{B}_{\mathcal{N}_p})$ is pre-compact in $\mathcal{N}_q^0$. Therefore, $C_\varphi(\mathbf{B}_{\mathcal{N}_p})$ can be covered by finitely many $\sqrt{\varepsilon}/2$-balls. That is, there exists a finite set $\{f_1, \dots, f_M\} \subset \mathbf{B}_{\mathcal{N}_p}$ such that for any $f \in \mathbf{B}_{\mathcal{N}_p}$, there is a number $j \in \{1, 2, \dots, M\}$ for which

$$\|C_\varphi(f) - C_\varphi(f_j)\|_q^2 < \frac{\varepsilon}{4}. \tag{10.8}$$

On the other hand, since $\{f_1 \circ \varphi, \dots, f_M \circ \varphi\}$ is contained in $\mathcal{N}_q^0$, there exists $0 < r < 1$ such that for all $1 \leq j \leq M$ and $|a| > r$,

$$\int_{\mathbb{B}} |f_j(\varphi(z))|^2 (1 - |\Phi_a(z)|^2)^q \, dV(z) < \frac{\varepsilon}{4}. \tag{10.9}$$

For each $a \in \mathbb{B}$ with $|a| > r$ and $f \in \mathcal{N}_p$ with $\|f\|_p < 1$, choose $1 \leq j \leq M$ such that (10.8) holds. Combining with (10.9), we have

$$\int_{\mathbb{B}} |f(\varphi(z))|^2 (1 - |\Phi_a(z)|^2)^q \, dV(z)$$

$$\leq 2 \int_{\mathbb{B}} \left(|f(\varphi(z)) - f_j(\varphi(z))|^2 + |f_j(\varphi(z))|^2 \right) (1 - |\Phi_a(z)|^2)^q \, dV(z)$$

$$= 2\|C_\varphi(f) - C_\varphi(f_j)\|_q^2 + 2 \int_{\mathbb{B}} |f_j(\varphi(z))|^2 (1 - |\Phi_a(z)|^2)^q \, dV(z)$$

$$< 2\left(\frac{\varepsilon}{2} + \frac{\varepsilon}{2}\right) = \varepsilon.$$

This shows that for all $r < |a| < 1$,

$$\sup_{f \in \mathbf{B}_{\mathcal{N}_p}} \int_{\mathbb{B}} |f(\varphi(z))|^2 (1 - |\Phi_a(z)|^2)^q \, dV(z) \leq \varepsilon,$$

which implies (10.6). This finishes the proof of part (i).

Now let $\{A_k\}$ be a decreasing sequence of measurable subsets of $\mathbb{B}$ whose intersection is empty. Since $\{f_1 \circ \varphi, \dots, f_M \circ \varphi\}$ is contained in $\mathcal{N}_q^0$, Lemma 9.3.3 shows that there exists an integer k_ε such that for any $k \geq k_\varepsilon$ and any $1 \leq j \leq M$,

$$\sup_{a\in\mathbb{B}} \int_{A_k} |f_j(\varphi(z))|^2 (1 - |\Phi_a(z)|^2)^q \, dV(z) < \frac{\varepsilon}{4}. \tag{10.10}$$

Inequalities (10.8) and (10.10) together give

$$\sup_{a\in\mathbb{B}} \int_{A_k} |f(\varphi(z))|^2 (1 - |\Phi_a(z)|^2)^q \, dV(z)$$

$$\leq 2 \sup_{a\in\mathbb{B}} \int_{A_k} \left(|f(\varphi(z)) - f_j(\varphi(z))|^2 + |f_j(\varphi(z))|^2 \right) (1 - |\Phi_a(z)|^2)^q \, dV(z)$$

$$\leq 2\|C_\varphi(f) - C_\varphi(f_j)\|_q^2 + 2 \sup_{a\in\mathbb{B}} \int_{A_k} |f_j(\varphi(z))|^2 (1 - |\Phi_a(z)|^2)^q \, dV(z)$$

$$< 2\left(\frac{\varepsilon}{4} + \frac{\varepsilon}{4}\right) = \varepsilon,$$

for any $k \geq k_\varepsilon$ and $f \in \mathbf{B}_{\mathcal{N}_p}$. The limit (10.7) then follows. $\qquad\square$

We are now ready to give a complete characterization of the compactness of C_φ acting from $\mathcal{N}_p$ into $\mathcal{N}_q$.

Theorem 10.3.2 *Let p, q be positive numbers and φ a holomorphic self-map of $\mathbb{B}$ such that the composition operator $C_\varphi : \mathcal{N}_p \longrightarrow \mathcal{N}_q$ is bounded. Let $E_1 \subset E_2 \subset \cdots \subset \mathbb{B}$ be an increasing sequence of measurable sets such that $\cup_{k\geq 1} E_k = \mathbb{B}$ and for each k, the closure $\overline{\varphi(E_k)}$ is compact in $\mathbb{B}$. Then $C_\varphi : \mathcal{N}_p \longrightarrow \mathcal{N}_q$ is compact if and only if*

$$\lim_{k\to\infty} \sup_{f\in\mathbf{B}_{\mathcal{N}_p}} \left[\sup_{a\in\mathbb{B}} \int_{\mathbb{B}\setminus E_k} |f(\varphi(z))|^2 (1 - |\Phi_a(z)|^2)^q \, dV(z) \right] = 0. \tag{10.11}$$

Proof *Necessity* Set $A_k = \mathbb{B}\setminus E_k$ for all integers $k \geq 1$. Then $\{A_k\}_{k\geq 1}$ is a decreasing sequence of measurable subsets of $\mathbb{B}$ and

$$\bigcap_{k\geq 1} A_k = \mathbb{B}\setminus\left(\bigcup_{k\geq 1} E_k\right) = \emptyset.$$

If C_φ is a compact operator from $\mathcal{N}_p$ into $\mathcal{N}_q$, then (10.11) follows from Lemma 10.3.1(ii).
Sufficiency Suppose (10.11) holds. Take any bounded sequence $(f_m) \subset \mathcal{N}_p$ converging to zero uniformly on every compact subset of $\mathbb{B}$. Again, by the compactness criteria mentioned above, it suffices to show that the sequence $(\|f_m \circ \varphi\|_q)$ converges to zero as $m \to \infty$.

Let $\varepsilon > 0$ be given. By (10.11), there exists a positive integer k such that for any $m \in \mathbb{N}$,

$$\sup_{a\in\mathbb{B}} \int_{\mathbb{B}\setminus E_k} |f_m(\varphi(z))|^2 (1 - |\Phi_a(z)|^2)^q \, dV(z) < \frac{\varepsilon}{2}. \tag{10.12}$$

On the other hand, since (f_m) converges to zero uniformly on the compact set $\overline{\varphi(E_k)}$, for sufficiently large m, we have

$$\sup_{a\in\mathbb{B}} \int_{E_k} |f_m(\varphi(z))|^2(1-|\Phi_a(z)|^2)^q \, dV(z) \le \int_{E_k} |f_m(\varphi(z))|^2 \, dV(z)$$

$$\le \sup_{z\in E_k} |f_m(\varphi(z))|^2$$

$$= \sup_{w\in\varphi(E_k)} |f_m(w)|^2 < \frac{\varepsilon}{2}.$$

This estimate together with (10.12) immediately yields

$$\|C_\varphi f_m\|_p^2 = \sup_{a\in\mathbb{B}} \int_{\mathbb{B}} |f_m(\varphi(z))|^2(1-|\Phi_a(z)|^2)^q \, dV(z) < \varepsilon,$$

for sufficiently large integers m. Consequently, $\|C_\varphi f_m\|_q \to 0$ as desired. $\qquad\square$

By weakening condition (10.11) and adding an extra condition, we obtain an equivalent characterization for the compactness of C_φ as follows.

Theorem 10.3.3 *Let p,q be positive numbers and φ a holomorphic self-map of $\mathbb{B}$ such that the composition operator $C_\varphi : \mathcal{N}_p \longrightarrow \mathcal{N}_q$ is bounded. Let $E_1 \subset E_2 \subset \cdots \subset \mathbb{B}$ be an increasing sequence of measurable sets such that $\cup_{k\ge1} E_k = \mathbb{B}$ and for each k, the closure $\overline{\varphi(E_k)}$ is compact in $\mathbb{B}$. Then the following statements are equivalent.*

(i) C_φ is compact from $\mathcal{N}_p$ into $\mathcal{N}_q$.
(ii) C_φ is compact from $\mathcal{N}_p$ into $\mathcal{N}_q^0$.
(iii) The following two conditions are satisfied:

$$\lim_{k\to\infty} \left(\sup_{f\in\mathbf{B}_{\mathcal{N}_p}} \left[\int_{\mathbb{B}\backslash E_k} |f(\varphi(z))|^2(1-|z|^2)^q \, dV(z) \right] \right) = 0; \tag{10.13}$$

and

$$\lim_{|a|\to1^-} \sup_{f\in\mathbf{B}_{\mathcal{N}_p}} \int_{\mathbb{B}} |f(\varphi(z))|^2(1-|\Phi_a(z)|^2)^q \, dV(z) = 0. \tag{10.14}$$

Proof The equivalence of (i) and (ii) stems from the fact that $\mathcal{N}_q^0$ is a subspace of $\mathcal{N}_q$ and Lemma 10.3.1(i), which shows that whenever $C_\varphi : \mathcal{N}_p \longrightarrow \mathcal{N}_q$ is compact, the range of C_φ is actually contained in $\mathcal{N}_q^0$. We only need to prove the equivalence of (i) and (iii).

Suppose (i) holds, that is, $C_\varphi : \mathcal{N}_p \rightarrow \mathcal{N}_q$ is compact. Then (10.13) follows from Lemma 10.3.1(ii) with $A_k = \mathbb{B} \backslash E_k$ and $a = 0$. In addition, (10.14) follows from Lemma 10.3.1(i).

Now suppose (c) hold, that is, both (10.13) and (10.14) are satisfied. Take any bounded sequence $(f_m) \subset \mathcal{N}_p$ converging to zero uniformly on every compact subset of $\mathbb{B}$. We may assume that $\|f_m\|_p \leq 1$ for each $m \in \mathbb{N}$. To show C_φ is compact, it suffices to show that

$$\lim_{m \to \infty} \|C_\varphi(f_m)\|_q = 0.$$

Let $\varepsilon > 0$ be given. By (10.14), there exists $0 < \delta < 1$ such that

$$\sup_{\delta < |a| < 1} \left(\sup_{f \in \mathbf{B}_{\mathcal{N}_p}} \int_{\mathbb{B}} |f(\varphi(z))|^2 (1 - |\Phi_a(z)|^2)^q dV(z) \right) < \frac{\varepsilon}{3}. \tag{10.15}$$

By (10.13), there exists an integer $k \geq 1$ such that

$$\sup_{f \in \mathbf{B}_{\mathcal{N}_p}} \left[\int_{\mathbb{B} \backslash E_k} |f(\varphi(z))|^2 (1 - |z|^2)^q dV(z) \right] < \frac{(1 - \delta)^{2q} \cdot \varepsilon}{3}. \tag{10.16}$$

If $a \in \mathbb{B}$ with $|a| \leq \delta$, then

$$1 - |\Phi_a(z)|^2 = \frac{(1 - |a|^2)(1 - |z|^2)}{|1 - \langle z, a \rangle|^2} \leq \frac{1 - |z|^2}{(1 - \delta)^2}.$$

This together with (10.16) implies

$$\sup_{f \in \mathbf{B}_{\mathcal{N}_p}} \left(\sup_{|a| \leq \delta} \int_{\mathbb{B} \backslash E_k} |f(\varphi(z))|^2 (1 - |\Phi_a(z)|^2)^q dV(z) \right)$$

$$\leq \frac{1}{(1 - \delta)^{2q}} \sup_{f \in \mathbf{B}_{\mathcal{N}_p}} \int_{\mathbb{B} \backslash E_k} |f(\varphi(z))|^2 (1 - |z|^2)^q dV(z) < \frac{\varepsilon}{3}. \tag{10.17}$$

Since (f_m) converges to zero uniformly on the compact set $\overline{\varphi(E_k)}$, there exists $m_0 \in \mathbb{N}$ such that whenever $m > m_0$,

$$\sup_{a \in \mathbb{B}} \left(\int_{E_k} |f_m(\varphi(z))|^2 (1 - |\Phi_a(z)|^2)^q dV(z) \right) \leq \int_{E_k} |f_m(\varphi(z))|^2 dV(z)$$

$$\leq \sup_{w \in \varphi(E_k)} |f_m(w)| < \frac{\varepsilon}{3}. \tag{10.18}$$

For $m > m_0$, by (10.15), (10.16), and (10.18), we have

$$\|C_\varphi(f_m)\|_q^2 = \sup_{a\in\mathbb{B}} \int_\mathbb{B} |f_m(\varphi(z))|^2(1-|\Phi_a(z)|^2)^q dV(z)$$

$$\leq \sup_{\delta<|a|<1} \int_\mathbb{B} |f_m(\varphi(z))|^2(1-|\Phi_a(z)|^2)^q dV(z)$$

$$+ \sup_{|a|\leq\delta} \int_\mathbb{B} |f_m(\varphi(z))|^2(1-|\Phi_a(z)|^2)^q dV(z)$$

$$\leq \sup_{\delta<|a|<1} \int_\mathbb{B} |f_m(\varphi(z))|^2(1-|\Phi_a(z)|^2)^q dV(z)$$

$$+ \sup_{|a|\leq\delta} \int_{\mathbb{B}\setminus E_k} |f_m(\varphi(z))|^2(1-|\Phi_a(z)|^2)^q dV(z)$$

$$+ \sup_{|a|\leq\delta} \int_{E_k} |f_m(\varphi(z))|^2(1-|\Phi_a(z)|^2)^q dV(z)$$

$$< \varepsilon.$$

It follows that $\|C_\varphi f_m\|_q \to 0$ as required. $\square$

By using Theorem 10.3.2 with certain choices of the sets $\{E_k\}_{k\geq1}$, we obtain somewhat more concrete criteria for the compactness of C_φ.

Corollary 10.3.4 *Let p, q be positive numbers and φ a holomorphic self-map of $\mathbb{B}$ such that the composition operator $C_\varphi : \mathcal{N}_p \longrightarrow \mathcal{N}_q$ is bounded. Then the following statements are equivalent.*

(i) C_φ is a compact operator from $\mathcal{N}_p$ into $\mathcal{N}_q$.
(ii) $\lim\limits_{t\to1^-} \left(\sup_{a\in\mathbb{B},\, f\in\mathbf{B}_{\mathcal{N}_p}} \left[\int_{|z|>t} |f(\varphi(z))|^2(1-|\Phi_a(z)|^2)^q\, dV(z) \right] \right) = 0.$
(iii) $\lim\limits_{t\to1^-} \left(\sup_{a\in\mathbb{B},\, f\in\mathbf{B}_{\mathcal{N}_p}} \left[\int_{|\varphi(z)|>t} |f(\varphi(z))|^2(1-|\Phi_a(z)|^2)^q\, dV(z) \right] \right) = 0.$

Proof Observe that statement (ii), and, respectively, statement (iii), is equivalent to the statement that for any sequence $\{t_k\}_{k\geq1}$ of positive numbers increasing to 1, we have

$$\lim_{k\to\infty} \left(\sup_{a\in\mathbb{B},\, f\in\mathbf{B}_{\mathcal{N}_p}} \left[\int_{|z|>t_k} |f(z)|^2(1-|\Phi_a(z)|^2)^q\, dV(z) \right] \right) = 0,$$

and, respectively,

$$\lim_{k\to\infty} \left(\sup_{a\in\mathbb{B},\, f\in\mathbf{B}_{\mathcal{N}_p}} \left[\int_{|\varphi(z)|>t_k} |f(z)|^2(1-|\Phi_a(z)|^2)^q\, dV(z) \right] \right) = 0.$$

For each integer $k \geq 1$, define $E_k = \{z : |z| \leq t_k\}$ in the case of statement (ii) and $E_k = \{z : |\varphi(z)| \leq t_k\}$ in the case of statement (iii). Since $\{t_k\}_{k\geq 1}$ is increasing to 1, we see that $\{E_k\}_{k\geq 1}$ is an increasing sequence of measurable sets and $\bigcup_{k=1}^\infty E_k = \mathbb{B}$. Furthermore, it is clear that the set $\overline{\varphi(E_k)}$ is compact for each k. The equivalence of (i) and (ii) and the equivalence of (i) and (iii) now follow from Theorem 10.3.2.   □

Corollary 10.3.5 *Let φ be a holomorphic self-map of $\mathbb{B}$ such that $\|\varphi\|_\infty < 1$. Then $C_\varphi : \mathcal{N}_p \longrightarrow \mathcal{N}_q$ is compact for all $p, q > 0$.*

Proof By Corollary 10.2.3, the operator C_φ is bounded from $\mathcal{N}_p$ into $\mathcal{N}_q$. In addition, condition (iii) in Corollary 10.3.4 is clearly satisfied since the set $\{|\varphi(z)| > t\}$ is empty for all $\|\varphi\|_\infty < t < 1$. Consequently, C_φ is compact.   □

By changing the role of $\mathcal{N}_p$ to $\mathcal{N}_p^0$ in the proofs of Theorems 10.3.2 and 10.3.3, we immediately obtain the following result describing the compactness of composition operators acting between $\mathcal{N}_p^0$ and $\mathcal{N}_q$.

Theorem 10.3.6 *Let p, q be positive numbers and φ a holomorphic self-map of $\mathbb{B}$ such that the composition operator $C_\varphi : \mathcal{N}_p^0 \longrightarrow \mathcal{N}_q$ is bounded. Let $E_1 \subset E_2 \subset \cdots \subset \mathbb{B}$ be an increasing sequence of measurable sets such that $\bigcup_{k\geq 1} E_k = \mathbb{B}$ and for each k, the closure $\overline{\varphi(E_k)}$ is compact in $\mathbb{B}$. Then the following statements are equivalent.*

(i) C_φ is compact from $\mathcal{N}_p^0$ into $\mathcal{N}_q$.

(ii) C_φ is compact from $\mathcal{N}_p^0$ into $\mathcal{N}_q^0$.

(iii) $\lim_{k\to\infty} \sup_{f \in \mathbb{B}_{\mathcal{N}_p^0}} \left[\sup_{a\in\mathbb{B}} \int_{\mathbb{B}\setminus E_k} |f(\varphi(z))|^2 (1 - |\Phi_a(z)|^2)^q \, dV(z) \right] = 0.$

(iv) *The following two conditions are satisfied:*

$$\lim_{k\to\infty} \left(\sup_{f \in \mathbb{B}_{\mathcal{N}_p^0}} \left[\int_{\mathbb{B}\setminus E_k} |f(\varphi(z))|^2 (1 - |z|^2)^q \, dV(z) \right] \right) = 0; \qquad (10.19)$$

and

$$\lim_{|a|\to 1^-} \sup_{f \in \mathbb{B}_{\mathcal{N}_p^0}} \int_{\mathbb{B}} |f(\varphi(z))|^2 (1 - |\Phi_a(z)|^2)^q \, dV(z) = 0. \qquad (10.20)$$

Here, $\mathbb{B}_{\mathcal{N}_p^0} = \{f \in \mathcal{N}_p^0 : \|f\|_p \leq 1\}$ is the unit ball of $\mathcal{N}_p^0$.

10.4   A Sufficient Compactness Condition

Although Theorems 10.3.2, 10.3.3, and 10.3.6 offer several characterizations of the compactness of composition operators $C_\varphi : \mathcal{N}_p \longrightarrow \mathcal{N}_q$, the conditions are rather abstract and difficult to verify. We shall provide in this section a necessary condition and a sufficient condition for the compactness of C_φ directly in terms of φ. These conditions are more useful in applications.

Theorem 10.4.1 *Let $p, q \in (0, n]$ be two positive numbers and φ a holomorphic self-map of $\mathbb{B}$ such that $C_\varphi : \mathcal{N}_p \longrightarrow \mathcal{N}_q$ is bounded. If*

$$\lim_{t\to 1^-} \sup_{a\in\mathbb{B}} \int_{|\varphi(z)|>t} \frac{(1 - |\Phi_a(z)|^2)^q}{(1 - |\varphi(z)|^2)^{n+1}}\, dV(z) = 0, \tag{10.21}$$

then C_φ is compact. Conversely, if $C_\varphi : \mathcal{N}_p \to \mathcal{N}_q$ is compact, then for $0 < \varepsilon \le p + 1$,

$$\lim_{t\to 1^-} \sup_{a\in\mathbb{B}} \int_{|\varphi(z)|>t} \frac{(1 - |\Phi_a(z)|^2)^q}{(1 - |\varphi(z)|^2)^{p+1-\varepsilon}}\, dV(z) = 0. \tag{10.22}$$

Proof Suppose (10.21) holds. As in the proof of Theorem 10.2.2, there is a constant $C > 0$ such that for any $f \in \mathcal{N}_p$ with $\|f\|_p \le 1$,

$$|f(z)| \le C\|f\|_p (1 - |z|^2)^{-(n+1)/2} \le C(1 - |z|^2)^{-(n+1)/2}.$$

It implies that for any $0 < t < 1$ and any $a \in \mathbb{B}$,

$$\int_{|\varphi(z)|>t} |f(\varphi(z))|^2 (1 - |\Phi_a(z)|^2)^q \, dV(z) \le C^2 \int_{|\varphi(z)|>t} \frac{(1 - |\Phi_a(z)|^2)^q}{(1 - |z|^2)^{n+1}}\, dV(z).$$

Now (10.21) shows that statement (iii) in Corollary 10.3.4 is satisfied. As a result, C_φ is a compact operator from $\mathcal{N}_p$ into $\mathcal{N}_q$.

   Now suppose $C_\varphi : \mathcal{N}_p \longrightarrow \mathcal{N}_q$ is compact. Let $\mu > 0$ be given. By Corollary 10.3.4, there is a number $t_\mu \in (0, 1)$ such that for all $t_\mu < t < 1$,

$$\sup_{a\in\mathbb{B},\, f\in B\mathcal{N}_p} \int_{|\varphi(z)|>t} |f(\varphi(z))|^2 (1 - |\Phi_a(z)|^2)^q \, dV(z) \le \mu.$$

An application of Lemma 10.2.1 with $E = \{z \in \mathbb{B} : |\varphi(z)| > t\}$ yields

$$\sup_{a\in\mathbb{B}} \int_{|\varphi(z)|>t} \frac{(1 - |\Phi_a(z)|^2)^q}{(1 - |z|^2)^{p+1-\varepsilon}}\, dV(z) \lesssim \mu,$$

for all $t_\mu < t < 1$. Since μ is arbitrary, (10.22) follows. $\square$

As an application of Theorems 10.2.2 and 10.4.1, we mention the following result.

Corollary 10.4.2 *Suppose $k > 0, p, q, r \in (0, n), r \geq q, \varepsilon \in (0, q + 1)$ and φ is a holomorphic self-map of $\mathbb{B}$. The following statements hold:*

(i) If $C_\varphi : A^{-k(q+1-\varepsilon)} \longrightarrow A^{-k(n+1)}$ is a bounded operator, then $C_\varphi : \mathcal{N}_p \longrightarrow \mathcal{N}_r$ is a bounded operator.

(ii) If $C_\varphi : A^{-k(q+1-\varepsilon)} \longrightarrow A^{-k(n+1)}$ is a compact operator, then $C_\varphi : \mathcal{N}_p \longrightarrow \mathcal{N}_r$ is a compact operator.

Proof The boundedness of $C_\varphi : A^{-k(q+1-\varepsilon)} \longrightarrow A^{-k(n+1)}$ is equivalent to

$$M = \sup_{z \in \mathbb{B}} \frac{(1 - |z|^2)^{q+1-\varepsilon}}{(1 - |\varphi(z)|^2)^{n+1}} < \infty. \tag{10.23}$$

By Theorem 10.2.2, it suffices to show that

$$\sup_{a \in \mathbb{B}} \int_{\mathbb{B}} \frac{(1 - |\Phi_a(z)|^2)^r}{(1 - |\varphi(z)|^2)^{n+1}} \, dV(z) < \infty.$$

Indeed, we have

$$\sup_{a \in \mathbb{B}} \int_{\mathbb{B}} \frac{(1 - |\Phi_a(z)|^2)^r}{(1 - |\varphi(z)|^2)^{n+1}} \, dV(z) \leq M \sup_{a \in \mathbb{B}} \int_{\mathbb{B}} \frac{(1 - |\Phi_a(z)|^2)^r}{(1 - |z|^2)^{q+1-\varepsilon}} \, dV(z)$$

$$= M \sup_{a \in \mathbb{B}} (1 - |a|^2)^r \int_{\mathbb{B}} \frac{(1 - |z|^2)^{r-q-1+\varepsilon}}{|1 - \langle z, a \rangle|^{2r}} \, dV(z)$$

$$= M \sup_{a \in \mathbb{B}} (1 - |a|^2)^r \int_{\mathbb{B}} \frac{(1 - |z|^2)^{r-q-1+\varepsilon}}{|1 - \langle z, a \rangle|^{n+1+(r-q-1+\varepsilon)+(r+q-n-\varepsilon)}} \, dV(z).$$

Therefore,

$$\sup_{a \in \mathbb{B}} (1 - |a|^2)^r \int_{\mathbb{B}} \frac{(1 - |z|^2)^{r-q-1+\varepsilon}}{|1 - \langle z, a \rangle|^{n+1+(r-q-1+\varepsilon)+(r+q-n-\varepsilon)}} \, dV(z) < \infty,$$

which implies the desired result.

The proof of statement (ii) is similar to that of (i) by using the fact that $C_\varphi : A^{-k(q+1-\varepsilon)} \longrightarrow A^{-k(n+1)}$ is compact if and only if

$$\lim_{t \to 1^-} \sup_{|\varphi(z)| > t} \frac{(1 - |z|^2)^{q+1-\varepsilon}}{(1 - |\varphi(z)|^2)^{n+1}} = 0.$$

$\square$

Notes on Chapter 10

The results of this chapter are mainly from [44]. In fact, the one-dimensional case of these results was considered in [73]. A sufficient condition and a necessary condition for C_φ to be bounded on $\mathcal{N}_p$ of the unit disc were given there. These conditions involve the generalized Nevanlinna counting function introduced by Shapiro. The existence of a sequence (10.1) is proved in [81]. The estimate (10.2) is borrowed from [95].

$\mathcal{N}_p$-Type Functions with Hadamard Gaps in the Unit Ball $\mathbb{B}$

A function $f \in \mathrm{Hol}(\mathbb{B})$ written in the form

$$f(z) = \sum_{k=0}^{\infty} P_{n_k}(z),$$

where P_{n_k} is a homogeneous polynomial of degree n_k, is said to have *Hadamard gaps* (see, e.g., [89]) if, for some $c > 1$,

$$\frac{n_{k+1}}{n_k} \geq c$$

for all $k \geq 0$. Hadamard gap series on spaces of holomorphic functions in $\mathbb{D}$ or in $\mathbb{B}$ has been extensively studied. In this chapter, we study some aspects of Hadamard gap series in $\mathcal{N}_p$-spaces.

11.1 Some Estimates for Volume and Surface Integrals

In the sequel, we need some relationships between the volume measure dv on the unit ball $\mathbb{B}$, normalized so that $v(\mathbb{B}) = 1$, and the surface measure ds on the unit sphere $\mathbb{S}$, similarly normalized so that $s(\mathbb{S}) = 1$. The normalizing constants, namely, the actual volume of $\mathbb{B}$ and the actual surface area of $\mathbb{S}$, are in general not important. The following result is referred to as integration in polar coordinates.

Lemma 11.1.1 *The measures v and s are related by*

© The Author(s), under exclusive license to Springer Nature Switzerland AG 2023
L. H. Khoi, J. Mashreghi, *Theory of $\mathcal{N}_p$ Spaces*, Frontiers in Mathematics,
https://doi.org/10.1007/978-3-031-39704-2_11

$$\int_{\mathbb{B}} f(z)\, dv(z) = 2n \int_0^1 r^{2n-1}\, dr \int_{\mathbb{S}} f(r\zeta)\, ds(\zeta).$$

Proof Let $dV = dx_1 dy_1 \cdots dx_n dy_n$ be the actual Lebesgue measure in $\mathbb{C}^n$ before normalization, where each z_k is identified with $x_k + iy_k$ $(k = 1, 2, \ldots, n)$. Similarly, let dS be the actual surface measure on $\mathbb{S}$ before normalization. For $r > 0$, let $V(r)$ be the actual volume of the ball

$$|z_1|^2 + \cdots + |z_n|^2 < r^2$$

and $S(r)$ be the actual surface area of the sphere

$$|z_1|^2 + \cdots + |z_n|^2 = r^2.$$

Then the Euclidean volume of the solid determined by dS on $\mathbb{S}$, by $r > 0$, and $r + dr$, is given by

$$dV = \frac{dS}{S(1)}[V(r + dr) - V(r)].$$

Hence, by the change of variables $z_k = rw_k$ $(1 \le k \le n)$, we have

$$V(r) = \int_{|z_1|^2 + \cdots + |z_n|^2 < r^2} dV(z) = r^{2n} V(1),$$

from which it follows that

$$dV = \frac{V(1)}{S(1)}[(r + dr)^{2n} - r^{2n}]\, dS.$$

Omitting powers of rd with exponents greater than 1, we obtain

$$dV = \frac{V(1)}{S(1)} 2nr^{2n-1}\, dr\, dS$$

or, equivalently,

$$dv = 2nr^{2n-1}\, dr\, ds.$$

$\square$

The following estimates are needed too.

Lemma 11.1.2 *Let* $m = (m_1, \ldots, m_n)$ *be a multi-index of nonnegative integers and* $\alpha > -1$. *Then*

$$\int_{\mathbb{S}} |\zeta^m|^2 \, ds(\zeta) = \frac{(n-1)!m!}{(n-1+|m|)!} \tag{11.1}$$

and

$$\int_{\mathbb{B}} |z^m|^2 \, dv_\alpha(z) = \frac{m! \, \Gamma(n+\alpha+1)}{\Gamma(n+|m|+\alpha+1)}. \tag{11.2}$$

Proof Identifying $\mathbb{C}^n$ with $\mathbb{R}^{2n}$ and denoting by dV the usual Lebesgue measure on $\mathbb{C}^n$, we have $c \, dv = dV$, where c is the Euclidean volume of $\mathbb{B}$.

We evaluate the integral

$$I = \int_{\mathbb{C}^n} |z^m|^2 e^{-|z|^2} \, dV(z)$$

in two different ways.

First, we use Fubini's theorem to obtain

$$I = \prod_{k=1}^{n} \int_{\mathbb{R}^2} (x^2 + y^2)^{m_k} e^{-(x^2 + y^2)} \, dx \, dy$$

$$= \pi^n \prod_{k=1}^{n} \int_0^\infty r^{m_k} e^{-r} \, dr = \pi^n m!.$$

Next, we use Lemma 11.1.1 about integration in polar coordinates to get

$$I = 2nc \int_0^\infty r^{2|m|+2n-1} e^{-r^2} \, dr \int_{\mathbb{S}} |\zeta^m|^2 \, ds(\zeta)$$

$$= nc(|m|+n-1)! \int_{\mathbb{S}} |\zeta^m|^2 \, ds(\zeta).$$

Comparing the two results yields

$$\int_{\mathbb{S}} |\zeta^m|^2 \, ds(\zeta) = \frac{\pi^n m!}{nc(|m|+n-1)!}.$$

In particular, for $m = (0, \ldots, 0)$, we have

$$c = \frac{\pi^n}{n!}.$$

Consequently,

$$\int_S |\zeta^m|^2 \, ds(\zeta) = \frac{(n-1)!m!}{(|m|+n-1)!}.$$

Furthermore, again by Lemma 11.1.1, we have

$$\int_{\mathbb{B}} |z^m|^2 \, dv_\alpha(z) = 2nc_\alpha \int_0^1 r^{2|m|+2n-1}(1-r^2)^\alpha \, dr \int_S |\zeta^m|^2 \, ds(\zeta)$$

$$= nc_\alpha \int_0^1 r^{|m|+n-1}(1-r^2)^\alpha \, dr \cdot \frac{(n-1)!m!}{(|m|+n-1)!}.$$

Note that

$$\int_0^1 r^{|m|+n-1}(1-r^2)^\alpha \, dr = \frac{\Gamma(n+|m|)\Gamma(\alpha+1)}{\Gamma(n+|m|+\alpha+1)}.$$

Thus, (11.2) follows from (11.1).   □

We also need the following estimates (see (7.13) in Chap. 7).

$$\int_{\mathbb{B}} \frac{(1-|z|^2)^{p-2k}}{|1-\langle z,a\rangle|^{2p}} \, dV(z) \asymp \begin{cases} \text{bounded in } \mathbb{B}, & p < n+1-2k, \\ \log \frac{1}{1-|a|^2}, & p = n+1-2k, \\ (1-|a|^2)^{n+1-p-2k}, & p > n+1-2k, \end{cases} \quad (11.3)$$

where $0 < k \le \dfrac{n+1}{2}$. In any case, there exists a positive constant C such that

$$(1-|a|^2)^p \int_{\mathbb{B}} \frac{(1-|z|^2)^{p-2k}}{|1-\langle z,a\rangle|^{2p}} \, dV(z) \le C, \qquad (a \in \mathbb{B}). \quad (11.4)$$

11.2   Hadamard Gaps in $\mathcal{N}_p$-Spaces ($p \le n$)

In this section and the next one, we characterize the holomorphic functions with Hadamard gaps in $\mathcal{N}_p$-space. Here, our focus is on the case $p \in (0, n]$. Let us denote

$$M_k = \sup_{\xi \in S} |P_{n_k}(\xi)|$$

and

$$I_k = \left(\int_{\xi \in \mathbb{S}} |P_{n_k}(\xi)|^2 \, d\sigma(\xi) \right)^{1/2},$$

where $d\sigma$ is the normalized surface measure on $\mathbb{S}$, that is, $\sigma(\mathbb{S}) = 1$. Clearly, for each $k \geq 0$, M_k and I_k are well-defined.

Theorem 11.2.1 *Let $p \in (0, n]$, and let*

$$f(z) = \sum_{k=0}^{\infty} P_{n_k}(z)$$

be a series with Hadamard gaps. Consider the following statements.

(i)

$$\sum_{k=0}^{\infty} \left(\frac{1}{2^{k(1+p)}} \sum_{2^k \leq n < 2^{k+1}} |M_j|^2 \right) < \infty.$$

(ii)

$$f \in \mathcal{N}_p^0.$$

(iii)

$$f \in \mathcal{N}_p.$$

(iv)

$$\sum_{k=0}^{\infty} \left(\frac{1}{2^{k(1+p)}} \sum_{2^k \leq n < 2^{k+1}} |I_j|^2 \right) < \infty.$$

We have (i) $\implies$ (ii) $\implies$ (iii) $\implies$ (iv).

Proof (i) $\implies$ (ii). Suppose (i) holds. We first prove that $f \in \mathcal{N}_p$. Using the integration in polar coordinates described in Lemma 11.1.1, we have

$$\|f\|_p^2 = \sup_{a \in \mathbb{B}} \int_{\mathbb{B}} \left| \sum_{k=0}^{\infty} P_{n_k}(z) \right|^2 (1 - |\Phi_a(z)|^2)^p \, dV(z)$$

$$\leq \sup_{a\in\mathbb{B}} \int_{\mathbb{B}} \left(\sum_{k=0}^{\infty} |P_{n_k}(z)|\right)^2 \frac{(1-|a|^2)^p (1-|z|^2)^p}{|1-\langle z,a\rangle|^{2p}} \, dV(z)$$

$$= \sup_{a\in\mathbb{B}} \left\{ (1-|a|^2)^p \int_{\mathbb{B}} \left(\sum_{k=0}^{\infty} |P_{n_k}(z)|\right)^2 \right.$$

$$\left. \times \frac{(1-|z|^2)^p}{|1-\langle z,a\rangle|^{2p}} \, dV(z) \right\}$$

$$\leq 2n \sup_{a\in\mathbb{B}} \left\{ (1-|a|^2)^p \int_0^1 \left(\sum_{k=0}^{\infty} |P_{n_k}(r\xi)|\right)^2 \right.$$

$$\left. \times (1-r^2)^p \left(\int_{\mathbb{S}} \frac{1}{|1-\langle r\xi,a\rangle|^{2p}} \, d\sigma(\xi)\right) dr \right\}$$

$$= 2n \sup_{a\in\mathbb{B}} \left\{ (1-|a|^2)^p \int_0^1 \left(\sum_{k=0}^{\infty} |P_{n_k}(\xi) r^{n_k}|\right)^2 \right.$$

$$\left. \times (1-r^2)^p \left(\int_{\mathbb{S}} \frac{1}{|1-\langle r\xi,a\rangle|^{2p}} \, d\sigma(\xi)\right) dr \right\}$$

$$\leq 2n \sup_{a\in\mathbb{B}} \left\{ (1-|a|^2)^p \int_0^1 \left(\sum_{k=0}^{\infty} M_k r^{n_k}\right)^2 (1-r^2)^p \left(\int_{\mathbb{S}} \frac{1}{|1-\langle r\xi,a\rangle|^{2p}} \, d\sigma(\xi)\right) dr \right\}.$$

Applying (11.3), for each $a \in \mathbb{B}$ and $r \in [0,1]$, we have

$$\int_{\mathbb{S}} \frac{1}{|1-\langle r\xi,a\rangle|^{2p}} \, d\sigma(\xi) = \int_{\mathbb{S}} \frac{1}{|1-\langle \xi,ar\rangle|^{2p}} \, d\sigma(\xi) = \int_{\mathbb{S}} \frac{1}{|1-\langle ar,\xi\rangle|^{n+(2p-n)}} \, d\sigma(\xi)$$

$$= \begin{cases} \text{bounded in } \mathbb{B}, & 0 < p < \frac{n}{2}, \\ \log\frac{1}{1-r^2|a|^2} \leq \log\frac{1}{1-|a|^2}, & p = \frac{n}{2}, \\ (1-r^2|a|^2)^{n-2p} \leq (1-|a|^2)^{n-2p}, & \frac{n}{2} < p < n. \end{cases}$$

By (11.4), for all cases of p, we have

$$(1-|a|^2)^p \int_{\mathbb{S}} \frac{1}{|1-\langle r\xi,a\rangle|^{2p}} \, d\sigma(\xi) \leq M, \quad a \in \mathbb{B},$$

where $M > 0$ is independent of both a and r. Thus, applying Theorem 3.5.1, we get

$$\|f\|_{\mathcal{N}_p}^2 \leq 2nM \int_0^1 \left(\sum_{k=0}^{\infty} M_k r^{n_k}\right)^2 (1-r^2)^p \, dr$$

$$\asymp 2nM \sum_{k=0}^{\infty} \frac{1}{2^{k(1+p)}} \left(\sum_{2^k \leq n_j < 2^{k+1}} M_j\right)^2.$$

Since f is in the Hadamard gap class, there exists a constant $c > 1$ such that $n_{j+1} \geq cn_j$ for all $j \geq 0$. Let us denote

$$N_{a,b}(n) = \sharp\{n : a \leq n < b\},$$

where $0 < a < b$ and $n \geq 0$. Then the maximum number of n_j between 2^k and 2^{k+1} is less than or equal to $[\log_c 2] + 1$, for all $k \geq 0$, i.e.,

$$N_{2^k, 2^{k+1}}(n_j) \leq [\log_c 2] + 1 \quad (k = 0, 1, 2, \ldots).$$

Note that, for every $k \geq 0$, by the Cauchy–Schwarz inequality,

$$\left(\sum_{2^k \leq n_j < 2^{k+1}} M_j\right)^2 \leq ([\log_c 2] + 1) \left(\sum_{2^k \leq n_j < 2^{k+1}} M_j^2\right).$$

Hence,

$$\|f\|_{\mathcal{N}_p}^2 \lesssim \sum_{k=0}^{\infty} \left(\frac{1}{2^{k(1+p)}} \sum_{2^k \leq n_j < 2^{k+1}} M_j^2\right) < \infty, \tag{11.5}$$

which shows that $f \in \mathcal{N}_p$.

Next we prove that $f \in \mathcal{N}_p^0$. For this, put

$$A = \int_{\mathbb{B}} (1 - |\Phi_a(z)|^2)^p \, dV(z).$$

We need to show that $A \to 0$ as $|a| \to 1^-$. Indeed, we have

$$\int_{\mathbb{B}} (1 - |\Phi_a(z)|^2)^p \, dV(z) = \int_{\mathbb{B}} \frac{(1 - |a|^2)^p (1 - |z|^2)^p}{|1 - \langle z, a\rangle|^{2p}} \, dV(z)$$

$$= (1 - |a|^2)^p \int_{\mathbb{B}} \frac{(1 - |z|^2)^p}{|1 - \langle z, a\rangle|^{n+1+p+p-n-1}} \, dV(z),$$

and hence, by (11.3),

$$\int_{\mathbb{B}} \frac{(1-|z|^2)^p}{|1-\langle z, a\rangle|^{n+1+p+p-n-1}} \, dV(z) \asymp \begin{cases} \text{bounded in } \mathbb{B}, & 0 < p < n+1, \\ \log \frac{1}{1-|a|^2}, & p = n+1, \\ (1-|a|^2)^{1+n-p}, & p > n+1. \end{cases}$$

In all cases, $A \to 0$, as $a \to 1^-$.
    Now put

$$f_m(z) = \sum_{k=0}^{m} P_{n_k}(z), \quad m\mathbb{N} \quad \text{and} \quad K_m = \max\{M_0, M_1, \dots, M_m\}.$$

Note that for each $a \in \mathbb{B}$,

$$\int_{\mathbb{B}} |f_m(z)|^2 (1-|\Phi_a(z)|^2)^p \, dV(z)$$

$$\leq \int_{\mathbb{B}} \left(\sum_{k=0}^{m} |P_{n_k}(z)|\right)^2 (1-|\Phi_a(z)|^2)^p \, dV(z)$$

$$\leq m^2 K_m^2 \int_{\mathbb{B}} (1-|\Phi_a(z)|^2)^p \, dV(z),$$

which tends to 0 as $|a| \to 1^{-1}$. This shows that $f \in \mathcal{N}_p^0$. Moreover, since $\mathcal{N}_p^0$ is closed and the set of all polynomials is dense in $\mathcal{N}_p^0$, it suffices to show that $\|f_m - f\|_{\mathcal{N}_p} \to 0$ as $m \to \infty$. We have

$$\|f_m - f\|_{\mathcal{N}_p}^2 \lesssim \sum_{k=m'}^{\infty} \sum_{k=0}^{\infty} \left(\frac{1}{2^{k(1+p)}} \sum_{2^k \leq n_j < 2^{k+1}} M_j^2\right), \tag{11.6}$$

where $m' = \left[\frac{m+1}{[\log_c 2]+1}\right]$. Then from condition (i) and (11.6), the desired result follows.
$(ii) \implies (iii)$. This is trivial.
$(iii) \implies (iv)$. Suppose $f \in \mathcal{N}_p$. Then we have

$$\|f\|_{\mathcal{N}_p}^2 = \sup_{a \in \mathbb{B}} \int_{\mathbb{B}} \left|\sum_{k=0}^{\infty} P_{n_k}(z)\right|^2 (1-|\Phi_a(z)|^2)^p \, dV(z)$$

$$\geq \int_{\mathbb{B}} \left|\sum_{k=0}^{\infty} P_{n_k}(z)\right|^2 (1-|z|^2)^p \, dV(z)$$

$$\asymp \int_{\mathbb{S}} \sum_{k=0}^{\infty} \left(\frac{1}{2^{k(1+p)}} \sum_{2^k \le n_j < 2^{k+1}} |P_{n_k}(\xi)|^2 \right) d\sigma(\xi)$$

$$= \sum_{k=0}^{\infty} \left(\frac{1}{2^{k(1+p)}} \sum_{2^k \le n_j < 2^{k+1}} I_j^2 \right),$$

which completes the proof of the theorem.   □

Generally, when $n > 1$, the above conditions in Theorem 11.2.1 are not equivalent. As a matter of fact, we show that $(iv) \nRightarrow (i)$. Put

$$f(z) = \sum_{k=0}^{\infty} 2^{\frac{k(p+1)}{2}} z_1^{2^k}, \quad z = (z_1, \dots, z_n) \in \mathbb{B}.$$

Since $M_k = 2^{\frac{k(p+1)}{2}}$, we have

$$\sum_{k=0}^{\infty} \left(\frac{1}{2^{k(1+p)}} \sum_{2^k \le n_j < 2^{k+1}} M_k^2 \right) = \infty.$$

On the other hand, for each $k \ge 0$, we have

$$I_k^2 = 2^{k(p+1)} \int_{z \in \mathbb{S}} |z_1^{2^k}|^2 d\sigma(z) = 2^{k(p+1)} \cdot \frac{(n-1)!(2^k)!}{(n-1+2^k)!} \asymp \frac{2^{k(p+1)}}{2^{k(n-1)}},$$

which implies

$$\sum_{k=0}^{\infty} \left(\frac{1}{2^{k(p+1)}} \sum_{2^k \le n_j < 2^{k+1}} I_k^2 \right) \asymp \sum_{k=0}^{\infty} \frac{1}{2^{k(n-1)}} < \infty.$$

Now, we consider some special cases where all the conditions in Theorem 11.2.1 are equivalent. As already mentioned in (9.21) of Chap. 9, there is a sequence of homogeneous polynomial $\{T_k\}_{k \in \mathbb{N}}$ satisfying $\deg(T_k) = k$,

$$\sup_{\xi \in \mathbb{S}} |T_k(\xi)| = 1 \quad \text{and} \quad \int_{\xi \in \mathbb{S}} |T_k(\xi)|^2 d\sigma(\xi) \ge \frac{\pi}{2^{2n}}. \tag{11.7}$$

An immediate corollary of the theorem above is stated as follows.

Corollary 11.2.2 *Let $p \in (0, n]$, and let*

$$f(z) = \sum_{k=0}^{\infty} a_k T_{n_k}(z)$$

be a series with Hadamard gaps, where $a_k \in \mathbb{C}$, $k \geq 0$. Then the following statements are equivalent:

(i)

$$\sum_{k=0}^{\infty} \left(\frac{1}{2^{k(1+p)}} \sum_{2^k \leq n < 2^{k+1}} |a_j|^2 \right) < \infty.$$

(ii)

$$f \in \mathcal{N}_p^0.$$

(iii)

$$f \in \mathcal{N}_p.$$

Proof The desired result follows from the fact that for each $k \geq 0$, $M_k \asymp I_k$.   □

The special case $n = 1$ leads to the following corollary describing the functions in $\mathcal{N}_p(\mathbb{D})$ with Hadamard gaps.

Corollary 11.2.3 *Let $p \in (0, 1]$, and let*

$$f(z) = \sum_{k=0}^{\infty} b_k T_{n_k}(z)$$

be a series with Hadamard gaps, where $b_k \in \mathbb{C}$, $k \geq 0$. Then the following conditions are equivalent.

(i) $f \in \mathcal{N}_p(\mathbb{D})$.
(ii) $f \in \mathcal{N}_p^0(\mathbb{D})$.
(iii)

$$\sum_{k=0}^{\infty} \left(\frac{1}{2^{k(1+p)}} \sum_{2^k \leq n < 2^{k+1}} |b_j|^2 \right) < \infty.$$

Proof The desired result follows from the fact that when $n = 1$, $M_j = I_j = |b_j|$.   □

11.3 Hadamard Gaps in $\mathcal{N}_p$-Spaces $(p > n)$

Recall that when $p > n$, all $\mathcal{N}_p$-spaces coincide with $A^{-\frac{n+1}{2}}$. In this section, we consider a more general question about the Hadamard gap series in A^{-l} for any $l > 0$.

Theorem 11.3.1 *Let $l > 0$ and*

$$f(z) = \sum_{k=0}^{\infty} P_{n_k}(z)$$

be a series with Hadamard gaps, where P_{n_k} is a homogeneous polynomial of degree n_k. The following assertions hold.

(i) $f \in A^{-l}$ if and only if

$$\sup_{k \geq 1} \frac{M_k}{M_\ell} < \infty.$$

(ii) $f \in A_0^{-\ell}$ if and only if

$$\lim_{k \to \infty} \frac{M_k}{M_\ell} = 0.$$

Proof (i)-**Necessity** Suppose that $f \in A^{-\ell}$. Fix $\xi \in \mathbb{S}$ and put

$$f_\xi(w) \sum_{k=0}^{\infty} P_{n_k}(\xi) w^{n_k} = \sum_{k=0}^{\infty} P_{n_k}(w\xi), \qquad (\xi \in \mathbb{D}).$$

Since $f \in \text{Hol}(\mathbb{B})$, for a fixed $\xi \in \mathbb{S}$, the function $f_\xi(w)$ is holomorphic in $\mathbb{D}$. Thus, for any $r \in (0, 1)$, we have

$$M_k = \sup_{\xi \in \mathbb{S}} |P_{n_k}(\xi)|$$

$$= \sup_{\xi \in \mathbb{S}} \left| \frac{1}{2\pi i} \int_{|w|=r} \frac{f_\xi(w)}{w^{n_k+1}} \, dw \right|$$

$$= \frac{1}{2\pi} \sup_{\xi \in \mathbb{S}} \left| \int_{|w|=r} \frac{f(\xi w)}{w^{n_k+1}} \, dw \right|$$

$$\leq \frac{1}{2\pi} \sup_{\xi \in \mathbb{S}} \int_{|w|=r} \frac{|f(\xi w)|}{r^{n_k+1}} |dw|$$

$$\leq \frac{1}{2\pi} \sup_{\xi \in \mathbb{S}} \int_{|w|=r} \frac{|f(\xi w)|(1-\xi w|^2)^\ell}{r^{n_k+1}(1-r)^\ell} |dw|$$

$$\leq \frac{|f|_\ell}{r^{n_k+1}(1-r)^\ell}. \tag{11.8}$$

Letting $r = 1 - \frac{1}{n_k}$ in (11.8) gives

$$M_k \leq \frac{|f|_\ell n_k^\ell}{(1-\frac{1}{n_k})^{n_k}}.$$

This means that for each $k \geq 2$,

$$\frac{M_k}{n_k^\ell} \leq \frac{|f|_\ell n_k^\ell}{(1-\frac{1}{n_k})^{n_k}} \leq 4|f|_\ell.$$

Hence,

$$\sup_{k \geq 1} \frac{M_k}{n_k^\ell} \leq \max\left\{\frac{M_1}{n_1^\ell}, 4|f|_\ell\right\} < \infty.$$

(i)-**Sufficiency** Suppose that $\sup_{k \geq 1} M_k/M_\ell < \infty$. We have

$$|f(z)| = \left|\sum_{k=0}^{\infty} P_{n_k}\left(\frac{z}{|z|}\right)|z|^{n_k}\right| \leq \sum_{k=0}^{\infty} M_k |z|^{n_k} \lesssim \sum_{k=0}^{\infty} n_k^\ell |z|^{n_k}.$$

In this case,

$$\frac{|f(z)|}{1-|z|} \lesssim \left(\sum_{k=0}^{\infty} n_k^\ell |z|^{n_k}\right)\left(\sum_{s=0}^{\infty} |z|^s\right) = \sum_{t=0}^{\infty} \left(\sum_{n_j \leq \ell} n_j^\ell\right) |z|^t.$$

Note that

$$\lim_{k \to \infty} \frac{k^\ell k!}{\ell(\ell+1)\cdots(\ell+k)} = \Gamma(\ell), \quad \ell > 0.$$

Then we have

$$\sup_{k \in \mathbb{N}} \left(\frac{k^\ell k!}{(k+\ell)(k+\ell-1)\cdots(\ell+1)} \right) \leq M,$$

where $M > 0$ depends on ℓ. From this, it follows that for every $k \geq 0$,

$$\frac{k^\ell}{(-1)^k \binom{-\ell-1}{k}} = \frac{k^\ell k!}{(-1)^k (-\ell-1)(-\ell-2)\cdots(-\ell-k)}$$

$$= \frac{k^\ell k!}{(k+\ell)(k+\ell-1)\cdots(\ell+1)} \leq M. \tag{11.9}$$

Here,

$$\binom{\alpha}{k} = \frac{\alpha(\alpha-1)\cdots(\alpha-k)}{k!}, \qquad (\alpha \in \mathbb{R}).$$

Moreover, since f is in the Hadamard gap class, there exists a constant $c > 1$ such that $n_{j+1} \geq c n_j$ for all $j \geq 0$. Then

$$\frac{1}{k^\ell} \left(\sum_{n_j \leq k} n_j^\ell \right) \leq \sum_{m=0}^{\infty} \left(\frac{1}{c^\ell} \right)^m = \frac{c^\ell}{c^\ell - 1}. \tag{11.10}$$

Combining (11.9) and (11.10) yields

$$\frac{k^\ell}{(-1)^k \binom{-\ell-1}{k}} \cdot \frac{1}{k^\ell} \left(\sum_{n_j \leq k} n_j^\ell \right) \leq \frac{M c^\ell}{c^\ell - 1},$$

from which it follows that

$$\sum_{n_j \leq k} n_j^\ell \leq (-1)^k \binom{-\ell-1}{k} \frac{M c^\ell}{c^\ell - 1}. \tag{11.11}$$

Consequently, for any $z \in \mathbb{B}$, by (11.11), we have

$$\frac{|f(z)|}{1 - |z|} \lesssim \frac{M c^\ell}{c^\ell - 1} \cdot \sum_{t=0}^{\infty} (-1)^t \binom{-\ell-1}{t} |z|^t = \frac{M c^\ell}{c^\ell - 1} \cdot \frac{1}{(1 - |z|)^{\ell+1}},$$

which gives

$$|f(z)|(1 - |z|^2)^\ell \lesssim \frac{M c^\ell}{c^\ell - 1}.$$

This shows that $f \in A^{-\ell}$.

(ii)-**Necessity** Suppose that $f \in A_0^{-\ell}$. This means that for any $\varepsilon > 0$, there exists $\delta \in (0, 1)$, such that whenever $\delta < |z| < 1$, we have

$$|f(z)|(1 - |z|^2)^\ell < \varepsilon.$$

Take $N_0 \in \mathbb{N}$ with $\delta < 1 - \frac{1}{n_k} < 1$ whenever $k > N_0$. Then for all $k > N_0$ and $r = 1 - \frac{1}{n_k}$, applying the proof in part (i), we get

$$M_k \leq \frac{n_k^\ell}{(1 - \frac{1}{n_k})^{n_k}} \cdot \sup_{\delta < |z| < 1} |f(z)|(1 - |z|^2)^\ell < \frac{\varepsilon n_k^\ell}{(1 - \frac{1}{n_k})^{n_k}}.$$

This gives

$$\frac{M_k}{n_k^\ell} \leq \frac{\varepsilon}{(1 - \frac{1}{n_k})^{n_k}} \leq 4\varepsilon, \quad k > N_0.$$

Hence,

$$\lim_{k \to \infty} \frac{M_k}{n_k^\ell} = 0.$$

(ii)-**Sufficiency** Suppose that $\lim_{k \to \infty} \frac{M_k}{n_k^\ell} = 0$. This implies that $\sup_{k \geq 1} \frac{M_k}{n_k^\ell} < \infty$, and so by part (i), we have $f \in A^{-\ell}$. Then for any $\varepsilon > 0$ given, there exists $N_0 \in \mathbb{N}$ such that

$$\frac{M_m}{n_m^\ell} < \varepsilon, \quad (m > N_0).$$

For each $m \in \mathbb{N}$, we put

$$f_m(z) = \sum_{k=0}^{\infty} P_{n_k}(z).$$

Note that

$$|f_m(z)|(1 - |z|^2)^\ell \leq \left(\sum_{k=0}^{m} |P_{n_k}(z)| \right) (1 - |z|^2)^\ell$$

$$= \left(\sum_{k=0}^{m} \left| P_{n_k}(\frac{z}{|z|})|z|^{n_k} \right| \right) (1 - |z|^2)^\ell$$

$$\leq K(1 - |z|^2)^\ell \sum_{k=0}^{m} |z|^{n_k} \leq Km(1 - |z|^2)^\ell.$$

Here, $K = \max\{M_0, M_1, \ldots, M_m\}$. Thus,

$$\lim_{|z| \to 1^{-1}} |f_m(z)|(1 - |z|^2)^\ell = 0,$$

which means that $f_m \in A^{-\ell}$ for each $m \in \mathbb{N}$. Then it suffices to show that $|f_m - f| \to 0$ as $m \to \infty$. Indeed, for $m > N_0$, we have

$$|f_m(z) - f(z)| = \left| \sum_{k=m+1}^{\infty} P_{n_k}(z) \right| \leq \sum_{k=m+1}^{\infty} |M_k||z|^{n_k} \leq \varepsilon \sum_{k=m+1}^{\infty} |n_k|^\ell |z|^{n_k}.$$

Doing as in the proof of part (i), we obtain

$$\frac{|f_m(z) - f(z)|}{1 - |z|} \leq \varepsilon \left(\sum_{k=m+1}^{\infty} |n_k|^\ell |z|^{n_k} \right) \left(\sum_{s=0}^{\infty} |z|^s \right)$$

$$= \varepsilon \sum_{\ell=n_{m+1}}^{\infty} \left(\sum_{n_{m+1} \leq n_j < \ell} n_j^\ell \right) |z|^\ell$$

$$\leq \varepsilon \sum_{\ell=0}^{\infty} \left(\sum_{n_j \leq \ell} n_j^\ell \right) |z|^\ell \leq M' \frac{\varepsilon}{1 - |z|)^{\ell+1}}.$$

Here, M' is a positive constant which is independent of m. Hence, whenever $m > N_0$, we have $|f_m - f| \leq M'\varepsilon$, which implies that $f \in A_0^{-\ell}$.   □

11.4   A Characterization of H_μ^∞ and $H_{\mu,0}^\infty$

We recall from Sect. 6.4 of Chap. 6 that a positive continuous function μ on $[0, 1)$ is called *normal* if there exist positive α and β with $\alpha < \beta$, and $\delta \in (0, 1)$, such that

$$\begin{cases} \frac{\mu(r)}{(1-r)^\alpha} \text{ is decreasing on } [\delta, 1), \ \lim_{r \to 1} \frac{\mu(r)}{(1-r)^\alpha} = 0, \\ \text{and} \\ \frac{\mu(r)}{(1-r)^\beta} \text{ is increasing on } [\delta, 1), \ \lim_{r \to 1} \frac{\mu(r)}{(1-r)^\beta} = \infty. \end{cases} \tag{11.12}$$

A normal function $\mu : [0, 1) \rightarrow [0, \infty)$ is decreasing in a neighborhood of 1 and satisfies $\lim_{r \to 1^-} \mu(r) = 0$. The corresponding weighted-type space in the ball is defined as

$$H_\mu^\infty = \left\{ f \in \mathcal{O}(\mathbb{B}) : \|f\| = \sup_{z \in \mathbb{B}} |f(z)|\mu(|z|) < \infty \right\},$$

where μ is normal on $[0, 1)$. It is well-known that H_μ^∞ is a Banach space with the norm $\| \cdot \|$. The little weighted-type space, denoted by $H_{\mu,0}^\infty$, is the closed subspace of H_μ^∞ that consists of $f \in H_\mu^\infty$ satisfying

$$\lim_{|z| \to 1^-} |f(z)|\mu(|z|) = 0.$$

In the case $\mu(|z|) = (1 - |z|^2)^\alpha$, $\alpha > 0$, the induced spaces H_μ^∞ and $H_{\mu,0}^\infty$ are, respectively, denoted by H_α^∞ and $H_{\alpha,0}^\infty$.

Let $f(z) = \sum_{k=0}^\infty P_k(z)$ be a holomorphic function in $\mathbb{B}$, where $P_k(z)$ is a homogeneous polynomial of degree k. We put

$$M_k = \sup_{\xi \in \mathbb{S}} |P_k(\xi)|, \qquad k \geq 0.$$

The following results are about estimates for M_k of a function f belonging H_μ^∞ and $H_{\mu,0}^\infty$, respectively.

Lemma 11.4.1 *Let μ be a normal function on $[0, 1)$, and let*

$$f(z) = \sum_{k=0}^\infty P_k(z) \in Hol(\mathbb{B}).$$

Then the following assertions hold.

(i) *If $f \in H_\mu^\infty$, then*

$$\sup_{k \geq 0} M_k \mu(1 - \frac{1}{k}) < \infty.$$

(ii) *If $f \in H_{\mu,0}^\infty$, then*

$$\lim_{k \to \infty} M_k \mu(1 - \frac{1}{k}) = 0.$$

Proof

(i): Suppose that $f \in H_\mu^\infty$. Fix $\xi \in \mathbb{S}$, and denote

$$f_\xi(w) = \sum_{k=0}^\infty P_k(\xi)w^k = \sum_{k=0}^\infty P_k(\xi w), \quad (w \in \mathbb{D}).$$

Since $f \in H_\mu^\infty$, for a fixed $\xi \in \mathbb{S}$, the function $f_\xi \in H_\mu^\infty$. In this case, for every $r \in (0, 1)$, we have

$$M_k = \sup_{\xi \in \mathbb{S}} |P_k(\xi)|$$

$$= \sup_{\xi \in \mathbb{S}} \left| \frac{1}{2\pi i} \int_{|w|=r} \frac{f_\xi(w)}{w^{k+1}} \, dw \right|$$

$$= \frac{1}{2\pi} \sup_{\xi \in \mathbb{S}} \left| \int_{|w|=r} \frac{f(\xi w)}{w^{k+1}} \, dw \right|$$

$$\leq \frac{1}{2\pi} \sup_{\xi \in \mathbb{S}} \int_{|w|=r} \frac{|f(\xi w)|}{r^{k+1}} |dw|$$

$$\leq \frac{\|f\|}{r^k \mu(r)}.$$

In particular, for $r = 1 - \frac{1}{k}$, $k \geq 2$, we have

$$M_k \leq \frac{\|f\|}{(1 - \frac{1}{k})^k \mu(1 - \frac{1}{k})},$$

which gives

$$M_k \mu(1 - \frac{1}{k}) \leq \frac{\|f\|}{(1 - \frac{1}{k})^k} \leq 4\|f\|, \quad k \geq 2.$$

Hence,

$$\sup_{k \geq 1} M_k \mu(1 - \frac{1}{k}) \leq \max\{M_1 \mu(0), 4\|f\|\} < \infty.$$

(ii): Suppose that $f \in H_{\mu,0}^\infty$. For any $\epsilon > 0$, there exists $\delta \in (0, 1)$ such that $\mu(|z|)|f(z)| < \epsilon$ whenever $\delta < |z| < 1$. Fix $N_0 \in \mathbb{N}$ satisfying $\delta < 1 - \frac{1}{k} < 1$ for all $k > N_0$. Then, as in the proof of part (i), for $k > N_0$ and $r = 1 - \frac{1}{k}$, we have

$$M_k \leq \frac{1}{(1-\frac{1}{k})^k \mu(1-\frac{1}{k})} \cdot \sup_{\delta < |z| < 1} \mu(|z|) |f(z)| < \frac{\epsilon}{(1-\frac{1}{k})^k \mu(1-\frac{1}{k})},$$

which gives

$$M_k \mu(1-\frac{1}{k}) \leq \frac{\epsilon}{(1-\frac{1}{k})^k} \leq 4\epsilon, \quad k > N_0.$$

Hence,

$$\lim_{k\to\infty} M_k \mu(1-\frac{1}{k}) = 0.$$

$\square$

Theorem 11.4.2 *Let μ be a normal function on $[0, 1)$, $f(z) = \sum_{k=0}^{\infty} P_{n_k}(z)$ with Hadamard gaps, where $P_{n_k}(z)$ is a homogeneous polynomial of degree n_k. Then the following statements hold.*

(i) *$f \in H_\mu^\infty$ if and only if $\sup_{k \geq 0} M_{n_k} \mu(1-\frac{1}{n_k}) < \infty$.*

(ii) *$f \in H_{\mu,0}^\infty$ if and only if $\lim_{k\to\infty} M_{n_k} \mu(1-\frac{1}{n_k}) = 0$.*

Proof By Lemma 11.4.1, it suffices to prove the sufficiency of both parts.

(i) Suppose $\sup_{k \geq 0} M_{n_k} \mu(1-\frac{1}{n_k}) < \infty$. Note that

$$|f(z)| = \left| \sum_{k=0}^{\infty} P_{n_k}\left(\frac{z}{|z|}\right) |z|^{n_k} \right| \leq \sum_{k=0}^{\infty} M_k |z|^{n_k} \lesssim \sum_{k=0}^{\infty} \frac{|z|^{n_k}}{\mu(1-\frac{1}{n_k})}.$$

By the proof of Theorem 6.6.1, we have

$$\frac{|f(z)|}{1-|z|} \lesssim \sum_{m=1}^{\infty} \left(\sum_{n_k \leq m} \frac{1}{\mu(1-\frac{1}{n_k})} \right) |z|^m \lesssim \sum_{m=1}^{\infty} \frac{|z|^m}{\mu(1-\frac{1}{m})} \lesssim \frac{1}{(1-|z|)\mu(|z|)},$$

which implies that $f \in H_\mu^\infty$.

(ii) Suppose $\lim_{k\to\infty} M_{n_k} \mu(1-\frac{1}{n_k}) = 0$. In this case, $\sup_{k \geq 1} M_{n_k} \mu(1-\frac{1}{n_k}) < \infty$. Then by part (i), $f \in H_m u^\infty$.

Now for any $\varepsilon > 0$, there exists $N_0 \in \mathbb{N}$ such that

$$M_{n_m} \mu \left(1 - \frac{1}{n_m}\right) < \varepsilon \quad \text{whenever } m > N_0.$$

For each $m \in \mathbb{N}$, put

$$S_m(z) = \sum_{k=0}^{m} P_{n_k}(z) \in H_{\mu,0}^\infty.$$

To show $f \in H_{\mu,0}^\infty$, it suffices to prove that $\|S_m - f\| \to 0$ as $m \to \infty$. We have

$$|S_m(z) - f(z)| = \left| \sum_{k=m+1}^{\infty} P_{n_k}(z) \right| \le \sum_{k=m+1}^{\infty} M_{n_k} |z|^{n_k} \le \varepsilon \sum_{k=m+1}^{\infty} \frac{|z|^{n_k}}{\mu\left(1 - \frac{1}{n_k}\right)},$$

from which the desired result follows. $\qquad\qquad\qquad\qquad\qquad\qquad\qquad\qquad\quad \square$

11.5 The Growth Rate in H_μ^∞

As an application of Theorem 11.4.2, we provide an estimate of the growth rate of functions in H_μ^∞. To achieve this goal, we need some auxiliary results. For $\xi, \zeta \in \mathbb{S}$, we denote

$$d(\xi, \zeta) = (1 - |\langle \xi, \zeta \rangle|^2)^{1/2}.$$

Here, d satisfies the triangle inequality. We also write

$$E_\delta(\zeta) = \{\xi \in \mathbb{S} : d(\xi, \zeta) < \delta\}, \qquad (0 < \delta < 1).$$

We say that a set $\Gamma \subset \mathbb{S}$ is d-separated by δ if the sets $E_\delta(\zeta)$, where $\zeta \in \Gamma$, are pairwise disjoint.

Lemma 11.5.1 *For each $a > 0$, there exists a positive integer $N = N_n(a)$ with the following property: if $\delta > 0$ and if $\Gamma \subset \mathbb{S}$ is d-separated by $a\delta$, then Γ can be decomposed as*

$$\Gamma = \bigcup_{j=1}^{N} \Gamma_j$$

in such a way that each Γ_j is d-separated by δ.

Lemma 11.5.2 *Suppose $\Gamma \subset \mathbb{S}$ is d-separated by δ and k is a positive integer. If*

$$P(z) = \sum_{\zeta \in \Gamma} \langle z, \zeta \rangle^k, \qquad (z \in \mathbb{B}),$$

then

$$|P(z)| \leq 1 + \sum_{m=1}^{\infty} (m+2)^{2n-2} e^{-m^2 \delta^2 k/2}.$$

Theorem 11.5.3 *Let μ be a normal function on $[0, 1)$. Then there exists a positive integer $N = N(n)$, and there are functions $f_j \in H_\mu^\infty$ $(j = 1, \ldots, N)$ such that*

$$\sum_{j=1}^{N} |f_j(z)| \gtrsim \frac{1}{\mu(|z|)}, \qquad (z \in \mathbb{B}).$$

Proof First we note that the integer N in the statement of the theorem must be at least 2. Indeed, assume in contrary that $N = 1$, i.e., there exists $f \in H_\mu^\infty$ such that

$$|f(z)| \gtrsim \frac{1}{\mu(|z|)}, \qquad (z \in \mathbb{B}).$$

This shows that f has no zeros in $\mathbb{B}$, and hence there exists $g \in \mathrm{Hol}(\mathbb{B})$ such that $f = e^g$. Then the last inequality means

$$e^{\mathrm{Reg}(z)} \gtrsim \frac{1}{\mu(|z|)} \quad \Longleftrightarrow \quad \mathrm{Reg}(z) \gtrsim \log \frac{1}{\mu(|z|)}, \quad z \in \mathbb{B}.$$

For each $r \in (0, 1)$, integrating both sides of the inequality above on $r\mathbb{S} = \{z \in \mathbb{B} : |z| = r\}$, we have

$$\int_{r\mathbb{S}} \mathrm{Reg}(z) \, d\sigma \gtrsim \int_{r\mathbb{S}} \log(\frac{1}{\mu(r)}) \, d\sigma = \log(\frac{1}{\mu(r)}) \cdot \sigma(r\mathbb{S}).$$

Hence, by the mean value property,

$$\mathrm{Reg}(0) \gtrsim \log(\frac{1}{\mu(r)}), \quad \text{for all } r \in (0, 1),$$

which is impossible.

Now we prove the theorem by constructing the functions $f_j \in H_\mu^\infty$ satisfying the given property near the boundary only; then by adding a proper constant, we can extend it to the whole unit ball. By normality of μ on $[0, 1)$, there exist positive constants $\alpha < \beta$ and $\delta \in (0, 1)$ that satisfy (11.12). Let $0 < A < 1$ be a sufficiently small number for which

$$\sum_{m=1}^{\infty} (m+2)^{2n-2} e^{-m^2/2A^2} \le \frac{1}{27}. \tag{11.13}$$

Let $N = N_n(\frac{A}{2})$ be a positive constant from Lemma 11.5.1 (with $a = \frac{A}{2}$). Take a positive integer p large enough satisfying the following conditions:

$$1 - \frac{1}{p} \ge \delta,$$

$$\frac{1}{3} \le (1 - \frac{1}{p})^p \le \frac{1}{2},$$

$$\frac{1}{p^{\alpha N} - 1} \le \frac{1}{200},$$

$$\frac{p^{\beta N} \cdot 2^{-p^{N-\frac{1}{2}}}}{1 - p^{\beta N} \cdot 2^{-(p^{2N} - \frac{1}{2} - p^{N-\frac{1}{2}})}} \le \frac{1}{200}. \tag{11.14}$$

For every positive number $j \le N$, we put $\delta_{j,0}$ with

$$A^2 p^j \delta_{j,0}^2 = 1$$

and inductively define $\delta_{j,\nu}$ by

$$p^N \delta_{j,\nu}^2 = \delta_{j,\nu-1}^2, \quad \nu \ge 1.$$

In this case, we have

$$A^2 p^{\nu N+j} \delta_{j,\nu}^2 = 1. \tag{11.15}$$

For each fixed j, ν, denote by $\Gamma^{j,\nu}$ the maximal subset of $\mathbb{S}$ subject to the condition that $\Gamma^{j,\nu}$ is d-separated by $A\delta_{j,\nu}/2$. Then by Lemma 11.5.1, write

$$\Gamma^{j,\nu} = \bigcup_{\ell=1}^{N} \Gamma_{j,\nu N+\ell} \tag{11.16}$$

in such a way that each $\Gamma_{j,\nu N+\ell}$ is d-separated by $\delta j, \nu$. For each $i, j = 1, 2, \ldots, N$ and $\nu \ge 0$, set

$$P_{i,\nu N+j}(z) = \sum_{\xi \in \Gamma_{j,\nu N+\tau^i(j)}} \langle z, \xi \rangle^{p^{\nu N+j}},$$

where τ^i is the i-th iteration of the permutation τ on $\{1, 2, \ldots, N\}$ defined by

$$
\tau(j) = \begin{cases} j+1, & j < N, \\ 1, & j = N. \end{cases}
$$

By Lemma 11.5.2, (11.13), and (11.15), we have

$$
|P_{i,\nu N+j}(z)| \le 1 + \sum_{m=1}^{\infty} (m+2)^{2n-2} e^{-\frac{m^2 \delta_{j,\nu} p^{\nu N+j}}{2}}
$$

$$
\le 1 + \sum_{m=1}^{\infty} (m+2)^{2n-2} e^{-\frac{m^2}{2A^2}} \le 2, \qquad (z \in \mathbb{B}),
$$

for all $i, j = 1, 2, \ldots, N$ and $\nu \ge 0$. Now define

$$
g_{i,j}(z) = \sum_{\nu=0}^{\infty} \frac{P_{i,\nu N+j}(z)}{\mu(1 - \frac{1}{p^{\nu N+j}})}, \qquad (z \in \mathbb{B}).
$$

By Theorem 11.4.2, $g_{i,j} \in H_\mu^\infty$ for each $i, j = 1, 2, \ldots, N$. We show that for every $\nu \ge 0$, $1 \le j \le N$, and $z \in \mathbb{B}$ with

$$
1 - \frac{1}{p^{\nu N+j}} \le |z| \le 1 - 1 - \frac{1}{p^{\nu N+j+\frac{1}{2}}}, \tag{11.17}
$$

there exists $i \in \{1, 2, \ldots, N\}$ such that $|g_{i,j}(z)| \ge \frac{C}{\mu(|z|)}$, where C is some positive constant independent of the choice of i, j, and z.

Fix ν, j, and $z = \eta|z|$ ($\eta \in \mathbb{S}$) for which (11.17) holds. Since d-balls centered at points of $\Gamma^{j,\nu}$ of radius $A\delta_{j,\nu}$ cover $\mathbb{S}$, by maximality, there exists some $\xi \in \Gamma^{j,\nu}$ such that $\eta \in E_{A\delta_{j,\nu}}(\zeta)$. Note that by (11.16), $\zeta \in \Gamma_{j,\nu N+\ell}$ for some $1 \le \ell \le N$. We now estimate $|g_{i,j}(z)|$, as follows.

$$
|g_{i,j}(z)| = \left| \sum_{k=0}^{\infty} \frac{P_{i,kN+j}(z)}{\mu(1 - \frac{1}{p^{kN+j}})} \right|
$$

$$
\ge \left| \frac{P_{i,\nu N+j}(z)}{\mu(1 - \frac{1}{p^{\nu N+j}})} \right| - \left| \sum_{k \ne \nu} \frac{P_{i,kN+j}(z)}{\mu(1 - \frac{1}{p^{kN+j}})} \right|
$$

$$
= \left| \frac{|z|^{p^{\nu N+j}} P_{i,\nu N+j}(\eta)}{\mu(1 - \frac{1}{p^{\nu N+j}})} - \left| \sum_{k \ne \nu} \frac{|z|^{kN+j} P_{i,kN+j}(\eta)}{\mu(1 - \frac{1}{p^{kN+j}})} \right| \right|
$$

$$\geq \frac{|z|^{p^{\nu N+j}} P_{i,\nu N+j}(\eta)}{\mu(1 - \frac{1}{p^{\nu N+j}})} - 2\sum_{k=0}^{\nu-1} \frac{|z|^{p^{kN+j}}}{\mu(1 - \frac{1}{p^{kN+j}})} - 2\sum_{k=\nu+1}^{\infty} \frac{|z|^{p^{kN+j}}}{\mu(1 - \frac{1}{p^{kN+j}})}$$

$$= J_1 - J_2 - J_3.$$

We estimate the above J's separately.

The estimation of J_1: By (11.14) and (11.17), we have

$$|z|^{p^{\nu N+j}} \geq (1 - \frac{1}{p^{\nu N+j}})^{p^{\nu N+j}} \geq \frac{1}{3},$$

which implies

$$J_1 = \frac{|z|^{p^{\nu N+j}} P_{i,\nu N+j}(\eta)}{\mu(1 - \frac{1}{p^{\nu N+j}})}$$

$$\geq \frac{P_{i,\nu N+j}(\eta)}{3\mu(1 - \frac{1}{p^{\nu N+j}})}$$

$$\geq \frac{|\langle \eta, \zeta \rangle|^{p^{\nu N+j}} - \sum_{\xi \in \Gamma_{j,\nu N+\tau^i(j),\xi \neq \zeta}} |\langle \eta, \zeta \rangle|^{p^{\nu N+j}}}{3\mu(1 - \frac{1}{p^{\nu N+j}})}$$

$$\geq \frac{2}{27\mu(1 - \frac{1}{p^{\nu N+j}})}.$$

The estimation of J_2: By normality of μ, with every $s \in \mathbb{N}$, we have

$$\frac{(1 - \frac{1}{p^{sN+j}})^\alpha}{(1 - \frac{1}{p^{(s+1)N+j}})^\alpha} \leq \frac{\mu(1 - \frac{1}{p^{sN+j}})}{\mu(1 - \frac{1}{p^{(s+1)N+j}})} \leq \frac{(1 - \frac{1}{p^{sN+j}})^\beta}{(1 - \frac{1}{p^{(s+1)N+j}})^\beta}.$$

This gives

$$1 < p^{N\alpha} \leq \frac{\mu(1 - \frac{1}{p^{sN+j}})}{\mu(1 - \frac{1}{p^{(s+1)N+j}})} \leq p^{N\beta}.$$

Combining this and (11.14) yields

$$J_2 = 2\sum_{k=0}^{\nu-1} \frac{|z|^{p^{kN+j}}}{\mu(1 - \frac{1}{p^{kN+j}})}$$

$$\leq 2 \sum_{k=0}^{v-1} \frac{1}{\mu(1 - \frac{1}{p^{kN+j}})}$$

$$= \frac{2}{\mu(1 - \frac{1}{p^{vN+j}})} \sum_{k=0}^{v-1} \left[\frac{\mu(1 - \frac{1}{p^{vN+j}})}{\mu(1 - \frac{1}{p^{(v-1)N+j}})} \cdot \frac{\mu(1 - \frac{1}{p^{(v-1)N+j}})}{\mu(1 - \frac{1}{p^{(v-2)N+j}})} \cdots \frac{\mu(1 - \frac{1}{p^{(k+1)N+j}})}{\mu(1 - \frac{1}{p^{kN+j}})} \right]$$

$$\leq \frac{2}{\mu(1 - \frac{1}{p^{vN+j}})} \sum_{k=0}^{v-1} \frac{1}{p^{\alpha N(v-k)}}$$

$$\leq \frac{2}{\mu(1 - \frac{1}{p^{vN+j}})} \cdot \frac{1}{p^{\alpha N} - 1}$$

$$\leq \frac{1}{100\mu(1 - \frac{1}{p^{vN+j}})}.$$

The estimation of J_3: Since

$$J_3 = 2 \sum_{k=v+1}^{\infty} \frac{|z|^{p^{kN+j}}}{\mu(1 - \frac{1}{p^{kN+j}})}$$

$$= \frac{2|z|^{p^{(v+1)N+j}}}{\mu(1 - \frac{1}{p^{vN+j}})} \sum_{k=v+1}^{\infty} \left[\frac{\mu(1 - \frac{1}{p^{vN+j}})}{\mu(1 - \frac{1}{p^{kN+j}})} |z|^{p^{kN+j} - p^{(v+1)N+j}} \right],$$

using (11.14) and (11.17) as for J_1, we have

$$\sum_{k=v+1}^{\infty} \left[\frac{\mu(1 - \frac{1}{p^{vN+j}})}{\mu(1 - \frac{1}{p^{kN+j}})} |z|^{p^{kN+j} - p^{(v+1)N+j}} \right]$$

$$= \sum_{k=v+1}^{\infty} \left[\frac{\mu(1 - \frac{1}{p^{vN+j}})}{\mu(1 - \frac{1}{p^{(v+1)N+j}})} \cdots \frac{\mu(1 - \frac{1}{p^{(k-1)N+j}})}{\mu(1 - \frac{1}{p^{kN+j}})} |z|^{p^{kN+j} - p^{(v+1)N+j}} \right]$$

$$\leq \sum_{k=v+1}^{\infty} p^{\beta N(k-v)} |z|^{p^{kN+j} - p^{(v+1)N+j}}$$

$$= \sum_{k=v+1}^{\infty} p^{\beta N} p^{\beta N(k-v-1)} |z|^{p^j p^{kN} - p^{(v+1)N}}$$

$$= \sum_{k=0}^{\infty} p^{\beta N} p^{\beta Ns} |z|^{p^{j+(v+1)N}(p^{sN} - 1)}.$$

Here, $s = k - v - 1$, and we used $p^{sN-1} \geq s(p^N - 1)$. Thus,

$$J_3 \leq \sum_{k=0}^{\infty} p^{\beta N} p^{\beta Ns} |z|^{p^{j+(v+1)N}} (p^{N-1})s$$

$$= \sum_{k=0}^{\infty} p^{\beta N} \left(p^{\beta N} |z|^{p^{j+(v+2)N} - p^{j+(v+1)N}} \right)^s.$$

Hence,

$$J_3 \leq \frac{2|z|^{p^{(v+1)N+j}}}{\mu(1 - \frac{1}{p^{vN+j}})} \sum_{k=0}^{\infty} p^{\beta N} \left(p^{\beta N} |z|^{p^{j+(v+2)N} - p^{j+(v+1)N}} \right)^s$$

$$= \frac{2}{\mu(1 - \frac{1}{p^{vN+j}})} \cdot \frac{p^{\beta N} (|z|^{p^{vN+j}})^{p^N}}{1 - p^{\beta N} |z|^{p^{vN+j}(p^{2N} - p^N)}}$$

$$\leq \frac{2}{\mu(1 - \frac{1}{p^{vN+j}})} \cdot \frac{p^{\beta N} \cdot 2^{-p^{N-\frac{1}{2}}}}{1 - p^{\beta N} \cdot 2^{-(p^{2N-\frac{1}{2}} - p^{N-\frac{1}{2}})}}$$

$$\leq \frac{1}{100\mu(1 - \frac{1}{p^{vN+j}})}.$$

Now combining all these estimates for J_1, J_2, and J_3, we obtain

$$|g_{i,j}(z)| \geq J_1 - J_2 - J_3$$

$$\geq \left(\frac{2}{27} - \frac{1}{100} - \frac{1}{100} \right) \frac{1}{\mu(1 - \frac{1}{p^{vN+j}})}$$

$$> \frac{1}{20\mu(1 - \frac{1}{p^{vN+j}})}$$

$$= \frac{1}{20\mu(1 - \frac{1}{p^{vN+j+\frac{1}{2}}})} \cdot \frac{\mu(1 - \frac{1}{p^{vN+j+\frac{1}{2}}})}{\mu(1 - \frac{1}{p^{vN+j}})}$$

$$\geq \frac{1}{20 p^{\frac{\beta}{2}} \mu(1 - \frac{1}{p^{vN+j+\frac{1}{2}}})}$$

$$\geq \frac{1}{20 p^{\frac{\beta}{2}} \mu(|z|)},$$

for all $z \in \mathbb{B}$ with $1 - \frac{1}{p^k} \leq |z| \leq 1 - \frac{1}{p^{k+\frac{1}{2}}}$, $k \geq 1$. Then take a sequence (q_k) of positive integers satisfying $0 \leq q_k - p^{k+\frac{1}{2}} < 1$, and for each $j = 1, 2, \ldots, N$, take a sequence $(\varepsilon_{j,v})$ satisfying $A^2 q_{vN+j} \varepsilon_{j,v}^2 = 1$.

Choose a sequence of subsets $\Delta_{j,v} \subset \mathbb{S}$ with the following property: for each nonnegative integer v, the set $\cup_{\ell=1}^{N} \Delta_{j,vN+\ell}$ is a maximal subset of $\mathbb{S}$ which is d-separated by $A\varepsilon_{j,v}/2$ and each $\Delta_{j,vN+\ell}$. For each $i, j = 1, 2, \ldots, N$ and $v \geq 0$, we set

$$Q_{i,vN+j}(z) = \sum_{\xi \in \Delta_{j,vN+\tau^i(j)}} \langle z, \xi \rangle^{q_{vN+j}}$$

and define

$$h_{i,j}(z) = \sum_{v=0}^{\infty} \frac{Q_{i,vN+j}(z)}{\mu(1 - \frac{1}{q_{vN+j}})}.$$

It is clear that $h_{i,j}$ is in the Hadamard gap series, since for each $v \geq 0$, we have

$$\frac{q_{vN+j}}{q_{(v-1)N+j}} \geq \frac{p^{vN+\frac{1}{2}}}{p^{(v-1)N+\frac{1}{2}} + 1} \geq \frac{p^N}{2} > 1.$$

Furthermore, the homogeneous polynomials $Q_{i,vN+j}$ are uniformly bounded by 2 as above. Then each $h_{i,j} \in H_\mu^\infty$ by Theorem 11.4.2. Moreover, by some modifications of the previous arguments, it shows that for each $v \geq 0$, $1 \leq j \leq N$, and $z \in \mathbb{B}$ with

$$1 - \frac{1}{p^{vN+j+\frac{1}{2}}} \leq |z| \leq 1 - \frac{1}{p^{vN+j+1}},$$

there exists $i \in \{1, 2, \ldots, N\}$ such that $|h_{i,j}(z)| \geq \frac{C_p}{\mu(|z|)}$, where C_p is some positive constant independent of the choice of i, j, and z. Consequently,

$$\sum_{i=1}^{N} \sum_{j=1}^{N} |h_{i,j}(z)| \geq \frac{C_p}{\mu(|z|)},$$

for all $1 - \frac{1}{p^{k+\frac{1}{2}}} \leq |z| \leq 1 - \frac{1}{p^{k+1}}$, $k \geq 1$. Finally, we arrive to the following:

$$\sum_{i=1}^{N} \sum_{j=1}^{N} |g_{i,j}(z)| + |h_{i,j}(z)| \geq \frac{C}{\mu(|z|)},$$

for all $z \in \mathbb{B}$ close enough to the boundary and for some positive constant C. □

Corollary 11.5.4 *There exist some positive integer N and functions $f_j \in H_\alpha^\infty$ ($j = 1, 2, \ldots, N$) such that*

$$\sum_{j=1}^{N} |f_j(z)| \gtrsim \frac{1}{(1 - |z|^2)^\alpha}, \qquad (z \in \mathbb{B}).$$

Notes on Chapter 11

Hadamard gap series on different spaces of holomorphic functions in higher dimensions, namely, in the unit ball $\mathbb{B}$, have been extensively studied, e.g., see [10, 59, 89, 107] and references therein. Lemma 11.1.1 is given in [108]. Corollary 11.2.3 is an extension of the results of [74]. The result in [74] is a particular case of the assertion (i) in Theorem 11.3.1. The main results in Sect. 11.4 are from papers [47, 48]. Lemmas 11.5.1 and 11.5.2 are, respectively, Lemmas 2.2 and 2.3 in [12]. Recently, in [55], some simplifications to the formulas in Theorem 11.2.1 have been discovered. Namely, condition (i) can be replaced by

$$\sum_{k=0}^{\infty} \frac{M_k^2}{n_k^{p+1}} < \infty,$$

and condition (iv) by

$$\sum_{k=0}^{\infty} \frac{L_k^2}{n_k^{p+1}} < \infty.$$

In this chapter, a new class, which is a generalization of $\mathcal{N}_p$ and $\mathcal{N}_p$-type spaces, is studied. Besides basic properties, several other topics are covered, including the distance between Bergman-type spaces and $\mathcal{N}(p, q, s)$-type spaces. The results and their proofs are mainly developed from the corresponding ones in the $\mathcal{N}_p$ spaces. Nevertheless, they have independent interests.

12.1 $\mathcal{N}(p, q, s)$ as a Functional Banach Space

Let

$$d\lambda(z) = \frac{dV(z)}{(1 - |z|^2)^{n+1}}.$$

This is an important Möbius-invariant measure, i.e.,

$$\int_{\mathbb{B}} f(z)\, d\lambda(z) = \int_{\mathbb{B}} f \circ \Phi(z)\, d\lambda(z),$$

for each $f \in L^1(\lambda)$ and an automorphism Φ of $\mathbb{B}$. Let $p, q, s > 0$. The $\mathcal{N}(p, q, s)$-space consists of holomorphic functions f on $\mathbb{B}$, for which

$$\|f\| = \left(\sup_{a \in \mathbb{B}} \int_{\mathbb{B}} |f(z)|^p (1 - |z|^2)^q (1 - |\Phi_a(z)|^2)^{ns}\, d\lambda(z) \right)^{1/p} < \infty.$$

The corresponding little space $\mathcal{N}^0(p, q, s)$, which is a subspace of $\mathcal{N}(p, q, s)$, includes those $f \in \mathcal{N}(p, q, s)$ with

$$\lim_{|a|\to 1} \int_{\mathbb{B}} |f(z)|^p (1 - |z|^2)^q (1 - |\Phi_a(z)|^2)^{ns} \, d\lambda(z) = 0.$$

When $p = 2$, $q = n + 1$, $\mathcal{N}(2, n + 1, s)$ coincides with the $\mathcal{N}_{ns}$-space. In particular, by Theorem 7.6.1 in Chap. 7, when $p = 2$, $q = n + 1$, and $s > 1$, we have

$$\mathcal{N}(2, n + 1, s) = A^{-\frac{n+1}{2}} (\mathbb{B}).$$

Similar results hold for the little space.

Recall that for $a \in \mathbb{B}$, Φ_a denotes the automorphism of $\mathbb{B}$. For $0 < R < 1$, we put

$$\mathbb{B}_R = \{z : |z| < R\}$$

and

$$D(a, R) = \Phi_a (\{z \in \mathbb{B} : |z| < R\}) = \{z \in \mathbb{B} : |\Phi_a(z)| < R\}.$$

First, we need the following result.

Lemma 12.1.1 *Let $p \geq 1$, and let $q, s > 0$. Then the point evaluation $K_z : f \mapsto f(z)$ is a continuous linear functional on $\mathcal{N}(p, q, s)$. Moreover, $\mathcal{N}(p, q, s) \subseteq A^{-\frac{q}{p}} (\mathbb{B})$.*

Proof For each $f \in \mathcal{N}(p, q, s)$ and $a_0 \in \mathbb{B}$, we have

$$\|f\|^p = \sup_{a \in \mathbb{B}} \int_{\mathbb{B}} |f(z)|^p (1 - |z|^2)^q (1 - |\Phi_a(z)|^2)^{ns} \, d\lambda(z)$$

$$\geq \int_{D(a_0, 1/2)} |f(z)|^p (1 - |z|^2)^q (1 - |\Phi_{a_0}(z)|^2)^{ns} \, d\lambda(z)$$

$$\geq C_{n,s} \int_{D(a_0, 1/2)} |f(z)|^p (1 - |z|^2)^q \, d\lambda(z)$$

$$= C_{n,s} \int_{\mathbb{B}_{1/2}} |f(\Phi_{a_0}(w))|^p (1 - |\Phi_{a_0}(w)|^2)^q \, d\lambda(w)$$

(change variable $z = \Phi_{a_0}(w)$)

$$= C_{n,s} \int_{\mathbb{B}_{1/2}} |f(\Phi_{a_0}(w))|^p \frac{(1 - |a_0|^2)^q (1 - |w|^2)^{q-n-1}}{|1 - \langle a_0, w \rangle|^{2q}} \, dV(w)$$

$$\geq C_{q,s,n} (1 - |a_0|^2)^q \int_{\mathbb{B}_{1/2}} |f(\Phi_{a_0}(w))|^p \, dV(w)$$

$$\geq C'_{q,s,n} (1 - |a_0|^2)^q |f(a_0)|^p \qquad \text{(because } |f \circ \Phi_{a_0}(\cdot)|^p \text{ is subharmonic),}$$

where $C_{n,s}$ is a constant depending on n and s and $C_{q,s,n}$ and $C'_{q,s,n}$ are some constants depending on q, s, and n. This shows that for any $z \in \mathbb{B}$,

$$|f(z)| \lesssim \frac{\|f\|}{(1 - |z|^2)^{\frac{q}{p}}},$$

which implies that the point evaluation is a continuous linear functional, as well as $\mathcal{N}(p, q, s) \subseteq A^{-\frac{q}{p}}(\mathbb{B})$.   □

Now we show that $\mathcal{N}(p, q, s)$ is a functional Banach space when $p \geq 1$ and $q, s > 0$.

Theorem 12.1.2 *Let $p \geq 1$ and $q, s > 0$. The $\mathcal{N}(p, q, s)$-space is a functional Banach space.*

Proof Since $\mathcal{N}(p, q, s)$ is a normed space with respect to the norm $\| \cdot \|$, it suffices to show the completeness of $\mathcal{N}(p, q, s)$. Let (f_m) be a Cauchy sequence in $\mathcal{N}(p, q, s)$. From this, by Lemma 12.1.1, it follows that (f_m) is a Cauchy sequence in the space Hol($\mathbb{B}$), and hence it converges to some $f \in$ Hol($\mathbb{B}$). We show that $f \in \mathcal{N}(p, q, s)$. Indeed, there exists a $\ell_0 \in \mathbb{N}$ such that for all $m, \ell \geq \ell_0$, $\|f_m - f_\ell\| \leq 1$. Take and fix an arbitrary $a \in \mathbb{B}$; by Fatou's lemma, we have

$$\int_{\mathbb{B}} |f(z) - f_{\ell_0}(z)|^p (1 - |z|^2)^q (1 - |\Phi_a(z)|^2)^{ns} \, d\lambda(z)$$

$$\leq \lim_{\ell \to \infty} \int_{\mathbb{B}} |f_\ell(z) - f_{\ell_0}(z)|^p (1 - |z|^2)^q (1 - |\Phi_a(z)|^2)^{ns} \, d\lambda(z)$$

$$\leq \lim_{\ell \to \infty} \|f_\ell - f_{\ell_0}\|^p \leq 1,$$

which implies that

$$\|f - f_{\ell_0}\| = \sup_{a \in \mathbb{B}} \int_{\mathbb{B}} |f(z) - f_{\ell_0}(z)|^p (1 - |z|^2)^q (1 - |\Phi_a(z)|^2)^{ns} \, d\lambda(z) \leq 1,$$

and hence $\|f\| \leq 1 + \|f_{\ell_0}\| < \infty$. By Lemma 12.1.1, the point evaluations on $\mathcal{N}(p, q, s)$ are continuous linear functionals. Therefore, $\mathcal{N}(p, q, s)$ is a functional Banach space.   □

Next, we show that when $p \geq 1$, $q > 0$ and $s > 1$, $\mathcal{N}(p, q, s) = A^{-\frac{q}{p}}(\mathbb{B})$. More precisely, we have the following result.

Theorem 12.1.3 *Let $p \geq 1$, $q, s > 0$. If $s > 1 - \frac{q - kp}{n}$, $k \in (0, q/p]$, then $A^{-k}(\mathbb{B}) \subseteq \mathcal{N}(p, q, s)$. In particular, when $s > 1$, $\mathcal{N}(p, q, s) = A^{-\frac{q}{p}}(\mathbb{B})$.*

Proof Suppose $p \geq 1$, $q > 0$, and $s > 1 - \frac{q-kp}{n}$ for some $k \in (0, q/p]$. Then $q + ns - n - 1 - kp > -1$, and hence, by formula (7.13) in Chap. 7, for each $a \in \mathbb{B}$, we have

$$\int_{\mathbb{B}} \frac{(1 - |z|^2)^{q+ns-n-1-pk}}{|1 - \langle a, z \rangle|^{2ns}} \, dV(z) \simeq \begin{cases} \text{bounded in } \mathbb{B}, & \text{if } ns + pk < q, \\ \log \frac{1}{1-|a|^2}, & \text{if } ns + pk = q, \\ (1 - |a|^2)^{q-ns-pk}, & \text{if } ns + pk > q, \end{cases}$$

which implies that there exists a positive constant C such that

$$\sup_{a \in \mathbb{B}} (1 - |a|^2)^{ns} \int_{\mathbb{B}} \frac{(1 - |z|^2)^{q+ns-n-1-pk}}{|1 - \langle a, z \rangle|^{2ns}} \, dV(z) \leq C. \tag{12.1}$$

Let $f \in A^{-k}(\mathbb{B})$. By (12.1), we have

$$\|f\|^p = \sup_{a \in \mathbb{B}} \int_{\mathbb{B}} |f(z)|^p (1 - |z|^2)^q (1 - |\Phi_a(z)|^2|)^{ns} \, d\lambda(z)$$

$$= \sup_{a \in \mathbb{B}} \int_{\mathbb{B}} |f(z)|^p (1 - |z|^2)^{pk} (1 - |z|^2)^{q-pk} (1 - |\Phi_a(z)|^2)^{ns} \, d\lambda(z)$$

$$\leq |f|_k^p \sup_{a \in \mathbb{B}} (1 - |a|^2)^{ns} \int_{\mathbb{B}} \frac{(1 - |z|^2)^{q+ns-n-1-pk}}{|1 - \langle a, z \rangle|^{2ns}} \, dV(z)$$

$$\leq C |f|_k^p,$$

which shows that $A^{-k}(\mathbb{B}) \subseteq \mathcal{N}(p, q, s)$. Furthermore, if $s > 1$, then taking $k = \frac{q}{p}$, we have $A^{-\frac{q}{p}}(\mathbb{B}) \subseteq \mathcal{N}(p, q, s)$. Combining this fact with Theorem 12.1.1, we get the desired result. $\square$

12.2   The Closure of Polynomials in $\mathcal{N}(p, q, s)$-Type Spaces

We start by showing that $\mathcal{N}^0(p, q, s)$ by itself is a Banach space.

Lemma 12.2.1 *Let $p \geq 1$ and $q, s > 0$. The little space $\mathcal{N}^0(p, q, s)$ is a closed subspace of $\mathcal{N}(p, q, s)$, and hence, it is a Banach space.*

Proof Since $\mathcal{N}^0(p, q, s)$ is a subspace of $\mathcal{N}(p, q, s)$, we show $\mathcal{N}^0(p, q, s)$ is complete. If (f_n) is a Cauchy sequence in $\mathcal{N}^0(p, q, s)$, then by Theorem 12.1.2, there exists a limit $f \in \mathcal{N}(p, q, s)$ of (f_n). For any $\varepsilon > 0$, we can find an $N \in \mathbb{N}$, such that

$$\|f - f_n\| < \left(\frac{\varepsilon}{2^p}\right)^{1/p}, \qquad (n \geq N).$$

Fix an $n_0 > N$. Since $f_{n_0} \in \mathcal{N}^0(p, q, s)$, there exists a $\delta \in (0, 1)$ such that whenever $\delta < |a| < 1$,

$$\sup_{\delta < |a| < 1} \int_{\mathbb{B}} |f_{n_0}(z)|^p (1 - |z|^2)^q (1 - |\Phi_a(z)|^2)^{ns} \, d\lambda(z) < \frac{\varepsilon}{2^p}.$$

Hence,

$$\sup_{\delta < |a| < 1} \int_{\mathbb{B}} |f(z)|^p (1 - |z|^2)^q (1 - |\Phi_a(z)|^2)^{ns} \, d\lambda(z)$$

$$\leq 2^{p-1} \|f - f_{n_0}\|^p$$

$$+ 2^{p-1} \sup_{\delta < |a| < 1} \int_{\mathbb{B}} |f_{n_0}(z)|^p (1 - |z|^2)^q (1 - |\Phi_a(z)|^2)^{ns} \, d\lambda(z) < \varepsilon,$$

which shows that $f \in \mathcal{N}^0(p, q, s)$. $\qquad\qquad\qquad\qquad\qquad\qquad\qquad\qquad\square$

In the next step, we show that $\mathcal{N}^0(p, q, s)$ is closed with respect to dilation.

Lemma 12.2.2 *Let $ns + q > n$ and $f \in \mathcal{N}(p, q, s)$. Then $f \in \mathcal{N}^0(p, q, s)$ if and only if $\|f_r - f\| \to 0$, $r \to 1$, where $f_r(z) = f(rz)$ ($z \in \mathbb{B}$).*

Proof **Necessity** Suppose $f \in \mathcal{N}^0(p, q, s)$. Then, for any $\varepsilon > 0$, there exists a $\delta > 0$, such that for $\delta < |a| < 1$, we have

$$\int_{\mathbb{B}} |f(z)|^p (1 - |z|^2)^q (1 - |\Phi_a(z)|^2)^{ns} \, d\lambda(z) < \frac{\varepsilon}{3 \cdot 2^{2n+p-1}}. \qquad (12.2)$$

Furthermore, by the Schwarz–Pick lemma,

$$|\Phi_{ra}(rz)| \leq |\Phi_a(z)|, \qquad r \in (0, 1) \text{ and } a, z \in \mathbb{B}. \qquad (12.3)$$

We fix a $\delta_0 \in (\delta, 1)$ and consider r satisfying $\max\left\{\frac{1}{2}, \frac{\delta}{\delta_0}\right\} < r < 1$. In this case, for all $a \in \mathbb{B}$ with $|a| \in (\delta_0, 1)$, by (12.2) and (12.3), we have

$$\int_{\mathbb{B}} |f(rz)|^p (1 - |z|^2)^q (1 - |\Phi_a(z)|^2)^{ns} \, d\lambda(z)$$

$$\leq \int_{\mathbb{B}} |f(rz)|^p (1 - |rz|^2)^q (1 - |\Phi_{ra}(rz)|^2)^{ns} \, d\lambda(z)$$

$$= \left(\frac{1}{r}\right)^{2n} \int_{\mathbb{B}} |f(w)|^p (1 - |w|^2)^q (1 - |\Phi_{ra}(w)|^2)^{ns} \, d\lambda(w)$$

$$\leq 4^n \int_{\mathbb{B}} |f(w)|^p (1 - |w|^2)^q (1 - |\Phi_{ra}(w)|^2)^{ns} \, d\lambda(w) < \frac{\varepsilon}{3 \cdot 2^{p-1}}.$$

On the other hand, since $ns + q > n$, we have $ns + q - n - 1 > -1$. Then f_r converges to f as $r \to 1$, in the norm topology of the weighted Bergman space $A^p_{ns+q-n-1}(\mathbb{B})$. This means for some $r_1 \in (0, 1)$, whenever $r_1 < r < 1$, we have

$$\int_{\mathbb{B}} |f(rz) - f(z)|^p (1 - |z|^2)^{ns+q-n-q} \, dV(z) < \frac{(1 - \delta_0)^{2ns} \, \varepsilon}{3}.$$

Consequently, for $|a| \leq \delta_0$ and $r_1 < r < 1$, we have

$$\sup_{|a| \leq \delta_0} \int_{\mathbb{B}} |f(rz) - f(z)|^p (1 - |z|^2)^q (1 - |\Phi_a(z)|^2)^{ns} \, d\lambda(z)$$

$$= \sup_{|a| \leq \delta_0} \left\{ (1 - |a|^2)^{ns} \int_{\mathbb{B}} |f(rz) - f(z)|^p \frac{(1 - |z|^2)^{ns+q-n-1}}{|1 - \langle a, z \rangle|^{2ns}} \, dV(z) \right\}$$

$$\leq \sup_{|a| \leq \delta_0} \int_{\mathbb{B}} |f(rz) - f(z)|^p \frac{(1 - |z|^2)^{ns+q-n-1}}{|1 - \langle a, z \rangle|^{2ns}} \, dV(z)$$

$$\leq \frac{1}{(1 - \delta_0)^{2ns}} \int_{\mathbb{B}} |f(rz) - f(z)|^p (1 - |z|^2)^{ns+q-n-1} \, dV(z) < \frac{\varepsilon}{3}.$$

Therefore, for all r with $\max \left\{ \frac{1}{2}, \frac{\delta}{\delta_0}, r_1 \right\} < r < 1$, we have

$$\|f_r - f\|^p = \sup_{a \in \mathbb{B}} \int_{\mathbb{B}} |f(rz) - f(z)|^p (1 - |z|^2)^q (1 - |\Phi_a(z)|^2)^{ns} \, d\lambda(z)$$

$$\leq \left(\sup_{|a| \leq \delta_0} + \sup_{\delta_0 < |a| < 1} \right) \int_{\mathbb{B}} |f(rz) - f(z)|^p (1 - |z|^2)^q (1 - |\Phi_a(z)|^2)^{ns} \, d\lambda(z)$$

$$\leq \frac{\varepsilon}{3} + 2^{p-1} \sup_{\delta_0 < |a| < 1} \int_{\mathbb{B}} (|f(rz)|^p + |f(z)|^p)(1 - |z|^2)^q (1 - |\Phi_a(z)|^2)^{ns} \, d\lambda(z)$$

$$< \frac{\varepsilon}{3} + 2^{p-1} \left(\frac{\varepsilon}{3 \cdot 2^{p-1}} + \frac{\varepsilon}{3 \cdot 2^{2n+p-1}} \right) < \frac{\varepsilon}{3} + \frac{\varepsilon}{3} + \frac{\varepsilon}{3} = \varepsilon,$$

which shows that $\|f_r - f\| \to 0$ as $r \to 1$.

Sufficiency First, we show that, for each $r \in (0, 1)$ and each $f \in \mathcal{N}(p, q, s)$, we have $f_r \in \mathcal{N}^0(p, q, s)$. By formula (7.13) in Chap. 7, we have

$$\int_{\mathbb{B}} \frac{(1 - |z|^2)^{q+ns-n-1}}{|1 - \langle a, z \rangle|^{2ns}} \, dV(z) \simeq \begin{cases} \text{bounded in } \mathbb{B}, & \text{if } ns < q, \\ \log \frac{1}{1-|a|^2}, & \text{if } ns = q, \\ (1 - |a|^2)^{q-ns}, & \text{if } ns > q, \end{cases}$$

which implies

$$\sup_{a \in \mathbb{B}} (1 - |a|^2)^{ns} \int_{\mathbb{B}} \frac{(1 - |z|^2)^{q+ns-n-1}}{|1 - \langle a, z \rangle|^{2ns}} \, dV(z) < \infty.$$

Then

$$\|f_r\|^p = \sup_{a \in \mathbb{B}} \int_{\mathbb{B}} |f_r(z)|^p (1 - |z|^2)^q (1 - |\Phi_a(z)|^2)^{ns} \, d\lambda(z)$$

$$\leq \left(\sup_{z \in \mathbb{B}} |f(rz)| \right)^p \cdot \sup_{a \in \mathbb{B}} \int_{\mathbb{B}} (1 - |z|^2)^q (1 - |\Phi_a(z)|^2)^{ns} \, d\lambda(z)$$

$$= \left(\sup_{z \in \mathbb{B}} |f(rz)| \right)^p \cdot \sup_{a \in \mathbb{B}} (1 - |a|^2)^{ns} \int_{\mathbb{B}} \frac{(1 - |z|^2)^{q+ns-n-1}}{|1 - \langle a, z \rangle|^{2ns}} \, dV(z) < \infty,$$

which shows that $f_r \in \mathcal{N}(p, q, s)$. Moreover, since $f(z)$ on $\{|z| \leq r\}$, for some $M > 0$, we have

$$\int_{\mathbb{B}} |f_r(z)|^p (1 - |z|^2)^q (1 - |\Phi_a(z)|^2)^{ns} d\lambda(z)$$

$$\leq M \int_{\mathbb{B}} (1 - |z|^2)^q (1 - |\Phi_a(z)|^2)^{ns} d\lambda(z).$$

Note that $ns + q > n$ and hence by the estimation on the term

$$\int_{\mathbb{B}} \frac{(1 - |z|^2)^{q+ns-n-1}}{|1 - \langle a, z \rangle|^{2ns}} dV(z)$$

in the above formula, we have

$$\int_{\mathbb{B}} (1 - |z|^2)^q (1 - |\Phi_a(z)|^2)^{ns} d\lambda(z) \to 0, \qquad \text{as } |a| \to 1.$$

This shows that $f_r \in \mathcal{N}^0(p, q, s)$ for all $r \in (0, 1)$.

Now suppose that $\|f_r - f\| \to 0$ as $r \to 1$. Then since $\mathcal{N}^0(p, q, s)$ is a closed subspace of $\mathcal{N}(p, q, s)$, we conclude that $f \in \mathcal{N}^0(p, q, s)$. $\qquad \square$

As a consequence of Lemma 12.2.2, we have the following result.

Theorem 12.2.3 *The set of polynomials is dense in* $\mathcal{N}^0(p, q, s)$, *whenever* $ns + q > n$.

Proof By Lemma 12.2.2, for each $f \in \mathcal{N}^0(p, q, s)$, we have

$$\lim_{r \to 1} \| f_r - f \| = 0.$$

Since each f_r can be uniformly approximated by polynomials, and moreover, as we have seen in the proof of Lemma 12.2.2, the sup-norm in $\mathbb{B}$ is stronger than the $\mathcal{N}(p, q, s)$-norm, every $f \in \mathcal{N}^0(p, q, s)$ can be approximated in the $\mathcal{N}(p, q, s)$-norm by polynomials.   □

However, in general, the space $\mathcal{N}(p, q, s)$ does not always contain all the polynomials. In fact, we have the following characterization.

Corollary 12.2.4 *Let* $p \geq 1$ *and* $q, s > 0$. *The space* $\mathcal{N}(p, q, s)$ *contains the set of polynomials if and only if* $ns + q > n$.

Proof **Necessity** Assume in contrary that $ns + q \leq n$. Then $\tau = n - ns - q \geq 0$. We show that the constant function $f(z) \equiv 1$ does not belong to $\mathcal{N}(p, q, s)$. Indeed, we have

$$\| f \|^p = \sup_{a \in \mathbb{B}} \int_{\mathbb{B}} (1 - |z|^2)^q (1 - |\Phi_a(z)|^2)^{ns} \, d\lambda(z)$$

$$= \sup_{a \in \mathbb{B}} (1 - |a|^2)^{ns} \int_{\mathbb{B}} \frac{(1 - |z|^2)^{q+ns-n-1}}{|1 - \langle a, z \rangle|^{2ns}} \, dV(z)$$

$$\geq \int_{\mathbb{B}} \frac{1}{(1 - |z|^2)^{\tau+1}} \, dV(z) \quad \text{(taking } a = 0\text{)}$$

$$\simeq \int_0^1 \frac{r^{2n-1}}{(1 - r^2)^{1+\tau}} \, dr \gtrsim \int_{1/2}^1 \frac{1}{(1 - r)^{1+\tau}} \, dr = \infty,$$

which is a contradiction.

Sufficiency Combining Lemma 12.2.2 and Theorem 12.2.3 yields the desired result.   □

12.3 The Space $\mathcal{N}_*(p, q, s)$

The invariant Green's function is defined as $G(z, a) = g(\Phi_a(z))$, where

$$g(z) = \frac{n+1}{2n} \int_{|z|}^{1} (1 - t^2)^{n-1} t^{-2n+1} dt.$$

The following property of g is important in the sequel.

Lemma 12.3.1 *Let $n \geq 2$ be an integer. Then there are positive constants C_1 and C_2 such that for all $z \in \mathbb{B} \backslash \{0\}$,*

$$C_1 (1 - |z|^2)^n |z|^{-2(n-1)} \leq g(z) \leq C_2 (1 - |z|^2)^n |z|^{-2(n-1)}.$$

Proof We have

$$\lim_{|z| \to 0} \frac{g(z)}{(1 - |z|^2)^n |z|^{-2(n-1)}} = \frac{n+1}{4n(n-1)}$$

and

$$\lim_{|z| \to 1} \frac{g(z)}{(1 - |z|^2)^n |z|^{-2(n-1)}} = \frac{n+1}{4n^2}.$$

From the above two identities and the continuity of $g(z)$, the desired result follows. $\quad\square$

For $p \geq 1$ and $q, s > 0$ and $n \geq 2$, define the $\mathcal{N}_*(p, q, s)$-type space and a corresponding little subspace by

$$\mathcal{N}_*(p, q, s) =$$

$$\left\{ f \in H(\mathbb{B}) : \|f\|_*^p = \sup_{a \in \mathbb{B}} \int_{\mathbb{B}} |f(z)|^p (1 - |z|^2)^q G^s(z, a) \, d\lambda(z) < \infty \right\}$$

and

$$\mathcal{N}_*^0(p, q, s) =$$

$$\left\{ f \in \mathcal{N}_*(p, q, s) : \lim_{|a| \to 1} \int_{\mathbb{B}} |f(z)|^p (1 - |z|^2)^q G^s(z, a) \, d\lambda(z) = 0 \right\}.$$

We note, by Lemma 12.3.1, that $\| \cdot \| \lesssim \| \cdot \|_*$, which means that $\mathcal{N}_*(p, q, s) \subseteq \mathcal{N}(p, q, s)$. Combining this fact with the proof of Lemma 12.1.1, Theorem 12.1.2, and Lemma 12.2.1, we have the following result.

Theorem 12.3.2 *For $p \geq 1$, and $q, s > 0$, and $n \geq 2$, the following assertions are true.*

(i) $\mathcal{N}_*(p, q, s)$ is a functional Banach space.

(ii) $\mathcal{N}_*^0(p,q,s)$ is a closed subspace of $\mathcal{N}_*(p,q,s)$.

(iii) $\mathcal{N}_*(p,q,s) \subseteq A^{-\frac{q}{p}}(\mathbb{B})$.

As in the $\mathcal{N}(p,q,s)$ space, in general, not every $\mathcal{N}_*(p,q,s)$ contains all the polynomials. For instance, by Corollary 12.2.4 and the fact that $\|\cdot\| \lesssim \|\cdot\|_*$, we see that in case $ns + q \leq n$, the function $f \equiv 1$ does not belong to $\mathcal{N}_*(p,q,s)$. However, we have the following result.

Lemma 12.3.3 *Let* $p \geq 1$, *let* $q, s > 0$, *and let* $n \geq 2$. *Then the space* $\mathcal{N}_*(p,q,s)$ *contains the set of all polynomials if and only if* $ns + q > n$ *and* $s < \frac{n}{n-1}$.

Proof **Necessity** Consider the constant function $f(z) \equiv 1$. By Lemma 12.3.1, we have

$$
\begin{aligned}
\|f\|_*^p &= \sup_{a \in \mathbb{B}} \int_{\mathbb{B}} (1 - |z|^2)^q G^s(z,a) \, d\lambda(z) \\
&\geq \int_{\mathbb{B}} (1 - |z|^2)^q G^s(z) \, d\lambda(z) \\
&\simeq \int_{\mathbb{B}} (1 - |z|^2)^q \frac{(1 - |z|^2)^{ns}}{|z|^{2(n-1)s}} \, d\lambda(z) \\
&= \int_{\mathbb{B}} \frac{(1 - |z|^2)^{q+ns-n-1}}{|z|^{2(n-1)s}} \, dV(z) \\
&= I_1 + I_2,
\end{aligned}
$$

where

$$
I_1 = \int_{\mathbb{B}_{1/2}} \frac{(1 - |z|^2)^{q+ns-n-1}}{|z|^{2(n-1)s}} \, dV(z)
$$

and

$$
I_2 = \int_{\mathbb{B} \setminus \mathbb{B}_{1/2}} \frac{(1 - |z|^2)^{q+ns-n-1}}{|z|^{2(n-1)s}} \, dV(z).
$$

There are now two cases to consider.

Case I: $ns + q \leq n$. In this case,

$$
I_2 \simeq \int_{\mathbb{B} \setminus \mathbb{B}_{1/2}} (1 - |z|^2)^{q+ns-n-1} \, dV(z) \simeq \int_{1/2}^1 \frac{1}{(1-r)^{n+1-q-ns}} \, dr = \infty,
$$

which is a contradiction.

Case II: $s \geq \frac{n}{n-1}$. In this case,

$$I_1 \simeq \int_{\mathbb{B}_{1/2}} \frac{1}{|z|^{2(n-1)s}} \, dV(z) \simeq \int_0^{1/2} \frac{1}{r^{2(n-1)s-2n+1}} \, dr = \infty,$$

which contradicts the assumption that $f \in \mathcal{N}_*(p, q, s)$.

Sufficiency Suppose $ns + q > n$ and $s < \frac{n}{n-1}$. Let P be a polynomial defined on $\mathbb{B}$. Then we have

$$\|P\|_*^p = \sup_{a \in \mathbb{B}} \int_{\mathbb{B}} |P(z)|^p (1 - |z|^2)^q G^s(z, a) \, d\lambda(z)$$

$$\lesssim \sup_{a \in \mathbb{B}} \int_{\mathbb{B}} (1 - |z|^2)^q G^s(z, a) \, d\lambda(z) \quad (P \text{ is bounded on } \mathbb{B})$$

$$\simeq \sup_{a \in \mathbb{B}} \int_{\mathbb{B}} (1 - |z|^2)^p \frac{(1 - |\Phi_a(z)|^2)^{ns}}{|\Phi_a(z)|^{2(n-1)s}} \, d\lambda(z)$$

$$= \sup_{a \in \mathbb{B}} \int_{\mathbb{B}} (1 - |\Phi_a(w)|^2)^q \frac{(1 - |w|^2)^{ns}}{|w|^{2(n-1)s}} \, d\lambda(w) \quad (\text{change variable } w = \Phi_a(z))$$

$$= \sup_{a \in \mathbb{B}} (1 - |a|^2)^q \int_{\mathbb{B}} \frac{(1 - |w|^2)^{q+ns-n-1}}{|w|^{2(n-1)s} |1 - \langle a, w \rangle|^{2q}} \, dV(w).$$

For each $a \in \mathbb{B}$, consider

$$\int_{\mathbb{B}} \frac{(1 - |w|^2)^{q+ns-n-1}}{|w|^{2(n-1)s} |1 - \langle a, w \rangle|^{2q}} \, dV(w) = I_{1,a} + I_{2,a},$$

where

$$I_{1,a} = \int_{\mathbb{B}_{1/2}} \frac{(1 - |w|^2)^{q+ns-n-1}}{|w|^{2(n-1)s} |1 - \langle a, w \rangle|^{2q}} \, dV(w)$$

and

$$I_{2,a} = \int_{\mathbb{B} \setminus \mathbb{B}_{1/2}} \frac{(1 - |w|^2)^{q+ns-n-1}}{|w|^{2(n-1)s} |1 - \langle a, w \rangle|^{2q}} \, dV(w).$$

For $I_{1,a}$, by the proof of necessary part, we have

$$I_{1,a} \simeq \int_{\mathbb{B}_{1/2}} \frac{1}{|w|^{2(n-1)s}} \, dV(w) < \infty$$

independent of the choice of a, which implies that $\sup_{a \in \mathbb{B}}(1 - |a|^2)^q I_{1,a} < \infty$.

For $I_{2,a}$, by formula (7.13) in Chap. 7, we have

$$
I_{2,a} \simeq \int_{\mathbb{B} \setminus \mathbb{B}_{1/2}} \frac{(1 - |w|^2)^{q+ns-n-1}}{|w|^{2(n-1)s} |1 - \langle a, w \rangle|^{2q}} \, dV(w)
$$

$$
\lesssim \int_{\mathbb{B}} \frac{(1 - |w|^2)^{q+ns-n-1}}{|1 - \langle a, w \rangle|^{2q}} \, dV(w)
$$

$$
\simeq \begin{cases} \text{bounded in } \mathbb{B}, & \text{if } ns > q \\ \log \frac{1}{1-|a|^2}, & \text{if } ns = q \\ (1 - |a|^2)^{ns-q}, & \text{if } ns < q, \end{cases}
$$

which shows that $\sup_{a \in \mathbb{B}}(1 - |a|^2)^q I_{2,a} < \infty$. Thus, $\|P\|_* < \infty$, and the proof is complete.

$\square$

Lemma 12.3.3 leads us to the following description of $\mathcal{N}(p, q, s)$-spaces by the invariant Green's function.

Theorem 12.3.4 *Let* $p \geq 1$, *let* $q, s > 0$, *and let* $n \geq 2$. *If* $s < \frac{n}{n-1}$, *then* $\mathcal{N}(p, q, s) = \mathcal{N}_*(p, q, s)$. *In particular, if* $1 < s < \frac{n}{n-1}$, *then*

$$
\mathcal{N}(p, q, s) = \mathcal{N}_*(p, q, s) = A^{-\frac{q}{p}}(\mathbb{B}).
$$

Proof It is clear $\mathcal{N}_*(p, q, s) \subseteq \mathcal{N}(p, q, s)$. We show that $\mathcal{N}(p, q, s) \subseteq \mathcal{N}_*(p, q, s)$, whenever $s < \frac{n}{n-1}$. Take $f \in \mathcal{N}_*(p, q, s)$. For each $a \in \mathbb{B}$, we have

$$
\int_{\mathbb{B}} |f(z)|^p (1 - |z|^2)^q G^s(z, a) \, d\lambda(z)
$$

$$
\simeq \int_{\mathbb{B}} |f(z)|^p (1 - |z|^2)^q \frac{(1 - |\Phi_a(z)|^2)^{ns}}{|\Phi_a(z)|^{2(n-1)s}} \, d\lambda(z)
$$

$$
= \int_{\mathbb{B}} |f(\Phi_a(w))|^p (1 - |\Phi_a(w)|^2)^q \frac{(1 - |w|^2)^{ns}}{|w|^{2(n-1)s}} \, d\lambda(w)
$$

(change variable $w = \Phi_a(z)$)

$$
= J_{1,a} + J_{2,a},
$$

where

$$J_{1,a} = \int_{\mathbb{B}_{1/2}} |f(\Phi_a(w))|^p (1 - |\Phi_a(w)|^2)^q \frac{(1 - |w|^2)^{ns}}{|w|^{2(n-1)s}} \, d\lambda(w)$$

and

$$J_{2,a} = \int_{\mathbb{B} \setminus \mathbb{B}_{1/2}} |f(\Phi_a(w))|^p (1 - |\Phi_a(w)|^2)^q \frac{(1 - |w|^2)^{ns}}{|w|^{2(n-1)s}} \, d\lambda(w).$$

For $J_{2,a}$, we have

$$J_{2,a} \lesssim \int_{\mathbb{B} \setminus \mathbb{B}_{1/2}} |f(\Phi_a(w))|^p (1 - |\Phi_a(w)|^2)^q (1 - |w|^2)^{ns} \, d\lambda(w)$$

$$\leq \int_{\mathbb{B}} |f(\Phi_a(w))|^p (1 - |\Phi_a(w)|^2)^q (1 - |w|^2)^{ns} \, d\lambda(w)$$

$$= \int_{\mathbb{B}} |f(z)|^p (1 - |z|^2)^q (1 - |\Phi_a(z)|^2)^{ns} \, d\lambda(z) \leq \|f\|^p.$$

For $J_{1,a}$, taking into account $s < \frac{n}{n-1}$, by Lemma 12.1.1, and the proof of Lemma 12.3.3, we have

$$J_{1,a} \leq |f|_{q/p}^p \int_{\mathbb{B}_{1/2}} \frac{(1 - |w|^2)^{ns-n-1}}{|w|^{2(n-1)s}} \, dV(w)$$

$$\lesssim \|f\|^p \int_{\mathbb{B}_{1/2}} \frac{1}{|w|^{2(n-1)s}} \, dV(w) \leq M \|f\|^p,$$

for some $M > 0$, which is independent of a choice of a. Combining the two estimates yields

$$\int_{\mathbb{B}} |f(z)|^p (1 - |z|^2)^q G^s(z, a) \, d\lambda(z) \lesssim \|f\|^p,$$

which gives $\|f\|_* \lesssim \|f\|$. In particular, when $1 < s < \frac{n}{n-1}$, from the above argument and Theorem 12.1.3, it follows that $\mathcal{N}(p, q, s) = \mathcal{N}_*(p, q, s) = A^{-\frac{q}{p}}(\mathbb{B})$. $\square$

The following result is a straightforward consequence of Theorem 12.3.4.

Corollary 12.3.5 *Let $n \geq 2$ and $0 < s < \frac{n}{n-1}$. Then*

$$\mathcal{N}_{ns} = \mathcal{N}(2, n+1, s) = \mathcal{N}_*(2, n+1, s).$$

Moreover, if $1 < s < \frac{n}{n-1}$, then $A^{-\frac{n+1}{2}} = \mathcal{N}(2, n+1, s) = \mathcal{N}_*(2, n+1, s).$

A question natural arises now. What happens when $s \geq \frac{n}{n-1}$? It turns out that in this case, all spaces $\mathcal{N}_*(p, q, s)$ are trivial. We end this section with this observation.

Theorem 12.3.6 *Let $p \geq 1$, let $q, s > 0$, and let $n \geq 2$. If $s \geq \frac{n}{n-1}$, then*

$$\mathcal{N}_*(p, q, s) = \{0\}.$$

Proof Assume in contrary that there exists a $a_0 \in \mathbb{B}$, such that $|f(a_0)| \geq \delta > 0$. Then for some $r > 0$,

$$|f(\Phi_a(w))| \geq \frac{\delta}{2}, \qquad (|w| < r).$$

Hence,

$$\|f\|_*^p = \sup_{a \in \mathbb{B}} \int_{\mathbb{B}} |f(z)|^p (1 - |z|^2)^q G^s(z, a) \, d\lambda(z)$$

$$\geq \int_{\mathbb{B}} |f(z)|^p (1 - |z|^2)^q G^s(z, a_0) \, d\lambda(z)$$

$$\gtrsim \int_{\mathbb{B}} |f(z)|^p (1 - |z|^2)^q \frac{(1 - |\Phi_{a_0}(z)|^2)^{ns}}{|\Phi_{a_0}(z)|^{2(n-1)s}} \, d\lambda(z)$$

$$= \int_{\mathbb{B}} |f(\Phi_{a_0}(w))|^p (1 - |\Phi_{a_0}(w)|^2)^q \frac{(1 - |w|^2)^{ns-n-1}}{|w|^{2(n-1)s}} \, dV(w)$$

(change variable $z = \Phi_a(w)$)

$$\geq \int_{|w|<r} |f(\Phi_{a_0}(w))|^p (1 - |\Phi_{a_0}(w)|^2)^q \frac{(1 - |w|^2)^{ns-n-1}}{|w|^{2(n-1)s}} \, dV(w)$$

$$\geq \left(\frac{\delta}{2}\right)^p (1 - |a_0|^2)^q \int_{|w|<r} \frac{(1 - |w|^2)^{ns+q-n-1}}{|w|^{2(n-1)s} |1 - \langle a, w \rangle|^{2q}} \, dV(w)$$

$$\gtrsim \int_{|w|<r} \frac{1}{|w|^{2(n-1)s}} \, dV(w) \gtrsim \int_0^r \frac{1}{t^{2(n-1)s-2n+1}} \, dt = \infty,$$

which is a contradiction.   □

By Theorems 12.3.4 and 12.3.6, it is clear that $\mathcal{N}_*(p, q, s)$-type spaces are a special case of $\mathcal{N}(p, q, s)$-type spaces. Therefore, in the sequel, we focus our attention on $\mathcal{N}(p, q, s)$-type spaces.

12.4   Hadamard Gaps in $\mathcal{N}(p, q, s)$-Type Spaces

In the previous chapter, problems with Hadamard gaps have been considered for $\mathcal{N}_p$ spaces in the unit ball. In this chapter, we study it for $\mathcal{N}(p, q, s)$ spaces. Namely, given a Hadamard gap series, we are interested in the following question: for $p \geq 1$ and $q, s > 0$, when does this Hadamard gap series belong to $\mathcal{N}(p, q, s)$? We note that a constant function has Hadamard gaps, and hence, by Corollary 12.2.4 and Lemma 12.3.3, we may assume that the condition $ns + q > n$, or equivalently $s > 1 - \frac{q}{n}$, holds. Moreover, if $s > 1$, then by Theorem 12.1.3, $\mathcal{N}(p, q, s) = A^{-\frac{q}{p}}(\mathbb{B})$, which was already studied in the previous chapter. Hence, it remains to study the case $s \leq 1$. We keep the notations

$$M_k = \sup_{\xi \in \mathbb{S}} |P_{n_k}(\xi)|$$

and

$$L_{k,p} = \left(\int_{\mathbb{S}} |P_{n_k}(\xi)|^p d\sigma(\xi) \right)^{1/p}, \qquad (p \geq 1),$$

where $d\sigma$ is the normalized surface measure on $\mathbb{S}$, that is, $\sigma(\mathbb{S}) = 1$. These quantities clearly are well-defined for each $k \geq 0$ and $p \geq 1$.

Theorem 12.4.1   *Let $p \geq 1$, let $q > 0$, and let $\max\left\{0, 1 - \frac{q}{n}\right\} < s \leq 1$. Suppose that*

$$f(z) = \sum_{k=0}^{\infty} P_{n_k}(z)$$

is a series with Hadamard gaps. Consider the following statements:

(i) $\displaystyle\sum_{k=0}^{\infty} \frac{1}{2^{k(ns+q-n)}} \left(\sum_{2^k \leq n_j < 2^{k+1}} M_j^p \right) < \infty.$

(ii) $f \in \mathcal{N}^0(p, q, s).$

(iii) $f \in \mathcal{N}(p, q, s).$

(iv) $\displaystyle\sum_{k=0}^{\infty} \frac{1}{2^{k(ns+q-n)}} \left(\sum_{2^k \leq n_j < 2^{k+1}} L_{j,p}^p \right) < \infty.$

Then (i) $\Longrightarrow$ (ii) $\Longrightarrow$ (iii) $\Longrightarrow$ (iv).

Proof (i) $\Longrightarrow$ (ii). Suppose that (i) holds. We first show that $f \in \mathcal{N}(p, q, s)$. For $f(z) = \sum_{k=0}^{\infty} P_{n_k}(z)$, by Lemma 11.1.1 about integration in polar coordinates, we have

$$\|f\|^p = \sup_{a \in \mathbb{B}} \int_{\mathbb{B}} \left| \sum_{k=0}^{\infty} P_{n_k}(z) \right|^p (1 - |z|^2)^q (1 - |\Phi_a(z)|^2)^{ns} \, d\lambda(z)$$

$$\leq \sup_{a \in \mathbb{B}} \int_{\mathbb{B}} \left(\sum_{k=0}^{\infty} |P_{n_k}(z)| \right)^p (1 - |z|^2)^q (1 - |\Phi_a(z)|^2)^{ns} \, d\lambda(z)$$

$$= \sup_{a \in \mathbb{B}} \int_{\mathbb{B}} \left(\sum_{k=0}^{\infty} |P_{n_k}(z)| \right)^p \frac{(1 - |a|^2)^{ns} (1 - |z|^2)^{ns+q-n-1}}{|1 - \langle a, z \rangle|^{2ns}} \, dV(z)$$

$$\lesssim \sup_{a \in \mathbb{B}} \int_0^1 (1 - |r|^2)^{ns+q-n-1} \int_{\mathbb{S}} \frac{(1 - |a|^2)^{ns} \left(\sum_{n=0}^{\infty} |P_{n_k}(r\zeta)| \right)^p}{|1 - \langle a, r\zeta \rangle|^{2ns}} \, d\sigma(\zeta) \, dr.$$

Since P_{n_k} is homogeneous, we have

$$\sum_{k=0}^{\infty} |P_{n_k}(r\zeta)| = \sum_{k=0}^{\infty} |P_{n_k}(\zeta) r^{n_k}| \leq \sum_{k=0}^{\infty} M_{n_k} r^{n_k}, \qquad (\zeta \in \mathbb{S}). \tag{12.4}$$

Moreover, for each $a \in \mathbb{B}$, by formula (7.13) in Chap. 7, we also have

$$\int_{\mathbb{S}} \frac{1}{|1 - \langle r\zeta, a \rangle|^{2ns}} d\sigma(\zeta) = \int_{\mathbb{S}} \frac{1}{|1 - \langle \zeta, ra \rangle|^{2ns}} d\sigma(\zeta)$$

$$\simeq \begin{cases} \text{bounded in } \mathbb{B}, & \text{if } s < \frac{1}{2}, \\ \log \frac{1}{1 - r^2 |a|^2}, & \text{if } s = \frac{1}{2}, \\ (1 - r^2 |a|^2)^{n-2ns}, & \text{if } \frac{1}{2} < s < 1, \end{cases}$$

$$\leq \begin{cases} \text{bounded in } \mathbb{B}, & \text{if } s < \frac{1}{2}, \\ \log \frac{1}{1 - |a|^2}, & \text{if } s = \frac{1}{2}, \\ (1 - |a|^2)^{n-2ns}, & \text{if } \frac{1}{2} < s < 1, \end{cases}$$

which implies that there exists a constant $C > 0$ such that

$$\sup_{a \in \mathbb{B}} (1 - |a|^2)^{ns} \int_{\mathbb{S}} \frac{1}{|1 - \langle r\zeta, a \rangle|^{2ns}} \, dV(z) \leq C. \tag{12.5}$$

Combining (12.4) and (12.5), by formula (9.23), we have

$$\|f\|^p \lesssim \int_0^1 \left(\sum_{k=0}^{\infty} M_{n_k} r^{n_k} \right)^p (1 - |r|^2)^{ns+q-n-1} \, dr$$

$$\simeq \sum_{k=0}^{\infty} \frac{1}{2^{k(ns+q-n)}} \Big(\sum_{2^k \leq n_j < 2^{k+1}} M_j \Big)^p.$$

Since f is in the Hadamard gap class, there exists a constant $c > 1$ such that $n_{j+1} \leq cn_j$ for all $j \geq 0$. Hence, the maximum number of n_j's between 2^k and 2^{k+1} is less or equal to $[\log_c 2] + 1$ for $k \geq 0$. Moreover, for every $k \geq 0$, the Hölder inequality gives

$$\Big(\sum_{2^k \leq n_j < 2^{k+1}} M_j \Big)^p \leq ([\log_c 2] + 1)^{p-1} \Big(\sum_{2^k \leq n_j < 2^{k+1}} M_j^p \Big).$$

Consequently, since (i) holds, we obtain

$$\|f\|^p \lesssim \sum_{k=0}^{\infty} \frac{1}{2^{k(ns+q-n)}} \Big(\sum_{2^k \leq n_j < 2^{k+1}} M_j^p \Big) < \infty, \tag{12.6}$$

which shows that $f \in \mathcal{N}(p, q, s)$.

Now we prove that $f \in \mathcal{N}^0(p, q, s)$. For each $m \in \mathbb{N}$, put

$$f_m(z) = \sum_{k=0}^{m} P_{n_k}(z), m \in \mathbb{N},$$

which is bounded in $\overline{\mathbb{B}}$. By Lemma 12.2.2, $f_m \in \mathcal{N}^0(p, q, s)$. Furthermore, $\mathcal{N}^0(p, q, s)$ is closed, and by Theorem 12.2.3, the set of all polynomials is dense in $\mathcal{N}^0(p, q, s)$. So it remains to show that $\|f_m - f\| \to 0$ as $\to \infty$. Indeed, by (12.6),

$$\|f_m - f\|^p \lesssim \sum_{k=m'}^{\infty} \Big(\frac{1}{2^{k(ns+q-n)}} \sum_{2^k \leq n_j < 2^{k+1}} M_j^p \Big), \tag{12.7}$$

where $m' = \left[\frac{m+1}{[\log_c 2]+1} \right]$. Hence, combining condition (i) and (12.7) yields the desired result.

$(ii) \implies (iii)$. It is obvious.

$(iii) \implies (iv)$. Suppose $f \in \mathcal{N}(p, q, s)$. As the proof of Theorem 9.6.2, we have

$$\|f\|^p = \sup_{a \in \mathbb{B}} \int_{\mathbb{B}} \Big| \sum_{k=0}^{\infty} P_{n_k}(z) \Big|^p (1 - |z|^2)^q (1 - |\Phi_a(z)|^2)^{ns} \, d\lambda(z)$$

$$\geq \int_{\mathbb{B}} \Big| \sum_{k=0}^{\infty} P_{n_k}(z) \Big|^p (1 - |z|^2)^{q+ns-n-1} \, dV(z)$$

$$\simeq \int_{\mathbb{S}} \left(\sum_{k=0}^{\infty} \frac{1}{2^{k(ns+q-n)}} \sum_{2^k \leq n_j < 2^{k+1}} |P_{n_k}(\zeta)|^p \right) d\sigma(\zeta)$$

$$= \sum_{k=0}^{\infty} \frac{1}{2^{k(ns+q-n)}} \left(\sum_{2^k \leq n_j < 2^{k+1}} L_j^p \right),$$

which gives (iv).   □

Note that in general, the part (iv) does not imply condition (i). A counterexample is given below.

Example 12.4.2 *Let*

$$f(z) = \sum_{k=0}^{\infty} 2^{\frac{k(p+1)}{2}} z_1^{2^k}, \quad z = (z_1, z_2, \ldots, z_n) \in \mathbb{B}.$$

In this case, $M_k = 2^{\frac{k(p+1)}{2}}$, and hence,

$$\sum_{k=0}^{\infty} \left(\frac{1}{2^{k(p+1)}} \sum_{2^k \leq n_j < 2^{k+1}} M_k^2 \right) = \infty.$$

On the other hand, for each $k \geq 0$,

$$L_k^2 = 2^{k(p+1)} \int_{z \in \mathbb{S}} |z_1^{2^k}|^2 \, d\sigma(z) = 2^{k(p+1)} \frac{(n-1)!(2^k)!}{(n-1+2^k)!} \lesssim \frac{2^{k(p+1)}}{2^{k(n-1)}},$$

which gives

$$\sum_{k=0}^{\infty} \left(\frac{1}{2^{k(p+1)}} \sum_{2^k \leq n_j < 2^{k+1}} L_k^2 \right) \lesssim \sum_{k=0}^{\infty} \frac{1}{2^{k(n-1)}} < \infty.$$

Despite the above example, there are some special cases for which all the conditions in Theorem 12.4.1 are equivalent. Once more, we make use of the sequence of homogeneous polynomials $\{P_k\}_{k \in \mathbb{N}}$ constructed in Chap. 9 (see (9.21)) which satisfy $\deg(P_k) = k$,

$$\sup_{\zeta \in \mathbb{S}} |P_k(\zeta)| = 1 \quad \text{and} \quad \int_{\mathbb{S}} |P_k(\zeta)|^p \, d\sigma(\zeta) \geq C(p, n),$$

where $C(p, n)$ is a positive constant depending on p and n.

Theorem 12.4.3 *Let $p \geq 1$, let $q > 0$, and let $\max\left\{0, 1 - \frac{q}{n}\right\} < s \leq 1$. Suppose that*

$$f(z) = \sum_{k=0}^{\infty} a_k P_{n_k}(z)$$

is a series with Hadamard gaps, where $a_k \in \mathbb{C}$ $(k \geq 0)$. Then the following statements are equivalent.

(i) $\displaystyle\sum_{k=0}^{\infty} \frac{1}{2^{k(ns+q-n)}} \left(\sum_{2^k \leq n_j < 2^{k+1}} |a_j|^p \right) < \infty.$

(ii) $f \in \mathcal{N}^0(p, q, s).$

(iii) $f \in \mathcal{N}(p, q, s).$

Proof The equivalence of all three statements follows from the fact that $M_k \simeq L_{k,p}$ for every $k \geq 0$. ☐

Let $p \geq 1$, let $q > 0$, and let $\max\left\{0, 1 - \frac{q}{n}\right\} < s_1 < s_2 \leq 1$. It is clear that

$$\mathcal{N}(p, q, s_1) \subseteq \mathcal{N}(p, q, s_2) \subseteq A^{-\frac{q}{p}}(\mathbb{B}). \tag{12.8}$$

A consequence of Theorem 12.4.1 is that the inclusions in (12.8) are strict.

Corollary 12.4.4 *Let $p \geq 1$, let $q > 0$, and let $\max\left\{0, 1 - \frac{q}{n}\right\} < s_1 < s_2 \leq 1$. Then*

$$\mathcal{N}(p, q, s_1) \subsetneq \mathcal{N}(p, q, s_2) \subsetneq A^{-\frac{q}{p}}(\mathbb{B}).$$

Proof First, we prove that $\mathcal{N}(p, q, s_2) \subsetneq A^{-\frac{q}{p}}(\mathbb{B})$. Indeed, take the series

$$f_1(z) = \sum_{k=0}^{\infty} 2^{\frac{kq}{p}} P_{2^k}(z),$$

which is in $A^{-\frac{q}{p}}$. We have

$$\sum_{k=0}^{\infty} \frac{1}{2^{k(ns_2+q-n)}} \left(\sum_{2^k \leq n_j < 2^{k+1}} \left|2^{\frac{kq}{p}}\right|^p \right) = \sum_{k=0}^{\infty} \frac{1}{2^{k(ns_2-n)}} = \infty,$$

which, by Theorem 12.4.3, shows that $f_1 \notin \mathcal{N}(p, q, s_2)$.

Next, we show that $\mathcal{N}(p, q, s_1) \subsetneq \mathcal{N}(p, q, s_2)$. Consider the series

$$f_2(z) = \sum_{k=0}^{\infty} 2^{\frac{k(ns_1+q-n)}{p}} P_{2^k}(z).$$

By Theorem 12.4.3,

$$\sum_{k=0}^{\infty} \frac{1}{2^{k(ns_2+q-n)}} \left(\sum_{2^k \leq n_j < 2^{k+1}} \left| 2^{\frac{k(ns_1+q-n)}{p}} \right|^p \right) = \sum_{k=0}^{\infty} \frac{1}{2^{kn(s_2-s_1)}} < \infty,$$

which implies that $f_2 \in \mathcal{N}(p, q, s_2)$.

Finally,

$$\sum_{k=0}^{\infty} \frac{1}{2^{k(ns_1+q-n)}} \left(\sum_{2^k \leq n_j < 2^{k+1}} \left| 2^{\frac{k(ns_1+q-n)}{p}} \right|^p \right) = \sum_{k=0}^{\infty} 1 = \infty,$$

which, again by Theorem 12.4.3, implies $f_2 \notin \mathcal{N}(p, q, s_1)$.   □

We can easily see that, for any $k \in \left(0, \dfrac{q + ns - n}{p} \right)$,

$$A^{-k}(\mathbb{B}) \subsetneq \mathcal{N}(p, q, s).$$

Indeed, there is s' satisfying $k < \dfrac{q + ns' - n}{p} < \dfrac{q + ns - n}{p}$. By Corollary 12.4.4 and Theorem 12.4.3,

$$A^{-k}(\mathbb{B}) \subseteq \mathcal{N}(p, q, s') \subsetneq \mathcal{N}(p, q, s),$$

from which the desired result follows.

12.5   Carleson Measure and Embedding Relationship

The definitions of Carleson measure, p-Carleson measure, and vanishing p-Carleson measure have been given in Sect. 9.5. As for $\mathcal{N}_p$ spaces, there is a good relationship between functions in $\mathcal{N}(p, q, s)$, $\mathcal{N}^0(p, q, s)$, and Carleson measures.

Proposition 12.5.1 *Let $f \in \text{Hol}(\mathbb{B})$, and let $p \geq 1$, $q, s > 0$, and*

$$d\mu_{f,p,q,s}(z) = |f(z)|^p (1 - |z|^2)^{q+ns} \, d\lambda(z).$$

Then:

(i) $f \in \mathcal{N}(p, q, s)$ if and only if $d\mu_{f,p,q,s}$ is an (ns)-Carleson measure.

(ii) $f \in \mathcal{N}^0(p, q, s)$ if and only if $d\mu_{f,p,q,s}$ is a vanishing (ns)-Carleson measure.

Moreover,

$$\|f\|^p \simeq \sup_{0<r<1, \zeta \in \mathbb{S}} \frac{\mu_{f,p,q,s}(Q_r(\zeta))}{r^{ns}}$$

$$= \sup_{0<r<1, \zeta \in \mathbb{S}} \frac{1}{r^{ns}} \int_{Q_r(\zeta)} |f(z)|^p (1 - |z|^2)^{q+ns} \, d\lambda(z). \qquad (12.9)$$

Proof

(i): Note that for $f \in \mathcal{N}(p, q, s)$, we can write

$$\|f\|^p = \sup_{a \in \mathbb{B}} \int_{\mathbb{B}} |f(z)|^p \frac{(1 - |a|^2)^{ns}(1 - |z|^2)^{q+ns-n-1}}{|1 - \langle a, z \rangle|^{2ns}} \, dV(z)$$

$$= \sup_{a \in \mathbb{B}} \int_{\mathbb{B}} \left(\frac{1 - |a|^2}{|1 - \langle a, z \rangle|^2} \right)^{ns} d\mu_{f,p,q,s}(z).$$

From Lemma 9.5.1, statement (i) and Equation (12.9) follow.

(ii): This is a consequence of the "little-o version" of the same Lemma 9.5.1.

$\square$

Let f be a holomorphic function on $\mathbb{B}$. For each $k > 0$, we put

$$M_k(r, f) = \left(\int_{\mathbb{S}} |f(r\zeta)|^k \, d\sigma(\zeta) \right)^{1/k}, \qquad 0 \leq r < 1.$$

Using this quantity, we can rewrite the norm of $\| \cdot \|_{k,\rho}$ of the weighted Bergman space A_ρ^k with $k \geq 1$ and $\rho > -1$, as follows.

$$\|f\|_{k,\rho} \simeq \left(\int_0^1 r^{2n-1}(1 - r^2)^\rho M_k^k(r, f) \, dr \right)^{1/k}$$

$$\simeq \left(\int_0^1 r^{2n-1}(1 - r)^\rho M_k^k(r, f) \, dr \right)^{1/k}.$$

With the help of Proposition 12.5.1, we can establish the embedding relation between A_ρ^k and $\mathcal{N}(p, q, s)$ under some conditions on k and ρ. Note that, since the set of all polynomials belongs to A_ρ^k, by Corollary 12.2.4, it is natural to assume that $ns + q > n$.

First, we need some auxiliary results. Recall that the Carleson tube is

$$Q_r(\zeta) = \{z \in \mathbb{B} : d(z, \zeta) < r\}, \quad r > 0, \zeta \in \mathbb{S},$$

where

$$d(z, w) = |1 - \langle z, w \rangle|^{1/2}, \quad z, w \in \overline{\mathbb{B}}.$$

Actually, a Carleson tube behaves much like a Bergman metric ball. To see this, we define the Euclidean tube

$$Q(\zeta, r) \times (s, 1) = \{z \in \mathbb{B} : s < |z| < 1, \frac{z}{|z|} \in Q(\zeta, r) \qquad (12.10)$$

for $\zeta \in \mathbb{S}, 0 < r < \sqrt{2}, 0 < s < 1$.

Lemma 12.5.2 *Let* $0 < r < 1$ *and* $R > 0$. *Then there exists a constant* $k \in (0, 1)$ *(depending on R but not on r) such that*

$$D(a, R) \subset Q_r(\zeta) \subset Q(\zeta, r') \times (1 - r^2, 1), \quad \zeta \in \mathbb{S},$$

where

$$a = (1 - kr^2)\zeta, \quad r' = \sqrt{\frac{2r^2}{1 - r^2}}.$$

Proof First, let $\zeta \in \mathbb{S}$ and $z \in Q_r(\zeta)$. Since $z \neq 0$, we can write $z = \eta|z|$ for some $\eta \in \mathbb{S}$. Note that

$$1 - \langle z, \zeta \rangle = 1 - |z| + (1 - \langle \eta, \zeta \rangle)|z|;$$

by the triangle inequality, we have

$$|1 - \langle \eta, \zeta \rangle||z| < r^2 + (1 - |z|).$$

Also,

$$r^2 > |1 - \langle \eta, \zeta \rangle| \geq 1 - |z|.$$

Consequently, $1 - r^2 < |z| < 1$ and

$$|1 - \langle \eta, \zeta \rangle| < \frac{r^2 + 1 - |z|}{|z|} < \frac{2r^2}{1 - r^2},$$

which shows that

$$Q_r(\zeta) \subset Q(\zeta, r') \times (1 - r^2, 1).$$

Next, let $\zeta \in \mathbb{S}$ and $a = (1 - kr^2)\zeta$, where $k \in (0, 1)$ is some constant to be chosen later. If $z \in D(a, R)$, then $z = \phi_a(w)$ for some w with $w| < R'$, where $R' = \tanh(R) \in (0, 1)$. This gives

$$1 - \langle \eta, \zeta \rangle = \left\langle \phi_a(w), \frac{a}{1 - kr^2} \right\rangle = \frac{1 - \langle \phi_a(w), a \rangle - kr^2}{1 - kr^2}.$$

Note also that

$$1 - \langle \phi_a(w), a \rangle = \frac{1 - |a|^2}{1 - \langle w, a \rangle} \quad \text{and} \quad 1 - |a| = kr^2.$$

Then we have

$$1 - \langle \eta, \zeta \rangle = \frac{kr^1}{1 - kr^2} \left[\frac{1 + |a|}{1 - \langle w, a \rangle} - 1 \right].$$

Now we choose $k \in (0, 1)$ such that

$$\frac{kr^1}{1 - kr^2} \left| \frac{1 + |a|}{1 - \langle w, a \rangle} - 1 \right| < 1$$

for all $|w| < R'$; then $D(a, R) \subset Q_r(\zeta)$.   □

As a practical corollary of Lemma 12.5.2, we have the following result.

Corollary 12.5.3 *Let $\alpha > -1$. Then there exist positive constants c and C such that*

$$cr^{2(n+1+\alpha)} \leq v_\alpha(Q_r(\zeta)) \leq Cr^{2(n+1+\alpha)}, \quad \zeta \in \mathbb{S}, 0 \leq r \leq \sqrt{2}.$$

Proof By Lemma 12.5.2, we have

$$v_\alpha(D(a, R)) \sim (1 - |a|^2)^{n+1+\alpha} \sim r^{2(n+1+\alpha)} \quad \text{as } r \to 0^+.$$

Also, it follows from polar coordinates that

$$v_\alpha\big(Q(\zeta, r')\big) \times (1 - r^2, 1) \sim r^{2(n+1+\alpha)} \quad \text{as } r \to 0^+.$$

Since the desired result is trivial for r not close to 0, the proof is complete. □

We are ready to have the main result.

Theorem 12.5.4 *Let $p \geq 1$, let $q > 0$, and let $s > \max\left\{0, 1 - \frac{q}{n}\right\}$. Then the following assertions hold.*

(i) If $0 < s < 1$, then for $\max\{0, q - n\} < \rho < \dfrac{q + ns - n}{1 - s}$, we have

$$\|f\| \lesssim \|f\|_{\frac{p(n+\rho)}{q}, \rho - 1},$$

that is,

$$A_{\rho-1}^{\frac{p(n+\rho)}{q}} \subseteq \mathcal{N}(p, q, s).$$

(ii) If $s \geq 1$, then for $\rho > \max\{0, q - n\}$, we have

$$\|f\| \lesssim \|f\|_{\frac{p(n+\rho)}{q}, (\rho - 1)},$$

that is,

$$A_{\rho-1}^{\frac{p(n+\rho)}{q}} \subseteq \mathcal{N}(p, q, s).$$

Proof Since the proof of (ii) is a simple modification of (i), we just prove (i). Suppose $0 < s < 1$. Note that

$$\rho < \frac{q + ns - n}{1 - s} \implies (q + ns - n - \rho) \cdot \frac{n + \rho}{n + \rho - q} + \rho > 0.$$

By Corollary 12.5.3 and Hölder's inequality, for fixed $\zeta \in \mathbb{S}$ and $0 < r < 1$, we have

$$I_{r,\zeta} = \frac{1}{r^{ns}} \int_{Q_r(\zeta)} |f(z)|^p (1 - |z|^2)^{q+ns} \, d\lambda(z)$$

$$\simeq \frac{1}{r^{ns}} \int_{Q_r(\zeta)} |f(z)|^p (1 - |z|^2)^{q+ns-n-\rho} \, dV_{\rho-1}(z)$$

$$\leq \frac{1}{r^{ns}} \left(\int_{Q_r(\zeta)} |f(z)|^{\frac{p(n+\rho)}{q}} \, dV_{\rho-1}(z) \right)^{\frac{q}{n+\rho}}$$

$$\cdot \left(\int_{Q_r(s)} (1 - |z|^2)^{(q+ns-n-\rho)\cdot\frac{n+\rho}{n+\rho-q}} \, dV_{\rho-1}(z) \right)^{\frac{n+\rho-q}{n+\rho}}.$$

Hence,

$$I_{r,\varsigma} \lesssim \frac{\|f\|^p_{\frac{p(n+\rho)}{q},\rho-1}}{r^{ns}} \cdot \left(\int_{Q_r(s)} (1 - |z|^2)^{(q+ns-n-\rho)\cdot\frac{n+\rho}{n+\rho-q}+\rho-1} \, dV(z) \right)^{\frac{n+\rho-q}{n+\rho}}$$

$$\simeq \frac{\|f\|^p_{\frac{p(n+\rho)}{q},\rho-1}}{r^{ns}} \cdot \left(r^{n+1+(q+ns-n-\rho)\cdot\frac{n+\rho}{n+\rho-q}+\rho-1} \right)^{\frac{n+\rho-q}{n+\rho}}$$

$$= \|f\|^p_{\frac{p(n+\rho)}{q},\rho-1},$$

which, by Proposition 12.5.1, implies

$$\|f\|^p \simeq \sup_{\varsigma\in\mathbb{S},0<r<1} I_{r,\varsigma} \lesssim \|f\|^p_{\frac{p(n+\rho)}{q},\rho-1}.$$

The proof is complete.   □

Corollary 12.5.5 *Let $p \geq 1$, let $q > 0$, and let $s > \max\left\{0, 1 - \frac{q}{n}\right\}$. If $q > n$, then we have*

$$\|f\| \lesssim \|f\|_{p,q-n-1},$$

that is,

$$A^p_{q-n-1} \subseteq \mathcal{N}(p,q,s).$$

Proof Note that for fixed $\varsigma \in \mathbb{S}$ and $0 < r < 1$, if $z \in Q_r(\varsigma)$, then we have

$$r > |1 - \langle z, \varsigma\rangle| \geq 1 - |\langle z, \varsigma\rangle| \geq 1 - |z||\varsigma| = 1 - |z|. \tag{12.11}$$

For fixed $\varsigma \in \mathbb{S}$ and $0 < r < 1$, by (12.11),

$$I_{r,\varsigma} = \frac{1}{r^{ns}} \int_{Q_r(\varsigma)} |f(z)|^p (1 - |z|^2)^{q+ns} \, d\lambda(z)$$

$$\simeq \frac{1}{r^{ns}} \int_{Q_r(\varsigma)} |f(z)|^p (1 - |z|^2)^{ns} \, dV_{q-n-1}(z)$$

$$\leq \frac{1}{r^{ns}} \cdot \sup_{z\in Q_r(\varsigma)} (1 - |z|^2)^{ns} \cdot \int_{Q_r(\varsigma)} |f(z)|^p \, dV_{q-n-1}(z)$$

$$\lesssim \frac{1}{r^{ns}} \cdot \sup_{z \in Q_r(\zeta)} (1 - |z|)^{ns} \cdot \|f\|_{p,q-n-1}^p$$

$$\leq \|f\|_{p,q-n-1}^p,$$

which implies that $A_{q-n-1}^p \subseteq \mathcal{N}(p, q, s)$. $\square$

Corollary 12.5.6 *Let $p \geq 1$, let $q > 0$, and let $s > \max \left\{ 0, 1 - \frac{q}{n} \right\}$. Then*

$$\Lambda_{q+ns-n-1}^{\frac{p(q+ns)}{q}} \subseteq \mathcal{N}(p, q, s) \subseteq A_{q+ns-n-1}^p.$$

Proof Putting in Theorem 12.5.4 $\rho = q + ns - n$, we get

$$A_{q+ns-n-1}^{\frac{p(q+ns)}{q}} \subseteq \mathcal{N}(p, q, s).$$

Moreover, it is clear that $\| \cdot \|_{p,q+ns-n-1} \leq \| \cdot \|$, that is, $\mathcal{N}(p, q, s) \subset A_{q+ns-n-1}^p$. Hence, combining the last facts and Theorem 12.5.4 yields the desired result. $\square$

12.6   Embedding Relationship with Weighted Hardy Space

Recall that for $0 < p < \infty$ and $f \in \text{Hol}(\mathbb{B})$, the integral means are defined as follows:

$$M_p(f, f) = \left(\int_{\mathbb{S}} |f(r\zeta)|^p \, d\sigma(\zeta) \right)^{1/p},$$

where $0 \leq r < 1$. Note that $M_p(r, f)^p$ as a function of r is differentiable on $[0, 1)$. The general weighted Hardy space H_β^α ($\alpha > 0$ and $\beta \geq 0$) was defined as

$$H_\beta^\alpha = \left\{ f \in \text{Hol}(\mathbb{B}) : \|f\|_{H_\beta^\alpha} = \sup_{0 < r < 1} (1 - r)^\beta M_\alpha(r, f) < \infty \right\}.$$

Moreover, the little weighted Hardy space $H_{\beta,0}^\alpha$ is the space of all $f \in H_\beta^\alpha$ with

$$\lim_{r \to 1} (1 - r)^\beta M_\alpha(r, f) = 0.$$

Note that when $\alpha \geq 1$, H_β^α is a Banach space with the norm $\| \cdot \|_{H_\beta^\alpha}$ and $H_{\beta,0}^\alpha \subseteq H_\beta^\alpha$. As for Theorem 12.5.4, we have the following relationship between $\mathcal{N}(p, q, s)$ and weighted Hardy spaces.

Theorem 12.6.1 *Let $p \geq 1$, let $q > 0$, and let $s > \max\left\{0, 1 - \frac{q}{n}\right\}$. Then the following statements hold.*

(i) If $0 < s < 1$, then

$$H^{\alpha}_{\frac{q}{p} - \frac{n}{\alpha}} \subseteq \mathcal{N}(p, q, s),$$

where $\max\left\{p, \frac{np}{q}\right\} \leq \alpha < \frac{p}{1-s}$.

(ii) If $s \geq 1$, then

$$H^{\alpha}_{\frac{q}{p} - \frac{n}{\alpha}} \subseteq \mathcal{N}(p, q, s),$$

where $\alpha \geq \max\left\{p, \frac{np}{q}\right\}$. In particular, if $\alpha \geq \max\left\{p, \frac{np}{q}\right\}$, then

$$H^{\alpha}_{\frac{q}{p} - \frac{n}{\alpha}} \subseteq A^{-\frac{q}{p}}.$$

Proof Since the proof of (ii) is a simple modification of (i), we just prove (i). We keep in mind that, by (12.11), for fixed $\zeta \in \mathbb{S}$ and $0 < r < 1$, if $z \in Q_r(\zeta)$, then

$$1 - r < |z| < 1. \tag{12.12}$$

We consider two cases.

Case I: $\alpha = p$. In this case, $q \geq n$. For each $\zeta \in \mathbb{S}$ and $0 < r < 1$, by (12.12), we have

$$
\begin{aligned}
I_{r,\zeta} &= \frac{1}{r^{ns}} \int_{Q_r(\zeta)} |f(z)|^p (1 - |z|^2)^{q+ns} \, d\lambda(z) \\
&= \frac{1}{r^{ns}} \int_{\{z \in \mathbb{B}: |1 - \langle z, \zeta \rangle| < r\}} |f(z)|^p (1 - |z|^2)^{q+ns-n-1} \, dV(z) \\
&\lesssim \frac{1}{r^{ns}} \int_{1-r}^1 (1 - t^2)^{q+ns-n-1} \left(\int_{Q(\zeta,r)} |f(\gamma t)|^p \, d\sigma(\gamma) \right) dt \\
&= \frac{1}{r^{ns}} \int_{1-r}^1 \frac{(1 - t^2)^{q+ns-n-1}}{(1-t)^{q-n}} (1-t)^{q-n} M_p^p(t, f) \, dt \\
&\leq \|f\|_{H^p_{\frac{q-n}{p}}}^p \cdot \frac{1}{r^{ns}} \int_{1-r}^1 (1-t)^{ns-1} \, dt \simeq \|f\|_{H^p_{\frac{q-n}{p}}}^p.
\end{aligned}
$$

Thus,

$$\|f\|^p = \sup_{0 < r < 1, \zeta \in \mathbb{S}} I_{r,\zeta} \lesssim \|f\|^p_{H^p_{\frac{q-n}{p}}}.$$

Case II: $\alpha > p$. For each $\zeta \in \mathbb{S}$ and $0 < r < 1$, by Hölder's inequality, we have

$$\int_{Q(\zeta,r)} |f(\gamma t)|^p d\sigma(\gamma) \le \sigma(Q(\zeta,r))^{1-\frac{p}{\alpha}} \left(\int_{Q(\zeta,r)} |f(\gamma t)|^\alpha d\sigma(\gamma) \right)^{\frac{p}{\alpha}}$$

$$\lesssim \frac{r^{n-\frac{np}{\alpha}}(1-t)^{q-\frac{np}{\alpha}} M^p_\alpha(t, f)}{(1-t)^{q-\frac{np}{\alpha}}}$$

$$\le \frac{r^{n-\frac{np}{\alpha}} \|f\|^p_{H^\alpha_{\frac{q}{p}-\frac{n}{\alpha}}}}{(1-t)^{q-\frac{np}{\alpha}}}.$$

Consequently, by (12.12) and previous calculation, we obtain

$$I_{r,\zeta} \lesssim \frac{1}{r^{ns}} \int_{1-r}^1 (1-t^2)^{q+ns-n-1} \left(\int_{Q(\zeta,r)} |f(\gamma t)|^p d\sigma(\gamma) \right) dt$$

$$\lesssim r^{n-\frac{np}{\alpha}-ns} \|f\|^p_{H^\alpha_{\frac{q}{p}-\frac{n}{\alpha}}} \int_{1-r}^1 \frac{(1-t)^{q+ns-n-1}}{(1-t)^{q-\frac{np}{\alpha}}} dt$$

$$= r^{n-\frac{np}{\alpha}-ns} \|f\|^p_{H^\alpha_{\frac{q}{p}-\frac{n}{\alpha}}} \int_{1-r}^1 (1-t)^{ns+\frac{np}{\alpha}-n-1} dt \simeq \|f\|^p_{H^\alpha_{\frac{q}{p}-\frac{n}{\alpha}}}.$$

Hence,

$$\|f\|^p \simeq \sup_{0 < r < 1, \zeta \in \mathbb{S}} I_{r,\zeta} \lesssim \|f\|^p_{H^\alpha_{\frac{q}{p}-\frac{n}{\alpha}}}.$$

$\square$

The following result describes the behavior of Hadamard gap series in H^α_β.

Theorem 12.6.2 *Let $\alpha > 0$, let $\beta > 0$, and let*

$$f(z) = \sum_{k=1}^\infty P_{n_k}(z)$$

be a series with Hadamard gaps. Then the following statements are true.

(i) $f \in H^\alpha_\beta$ *if and only if* $\sup_{k \ge 1} \dfrac{L_{k,\alpha}}{n_k^\beta} < \infty.$

(ii) $f \in H_{\beta,0}^{\alpha}$ if and only if $\lim\limits_{k\to\infty} \dfrac{L_{k,\alpha}}{n_k^{\beta}} = 0.$

Here,

$$L_{k,\alpha} = \left(\int_{\mathbb{S}} |P_{n_k}(\zeta)|^{\alpha} d\sigma(\zeta) \right)^{\frac{1}{\alpha}}.$$

Proof (i)-**Necessity** Let $f \in H_{\beta}^{\alpha}$. Then for each $k \in \mathbb{N}$,

$$
\begin{aligned}
\int_{\mathbb{S}} |f(r\zeta)|^{\alpha} \, d\sigma(\zeta) &= \int_{\mathbb{S}} \left(\int_0^{2\pi} |f(r\zeta e^{i\theta})|^{\alpha} \frac{d\theta}{2\pi} \right) d\sigma(\zeta) \\
&= \int_{\mathbb{S}} \left(\int_0^{2\pi} \left| \sum_{k=0}^{\infty} P_{n_k}(r\zeta e^{i\theta}) \right|^{\alpha} \frac{d\theta}{2\pi} \right) d\sigma(\zeta) \\
&\simeq \int_{\mathbb{S}} \left(\sum_{k=0}^{\infty} |P_{n_k}(\zeta)|^2 r^{2n_k} \right)^{\alpha/2} d\sigma(\zeta) \\
&\geq r^{\alpha n_k} \int_{\mathbb{S}} |P_{n_k}(\zeta)|^{\alpha} \, d\sigma(\zeta). \qquad\qquad (12.13)
\end{aligned}
$$

Hence,

$$(1-r)^{\beta} r^{n_k} L_{k,\alpha} \leq (1-r)^{\beta} M_{\alpha}(r, f) \leq \|f\|_{H_{\beta}^{\alpha}}. \qquad (12.14)$$

Choosing $r = 1 - \frac{1}{n_k}$ and using the well-known inequality $(1+\frac{1}{m})^{m+1} \leq 4, m \in \mathbb{N}$, we obtain

$$\sup_{k\in\mathbb{N}} \frac{L_{k,\alpha}}{n_k^{\beta}} \leq C\|f\|_{H_{\beta}^{\alpha}},$$

as desired.

(i)-**Sufficiency** Suppose that $\sup\limits_{k\in\mathbb{N}} \dfrac{L_{k,\alpha}}{n_k^{\beta}} < \infty$. For a fixed $r \in (0, 1)$, we have

$$\frac{\sum_{k=0}^{\infty} r^{\alpha n_k} n_k^{\alpha\beta}}{1 - r^{\alpha}} = \left(\sum_{k=0}^{\infty} r^{\alpha n_k} n_k^{\alpha\beta} \right) \cdot \left(\sum_{s=0}^{\infty} r^{\alpha s} \right) = \sum_{t=0}^{\infty} \left(\sum_{n_j \leq t} n_j^{\alpha\beta} \right) r^{\alpha t}.$$

Since

$$\lim_{k \to \infty} \frac{k^{\alpha\beta} k!}{(\alpha\beta)(\alpha\beta + 1)\ldots(\alpha\beta + k)} = \Gamma(\alpha\beta), \qquad (\alpha\beta > 0),$$

we have

$$\sup_{k \in \mathbb{N}} \left(\frac{k^{\alpha\beta} k!}{(k + \alpha\beta)(k + \alpha\beta - 1)\ldots(\alpha\beta + 1)} \right) \leq M,$$

where M is some positive number depending on α and β. Hence, for each $k \geq 0$,

$$\frac{k^{\alpha\beta}}{(-1)^k \binom{-\alpha\beta-1}{k}} = \frac{k^{\alpha\beta} k!}{(-1)^k (-\alpha\beta - 1)(-\alpha\beta - 2)\ldots(-\alpha\beta - k)}$$

$$= \frac{k^{\alpha\beta} k!}{(k + \alpha\beta)(k + \alpha\beta - 1)\ldots(\alpha\beta + 1)} \leq M, \qquad (12.15)$$

where

$$\binom{\gamma}{k} = \frac{\gamma(\gamma - 1)\ldots(\gamma - k + 1)}{k!}, \gamma \in \mathbb{R}.$$

Furthermore, since f is in Hadamard gap class, there exists a constant $c > 1$ such that $n_{j+1} \geq c n_j$ for all $j \geq 0$. Hence,

$$\frac{1}{t^{\alpha\beta}} \left(\sum_{n_j \leq t} n_j^{\alpha\beta} \right) \leq \sum_{m=0}^{\infty} \left(\frac{1}{c^{\alpha\beta}} \right)^m = \frac{c^{\alpha\beta}}{c^{\alpha\beta} - 1}. \qquad (12.16)$$

Combining (12.15) and (12.16), we have

$$\frac{t^{\alpha\beta}}{(-1)^t \binom{-\alpha\beta-1}{t}} \cdot \frac{1}{t^{\alpha\beta}} \left(\sum_{n_j \leq t} n_j^{\alpha\beta} \right) \leq \frac{M c^{\alpha\beta}}{c^{\alpha\beta} - 1},$$

which implies

$$\sum_{n_j \leq t} n_j^{\alpha\beta} \leq (-1)^t \binom{-\alpha\beta - 1}{t} \frac{M c^{\alpha\beta}}{c^{\alpha\beta} - 1}. \qquad (12.17)$$

Hence, by (12.17), we have

$$\frac{\sum_{k=0}^{\infty} r^{\alpha n_k} n_k^{\alpha\beta}}{1 - r^{\alpha}} \lesssim \sum_{r=0}^{\infty} (-1)^t \binom{-\alpha\beta - 1}{t} r^{\alpha t} = \frac{1}{(1 - r^{\alpha})^{\alpha\beta+1}},$$

which implies for $\alpha, \beta > 0$,

$$(1 - r^{\alpha})^{\alpha\beta} \sum_{k=0}^{\infty} r^{\alpha n_k} n_k^{\alpha\beta} \lesssim 1. \tag{12.18}$$

Now, we need to consider two cases.

Case I: $\alpha \in (0, 2]$. From (12.13) and (12.18) and by using the known inequality

$$\left(\sum_{k=1}^{\infty} a_k\right)^q \leq \sum_{k=1}^{\infty} a_k^q,$$

where $a_k \geq 0$, $k \in \mathbb{N}$, $q \in [0, 1]$, we have that

$$\|f\|_{H_\beta^\alpha}^\alpha \simeq \sup_{0<r<1} (1 - r)^{\alpha\beta} \int_{\mathbb{S}} \left(\sum_{k=0}^{\infty} |P_{n_k}(\zeta)|^2 r^{2n_k}\right)^{\alpha/2} d\sigma(\zeta)$$

$$\leq \sup_{0<r<1} (1 - r)^{\alpha\beta} \int_{\mathbb{S}} \left(\sum_{k=0}^{\infty} |P_{n_k}(\zeta)|^\alpha r^{\alpha n_k}\right) d\sigma(\zeta)$$

$$= \sup_{0<r<1} (1 - r)^{\alpha\beta} \cdot \sum_{k=0}^{\infty} r^{\alpha n_k} L_{k,\alpha}^\alpha$$

$$\lesssim \sup_{0<r<1} (1 - r)^{\alpha\beta} \cdot \sum_{k=0}^{\infty} r^{\alpha n_k} n_k^{\alpha\beta} \lesssim \sup_{0<r<1} \left(\frac{1 - r}{1 - r^{\alpha}}\right)^{\alpha\beta} < \infty,$$

which implies the desired result.

Case II: $\alpha > 2$. For each $r \in (0, 1)$, by Minkowski's inequality and (12.18), we have

$$\left[\int_{\mathbb{S}} \left(\sum_{k=0}^{\infty} |P_{n_k}(\zeta)|^2 r^{2n_k}\right)^{\alpha/2} d\sigma(\zeta)\right]^{\frac{2}{\alpha}}$$

$$\leq \sum_{k=0}^{\infty} \left(\int_{\mathbb{S}} \left(|P_{n_k}(\zeta)|^2 r^{2n_k}\right)^{\frac{\alpha}{2}} d\sigma(\zeta)\right)^{\frac{2}{\alpha}} \tag{12.19}$$

$$= \sum_{k=0}^{\infty} r^{2n_k} L_{k,\alpha}^2 \lesssim \sum_{k=0}^{\infty} r^{2n_k} n_k^{2\beta}. \tag{12.20}$$

Thus, by (12.13) and (12.18),

$$\|f\|_{H_\beta^\alpha}^\alpha \simeq \sup_{0<r<1} (1-r)^{\alpha\beta} \int_{\mathbb{S}} \left(\sum_{k=0}^{\infty} |P_{n_k}(\zeta)|^2 r^{2n_k} \right)^{\alpha/2} d\sigma(\zeta)$$

$$\leq \sup_{0<r<1} (1-r)^{\alpha\beta} \left(\sum_{k=0}^{\infty} r^{2n_k} L_{k,\alpha}^2 \right)^{\frac{\alpha}{2}} \tag{12.21}$$

$$\lesssim \sup_{0<r<1} (1-r)^{\alpha\beta} \left(\sum_{k=0}^{\infty} r^{2n_k} n_k^{2\beta} \right)^{\frac{\alpha}{2}}$$

$$\lesssim \sup_{0<r<1} \left(\frac{1-r}{1-r^2} \right)^{\alpha\beta} < \infty,$$

and hence, $f \in H_\beta^\alpha$.

(ii)-Necessity Let $f \in H_{\beta,0}^\alpha$. Then for every $\varepsilon > 0$; there is a $\delta > 0$, such that

$$(1-r)^\beta M_\alpha(r, f) < \varepsilon \tag{12.22}$$

whenever $\delta < r < 1$. From the first inequality in (12.14) and (12.22), we have

$$(1-r)^\beta r^{n_k} L_{k,\alpha} < \varepsilon$$

for each $k \in \mathbb{N}$ and $r \in (\delta, 1)$. Choosing $r = 1 - \frac{1}{n_k}$, where $n_k > \frac{1}{1-\delta}$, we obtain

$$\frac{L_{k,\alpha}}{n_k^\beta} < 4\varepsilon.$$

From this and since ε is an arbitrary positive number, it follows that

$$\lim_{k\to\infty} \frac{L_{k,\alpha}}{n_k^\beta} = 0.$$

(ii)-Sufficiency Suppose $\lim_{k\to\infty} \frac{L_{k,\alpha}}{n_k^\beta} = 0$. Take and fix a $\varepsilon > 0$, there is a $k_0 \in \mathbb{N}$ such that

$$L_{k,\alpha} < \varepsilon^{\frac{1}{\alpha}} n_k^\beta, \quad \text{for} \quad (k \geq k_0).$$

Fix the k_0 chosen above. Then, there exists a $\delta > 0$, such that when $\delta < r < 1$,

$$(1 - r)^{\alpha\beta} \sum_{k=0}^{k_0} L_{k,\alpha}^{\alpha} < \varepsilon.$$

Again, we consider two cases as the proof in part (i).

Case I: $\alpha \in (0, 2]$. By (12.14) and the proof in part (1), for $r \in (\delta, 1)$, we have

$$(1 - r)^{\alpha\beta} M_{\alpha}^{\alpha}(r, f) \simeq (1 - r)^{\alpha\beta} \int_{\mathbb{S}} \left(\sum_{k=0}^{\infty} |P_{n_k}(\zeta)|^2 r^{2n_k} \right)^{\alpha/2} d\sigma(\zeta)$$

$$\leq (1 - r)^{\alpha\beta} \cdot \left[\left(\sum_{k=0}^{k_0} + \sum_{k=k_0+1}^{\infty} \right) r^{\alpha n_k} L_{k,\alpha}^{\alpha} \right]$$

$$\leq \varepsilon + (1 - r)^{\alpha\beta} \cdot \sum_{k=k_0+1}^{\infty} r^{\alpha n_k} L_{k,\alpha}^{\alpha}$$

$$\leq \varepsilon + \varepsilon(1 - r)^{\alpha\beta} \cdot \sum_{k=k_0+1}^{\infty} r^{\alpha n_k} n_k^{\alpha\beta}$$

$$\lesssim \varepsilon \cdot \left(1 + \sup_{0 < r < 1} \left(\frac{1 - r}{1 - r^2} \right)^{\alpha\beta} \right) \lesssim \varepsilon,$$

which implies

$$\lim_{r \to 1^-} (1 - r)^{\beta} M_{\alpha}(r, f) = 0$$

and hence $f \in H_{\beta,0}^{\alpha}$ as desired.

Case II: $\alpha > 2$. The implication for case $p \geq 2$ follows similarly, from (12.19), (12.21), and the known inequality

$$(a + b)^{\frac{\alpha}{2}} \leq 2^{\frac{\alpha}{2}-1} \left(a^{\frac{\alpha}{2}} + b^{\frac{\alpha}{2}} \right), \quad a, b \geq 0.$$

Hence, we omit the detail here.

$\square$

As a corollary of Theorem 12.6.2, we can show that when $\frac{q}{p} - \frac{n}{\alpha} > 0$, the inclusion in Theorem 12.6.1 is strict.

Corollary 12.6.3 *Let $p, q, s,$ and α satisfy the condition in Proposition 12.6.1. If*

$$\frac{q}{p} - \frac{n}{\alpha} > 0,$$

then the inclusion in Theorem 12.6.1 is strict.

Proof First, we take a sequence of homogeneous polynomials $\{P_k\}_{k \in \mathbb{N}}$ satisfying $\deg(P_k) = k,$

$$\sup_{\zeta \in \mathbb{S}} |P_k(\zeta)| = 1$$

and

$$\int_{\mathbb{S}} |P_k(\zeta)|^p d\sigma(\zeta) \geq C(p, n) > 0.$$

Note that since $p \leq \alpha$, by Hölder's inequality, we have

$$\int_{\mathbb{S}} |P_k(\zeta)|^p d\sigma(\zeta) \leq \left(\int_{\mathbb{S}} |P_k(\zeta)|^\alpha d\sigma(\zeta) \right)^{\frac{p}{\alpha}},$$

that is,

$$\int_{\mathbb{S}} |P_k(\zeta)|^\alpha d\sigma(\zeta) \geq \left(\int_{\mathbb{S}} |P_k(\zeta)|^p d\sigma(\zeta) \right)^{\frac{\alpha}{p}} \geq C^{\frac{\alpha}{p}}(p, n). \qquad (12.23)$$

Note that when $s > 1, \mathcal{N}(p, q, s) = A^{-\frac{q}{p}}$, and hence, we consider three cases.

Case I: $s > 1$. Consider the series $f_1(z) = \sum_{k=0}^{\infty} 2^{\frac{kq}{p}} P_{2^k}(z)$, which belongs to $A^{-\frac{q}{p}}$. On the other hand, for each $k \in \mathbb{N}$, by (12.23),

$$\frac{L_{k,\alpha}}{2^{k\left(\frac{q}{p} - \frac{n}{\alpha}\right)}} = \frac{\left(\int_{\mathbb{S}} \left| 2^{\frac{kq}{p}} P_{2^k}(z) \right|^\alpha d\sigma(\zeta) \right)^{\frac{1}{\alpha}}}{2^{k\left(\frac{q}{p} - \frac{n}{\alpha}\right)}} \geq C^{\frac{1}{p}}(p, n) 2^{\frac{kn}{\alpha}}.$$

Hence, we have

$$\sup_{k \geq 0} \frac{L_{k,\alpha}}{2^{k\left(\frac{q}{p} - \frac{n}{\alpha}\right)}} = \infty,$$

which implies that $f_1 \notin H^\alpha_{\frac{q}{p} - \frac{n}{\alpha}}$.

Case II: $s = 1$. Take any $\varepsilon \in \left(0, \frac{np}{\alpha}\right)$, and consider the series

$$f_2(z) = \sum_{k=0}^{\infty} 2^{\frac{k(q-\varepsilon)}{p}} P_{2^k}(z).$$

On the one hand, we have

$$\sum_{k=0}^{\infty} \frac{1}{2^{kq}} \cdot \left| 2^{\frac{k(q-\varepsilon)}{p}} \right|^p = \sum_{k=0}^{\infty} \frac{1}{2^{k\varepsilon}} < \infty,$$

which, by Theorem 12.4.3, implies that $f_2 \in \mathcal{N}(p, q, 1)$. However, for each $k \in \mathbb{N}$, by (12.23),

$$\frac{L_{k,\alpha}}{2^{k\left(\frac{q}{p} - \frac{n}{\alpha}\right)}} = \frac{\left(\int_{\mathbb{S}} \left| 2^{\frac{k(q-\varepsilon)}{p}} P_{2^k}(z) \right|^{\alpha} d\sigma(\zeta) \right)^{\frac{1}{\alpha}}}{2^{k\left(\frac{q}{p} - \frac{n}{\alpha}\right)}} \geq C^{\frac{1}{p}}(p, n) 2^{k\left(\frac{n}{\alpha} - \frac{\varepsilon}{p}\right)}.$$

Hence, we have

$$\sup_{k \geq 0} \frac{L_{k,\alpha}}{2^{k\left(\frac{q}{p} - \frac{n}{\alpha}\right)}} = \infty,$$

which implies that $f_2 \notin H^{\alpha}_{\frac{q}{p} - \frac{n}{\alpha}}$.

Case III: $0 < s < 1$. Since $\alpha < \frac{p}{1-s}$, we have $\frac{s-1}{p} + \frac{1}{\alpha} > 0$. Take any $\varepsilon \in \left(0, n(s-1) + \frac{np}{\alpha}\right)$, and consider the series $f_3(z) = \sum_{k=0}^{\infty} 2^{\frac{k(ns+q-n-\varepsilon)}{p}} P_{2^k}(z)$. On the one hand, we have

$$\sum_{k=0}^{\infty} \frac{1}{2^{k(ns+q-n)}} \left| 2^{\frac{k(ns+q-n-\varepsilon)}{p}} \right|^p = \sum_{k=0}^{\infty} \frac{1}{2^{k\varepsilon}} < \infty,$$

which, by Theorem 12.4.3, implies that $f_3 \in \mathcal{N}(p, q, s)$. However, for each $k \in \mathbb{N}$, by (12.23),

$$\frac{L_{k,\alpha}}{2^{k\left(\frac{q}{p} - \frac{n}{\alpha}\right)}} = \frac{\left(\int_{\mathbb{S}} \left| 2^{\frac{k(ns+q-n-\varepsilon)}{p}} P_{2^k}(z) \right|^{\alpha} d\sigma(\zeta) \right)^{\frac{1}{\alpha}}}{2^{k\left(\frac{q}{p} - \frac{n}{\alpha}\right)}} \geq C^{\frac{1}{p}}(p, n) 2^{k\left(\frac{n(s-1)}{p} + \frac{n}{\alpha} - \frac{\varepsilon}{p}\right)}.$$

Hence, we have

$$\sup_{k \geq 0} \frac{L_{k,\alpha}}{2^{k\left(\frac{q}{p} - \frac{n}{\alpha}\right)}} = \infty,$$

which implies that $f_3 \notin H^{\alpha}_{\frac{q}{p} - \frac{n}{\alpha}}$.

□

Notes on Chapter 12

The results of this chapter are mostly taken from [49–51]. The $\mathcal{N}(p, q, s)$ spaces are closely related to $\mathcal{F}(p, q, s)$ spaces. In fact, the study of $\mathcal{N}(p, q, s)$ spaces was inspired by the latter spaces. The $\mathcal{F}(p, q, s)$ spaces were first introduced by Zhao on the unit disc and later studied on the unit disc and the unit ball $\mathbb{B}$ of $\mathbb{C}^n$ by various authors. For recent developments on this topic, see [105, 106]. The invariant Green's function is introduced in [93, 96]. The important property of this function is given in [72]. Example 12.4.2 is given in [47]. Corollary 12.5.3 is from [108]. Theorem 12.6.2 stems from [59]. Recently, in [55], some simplifications to the formulas in Sect. 12.4 have been discovered. Namely, in Theorem 12.4.1, condition (i) can be replaced by

$$\sum_{k=0}^{\infty} n_k^{n-ns-q} M_k^p < \infty,$$

and condition (iv) by

$$\sum_{k=0}^{\infty} n_k^{n-ns-q} L_{k,p}^p < \infty.$$

Also, in Theorem 12.4.3, condition (i) can be changed to

$$\sum_{k=0}^{\infty} n_k^{n-ns-q} |a_k|^p < \infty.$$

Bibliography

1. Aleksandrov, A.B.: Proper holomorphic mappings from the ball to the polydisk. Dokl. Akad. Nauk SSSR **286**(1), 11–15 (1986)
2. Anderson, J.M.: Bloch functions: the basic theory. In: Operators and Function Theory (Lancaster, 1984). NATO Advanced Science Institutes Series C Mathematical and Physical Sciences, vol. 153, pp. 1–17. Reidel, Dordrecht (1985)
3. Anderson, J.M., Clunie, J., Pommerenke, C.: On Bloch functions and normal functions. J. Reine Angew. Math. **270**, 12–37 (1974)
4. Arcozzi, N., Rochberg, R., Sawyer, E.T., Wick, B.D.: The Dirichlet Space and Related Function Spaces. Mathematical Surveys and Monographs, vol. 239. American Mathematical Society, Providence (2019)
5. Aronszajn, N.: Theory of reproducing kernels. Trans. Am. Math. Soc. **68**, 337–404 (1950)
6. Aulaskari, R., Stegenga, D.A., Xiao, J.: Some subclasses of BMOA and their characterization in terms of Carleson measures. Rocky Mountain J. Math. **26**(2), 485–506 (1996)
7. Aulaskari, R., Xiao, J., Zhao, R.H.: On subspaces and subsets of BMOA and UBC. Analysis **15**(2), 101–121 (1995)
8. Bungart, L.: Boundary kernel functions for domains on complex manifolds. Pac. J. Math. **14**, 1151–1164 (1964)
9. Burbea, J., Li, S.-Y.: Weighted Hadamard products of holomorphic functions in the ball. Pac. J. Math. **168**(2), 235–270 (1995)
10. Choa, J.S.: Some properties of analytic functions on the unit ball with Hadamard gaps. Complex Variables Theory Appl. **29**(3), 277–285 (1996)
11. Choe, B.R., Koo, H., Park, I.: Compact differences of composition operators over polydisks. Integral Equ. Oper. Theory **73**(1), 57–91 (2012)
12. Choe, B.R., Rim, K.S.: Fractional derivatives of Bloch functions, growth rate, and interpolation. Acta Math. Hungar. **72**(1–2), 67–86 (1996)
13. Collingwood, E.F., Lohwater, A.J.: The Theory of Cluster Sets. Cambridge Tracts in Mathematics and Mathematical Physics, vol. 56. Cambridge University Press, Cambridge (1966)
14. Contreras, M.D., Hernandez-Diaz, A.G.: Weighted composition operators in weighted Banach spaces of analytic functions. J. Aust. Math. Soc. Ser. A **69**(1), 41–60 (2000)
15. Conway, J.B.: A Course in Functional Analysis. Graduate Texts in Mathematics, vol. 96, 2nd edn. Springer, New York (1990)
16. Conway, J.B.: A Course in Operator Theory. Graduate Studies in Mathematics, vol. 21. American Mathematical Society, Providence (2000)

© The Author(s), under exclusive license to Springer Nature Switzerland AG 2023
L. H. Khoi, J. Mashreghi, *Theory of $\mathcal{N}_p$ Spaces*, Frontiers in Mathematics,
https://doi.org/10.1007/978-3-031-39704-2

17. Cowen, C.C., MacCluer, B.D.: Composition Operators on Spaces of Analytic Functions. Studies in Advanced Mathematics. CRC Press, Boca Raton (1995)

18. Doan, M.L., Hu, B., Khoi, L.H., Queffélec, H.: Approximation numbers for composition operators on spaces of entire functions. Indag. Math. (N.S.) **28**(2), 294–305 (2017)

19. Doan, M.L., Khoi, L.H.: Hilbert spaces of entire functions and composition operators. Complex Anal. Oper. Theory **10**(1), 213–230 (2016)

20. Doan, M.L., Khoi, L.H., Le, T.: Composition operators on Hilbert spaces of entire functions of several variables. Integral Equ. Oper. Theory **88**(3), 301–330 (2017)

21. Duren, P., Schuster, A.: Bergman Spaces. Mathematical Surveys and Monographs, vol. 100. American Mathematical Society, Providence (2004)

22. Duren, P., Weir, R.: The pseudohyperbolic metric and Bergman spaces in the ball. Trans. Am. Math. Soc. **359**(1), 63–76 (2007)

23. Duren, P.L.: Theory of H^p Spaces. Pure and Applied Mathematics, vol. 38. Academic, New York/London (1970)

24. Duren, P.L., Romberg, B.W., Shields, A.L.: Linear functionals on H^p spaces with $0 < p < 1$. J. Reine Angew. Math. **238**, 32–60 (1969)

25. El-Fallah, O., Kellay, K., Mashreghi, J., Ransford, T.: A Primer on the Dirichlet Space. Cambridge Tracts in Mathematics, vol. 203. Cambridge University Press, Cambridge (2014)

26. Essén, M., Shea, D.F., Stanton, C.S.: A value-distribution criterion for the class $L \log L$, and some related questions. Ann. Inst. Fourier (Grenoble) **35**(4), 127–150 (1985)

27. Fatou, P.: Séries trigonométriques et séries de Taylor. Acta Math. **30**(1), 335–400 (1906)

28. Fefferman, C.: The Bergman kernel and biholomorphic mappings of pseudoconvex domains. Invent. Math. **26**, 1–65 (1974)

29. Fleming, R.J., Jamison, J.E.: Isometries on Banach Spaces: Function Spaces. Chapman & Hall/CRC Monographs and Surveys in Pure and Applied Mathematics, vol. 129. Chapman & Hall/CRC, Boca Raton (2003)

30. Folland, G.B.: The tangential Cauchy-Riemann complex on spheres. Trans. Am. Math. Soc. **171**, 83–133 (1972)

31. Fricain, E., Mashreghi, J.: The theory of $\mathcal{H}(b)$ spaces. Vol. 1. New Mathematical Monographs, vol. 20. Cambridge University Press, Cambridge (2016)

32. Fricain, E., Mashreghi, J.: The Theory of $\mathcal{H}(b)$ Spaces. Vol. 2. New Mathematical Monographs, vol. 21. Cambridge University Press, Cambridge (2016)

33. Fried, H.: On analytic functions with bounded characteristic. Bull. Am. Math. Soc. **52**, 694–699 (1946)

34. Galindo, P., Lindström, M., Stević, S.: Essential norm of operators into weighted-type spaces on the unit ball. Abstr. Appl. Anal. **Art. ID 939873**, 13 (2011)

35. Garcia, S.R., Mashreghi, J., Ross, W.T.: Introduction to Model Spaces and Their Operators. Cambridge Studies in Advanced Mathematics, vol. 148. Cambridge University Press, Cambridge (2016)

36. Garcia, S.R., Mashreghi, J., Ross, W.T.: Finite Blaschke products: a survey. In: Harmonic Analysis, Function Theory, Operator Theory, and Their Applications. Theta Series Advances in Mathematics, vol. 19, pp. 133–158. Theta, Bucharest (2017)

37. Garnett, J.B.: Bounded Analytic Functions. Pure and Applied Mathematics, vol. 96. Academic [Harcourt Brace Jovanovich, Publishers], New York/London (1981)

38. Garnett, J.B.: Bounded Analytic Functions. Graduate Texts in Mathematics, vol. 236, 1st edn. Springer, New York (2007)

39. Hardy, G.H., Littlewood, J.E.: A further note on the converse of Abel's theorem. Proc. Lond. Math. Soc. (2) **25**, 219–236 (1926)

40. Hedenmalm, H., Korenblum, B., Zhu, K.: Theory of Bergman Spaces. Graduate Texts in Mathematics, vol. 199. Springer, New York (2000)

41. Heller, K., MacCluer, B.D., Weir, R.J.: Compact differences of composition operators in several variables. Integral Equ. Oper. Theory **69**(2), 247–268 (2011)
42. Hu, B., Khoi, L.H.: Weighted-composition operators on $\mathcal{N}_p$-spaces in the ball. C. R. Math. Acad. Sci. Paris **351**(19–20), 719–723 (2013)
43. Hu, B., Khoi, L.H.: Compact difference of weighted composition operators on $\mathcal{N}_p$-spaces in the ball. Rev. Roumaine Math. Pures Appl. **60**(2), 101–116 (2015)
44. Hu, B., Khoi, L.H., Le, T.: Composition operators between $\mathcal{N}_p$-spaces in the ball. North-West. Eur. J. Math. **2**, 103–120, i (2016)
45. Hu, B., Khoi, L.H., Le, T.: Essential norms of weighted composition operators on $\mathcal{N}_p$-spaces in the ball. Vietnam J. Math. **44**(2), 431–439 (2016)
46. Hu, B., Khoi, L.H., Le, T.: On the structure of $\mathcal{N}_p$-spaces in the ball. Acta Math. Vietnam. **43**(3), 433–448 (2018)
47. Hu, B., Li, S.: $\mathcal{N}_p$-type functions with Hadamard gaps in the unit ball. Complex Var. Elliptic Equ. **61**(6), 843–853 (2016)
48. Hu, B., Li, S.: Hadamard gap series in weighted-type spaces on the unit ball. Ann. Funct. Anal. **8**(2), 259–269 (2017)
49. Hu, B., Li, S.: $\mathcal{N}(p, q, s)$-type spaces in the unit ball of $\mathbb{C}^n$ (I): basic properties and Hadamard gaps. Complex Var. Elliptic Equ. **65**(6), 956–976 (2020)
50. Hu, B., Li, S.: $\mathcal{N}(p, q, s)$-type spaces in the unit ball of $\mathbb{C}^n$ (II): Carleson measure and its application. Forum Math. **32**(1), 79–94 (2020)
51. Hu, B., Li, S.: $\mathcal{N}(p, q, s)$-type spaces in the unit ball of $\mathbb{C}^n$ (III): various characterizations. Publ. Math. Debrecen **97**(1–2), 41–61 (2020)
52. Hu, B., Li, S.: $\mathcal{N}(p, q, s)$-type spaces in the unit ball of $\mathbb{C}^n$ (V): Riemann-Stieltjes operators and multipliers. Bull. Sci. Math. **166**, 102929, 27 (2021)
53. Jiang, L., Ouyang, C.: Compact differences of composition operators on holomorphic function spaces in the unit ball. Acta Math. Sci. Ser. B Engl. Ed. **31**(5), 1679–1693 (2011)
54. Khoi, L.H.: On $\mathcal{N}_p$-spaces in the ball. In: Geometric Complex Analysis. Springer Proceedings in Mathematics and Statistics, vol. 246, pp. 219–233. Springer, Singapore (2018)
55. Khoi, L.H., Mashreghi, J., Nasri, M.: Intrinsic characterization of $\mathcal{N}_p$-spaces via the Hadamard gap class, 21pp. Preprint (2023)
56. Koosis, P.: Introduction to H_p Spaces. London Mathematical Society Lecture Note Series, vol. 40. Cambridge University Press, Cambridge/New York (1980). With an appendix on Wolff's proof of the corona theorem
57. Krantz, S.G.: Function Theory of Several Complex Variables The Wadsworth & Brooks/Cole Mathematics Series, 2nd edn. Wadsworth & Brooks/Cole Advanced Books & Software, Pacific Grove (1992)
58. Levin, B.Y.: Lectures on Entire Functions. Translations of Mathematical Monographs, vol. 150. American Mathematical Society, Providence (1996). In collaboration with and with a preface by Y. Lyubarskii, M. Sodin and V. Tkachenko, Translated from the Russian manuscript by Tkachenko
59. Li, S., Stević, S.: Weighted-Hardy functions with Hadamard gaps on the unit ball. Appl. Math. Comput. **212**(1), 229–233 (2009)
60. Lindström, M., Makhmutov, S., Taskinen, J.: The essential norm of a Bloch-to-Q_p composition operator. Can. Math. Bull. **47**(1), 49–59 (2004)
61. Lindström, M., Palmberg, N.: Spectra of composition operators on BMOA. Integral Equ. Oper. Theory **53**(1), 75–86 (2005)
62. Littlewood, J.E.: On inequalities in the theory of functions. Proc. Lond. Math. Soc. (2) **23**(7), 481–519 (1925)
63. Lou, Z., Wulan, H.: Characterisations of Bloch functions in the unit ball of $\mathbb{C}^n$. I. Bull. Aust. Math. Soc. **68**(2), 205–212 (2003)

64. MacCluer, B.D., Stroethoff, K., Zhao, R.: Schwarz-Pick type estimates. Complex Var. Theory Appl. **48**(8), 711–730 (2003)

65. Mashreghi, J.: The rate of increase of mean values of functions in Hardy spaces. J. Aust. Math. Soc. **86**(2), 199–204 (2009)

66. Mashreghi, J.: Representation theorems in Hardy Spaces. London Mathematical Society Student Texts, vol. 74. Cambridge University Press, Cambridge (2009)

67. Mashreghi, J.: Derivatives of Inner Functions. Fields Institute Monographs, vol. 31. Springer, New York; Fields Institute for Research in Mathematical Sciences, Toronto (2013)

68. Mateljević, M., Pavlović, M.: L^p-behavior of power series with positive coefficients and Hardy spaces. Proc. Am. Math. Soc. **87**(2), 309–316 (1983)

69. Mateljević, M., Pavlović, M.: An extension of the Forelli-Rudin projection theorem. Proc. Edinb. Math. Soc. (2) **36**(3), 375–389 (1993)

70. Mathews, J.H.: Coefficients of uniformly normal-Bloch functions. Yokohama Math. J. **21**, 29–31 (1973)

71. Miao, J.: A property of analytic functions with Hadamard gaps. Bull. Aust. Math. Soc. **45**(1), 105–112 (1992)

72. Ouyang, C.H., Yang, W.S., Zhao, R.H.: Characterizations of Bergman spaces and Bloch space in the unit ball of $\mathbf{C}^n$. Trans. Am. Math. Soc. **347**(11), 4301–4313 (1995)

73. Palmberg, N.: Composition operators acting on $\mathcal{N}_p$-spaces. Bull. Belg. Math. Soc. Simon Stevin **14**(3), 545–554 (2007)

74. Palmberg, N.: Weighted composition operators with closed range. Bull. Aust. Math. Soc. **75**(3), 331–354 (2007)

75. Pau, J., Peláez, J.A.: Multipliers of Möbius invariant Q_s spaces. Math. Z. **261**(3), 545–555 (2009)

76. Paulsen, V.I., Raghupathi, M.: An Introduction to the Theory of Reproducing Kernel Hilbert Spaces. Cambridge Studies in Advanced Mathematics, vol. 152. Cambridge University Press, Cambridge (2016)

77. Pietsch, A.: Operator Ideals. Mathematische Monographien [Mathematical Monographs], vol. 16. VEB Deutscher Verlag der Wissenschaften, Berlin (1978)

78. Ramey, W., Ullrich, D.: Bounded mean oscillation of Bloch pull-backs. Math. Ann. **291**(4), 591–606 (1991)

79. Rudin, W.: Function Theory in the Unit Ball of $\mathbf{C}^n$. Grundlehren der Mathematischen Wissenschaften [Fundamental Principles of Mathematical Science], vol. 241. Springer, New York/Berlin (1980)

80. Rutovitz, D.: Some parameters associated with finite-dimensional Banach spaces. J. Lond. Math. Soc. **40**, 241–255 (1965)

81. Ryll, J., Wojtaszczyk, P.: On homogeneous polynomials on a complex ball. Trans. Am. Math. Soc. **276**(1), 107–116 (1983)

82. Shapiro, J.H.: The essential norm of a composition operator. Ann. Math. (2) **125**(2), 375–404 (1987)

83. Shapiro, J.H.: Composition Operators and Classical Function Theory. Universitext: Tracts in Mathematics. Springer, New York (1993)

84. Shields, A.L.: Weighted shift operators and analytic function theory. In: Topics in Operator Theory. Mathematical Surveys, vol. 13, pp. 49–128. American Mathematical Society, Providence (1974)

85. Shields, A.L., Williams, D.L.: Bounded projections, duality, and multipliers in spaces of analytic functions. Trans. Am. Math. Soc. **162**, 287–302 (1971)

86. Smith, W.: Composition operators between Bergman and Hardy spaces. Trans. Am. Math. Soc. **348**(6), 2331–2348 (1996)

87. Stanton, C.S.: Counting functions and majorization for Jensen measures. Pac. J. Math. **125**(2), 459–468 (1986)

88. Stein, P.: On a Theorem of M. Riesz. J. Lond. Math. Soc. **8**(4), 242–247 (1933)

89. Stević, S.: A generalization of a result of Choa on analytic functions with Hadamard gaps. J. Kor. Math. Soc. **43**(3), 579–591 (2006)

90. Stević, S., Ueki, S.-I.: Weighted composition operators and integral-type operators between weighted Hardy spaces on the unit ball. Discret. Dyn. Nat. Soc. **Art. ID 952831**, 21 (2009)

91. Stochel, J., Stochel, J.B.O.: Composition operators on Hilbert spaces of entire functions with analytic symbols. J. Math. Anal. Appl. **454**(2), 1019–1066 (2017)

92. Stochel, J.B.: Representation of generalised creation and annihilation operators in Fock space. Univ. Iagel. Acta Math. **34**, 135–148 (1997)

93. Stoll, M.: A characterization of Hardy spaces on the unit ball of $\mathbf{C}^n$. J. Lond. Math. Soc. (2) **48**(1), 126–136 (1993)

94. Tong, C.-Z.: Compact differences of weighted composition operators on $H^\infty(B_N)$. J. Comput. Anal. Appl. **14**(1), 32–41 (2012)

95. Ueki, S.-I.: Weighted composition operators acting between the $\mathcal{N}_p$-space and the weighted-type space H_α^∞. Indag. Math. (N.S.) **23**(3), 243–255 (2012)

96. Ullrich, D.: Radial limits of M-subharmonic functions. Trans. Am. Math. Soc. **292**(2), 501–518 (1985)

97. Ullrich, D.C.: A Bloch function in the ball with no radial limits. Bull. Lond. Math. Soc. **20**(4), 337–341 (1988)

98. Ullrich, D.C.: Radial divergence in BMOA. Proc. Lond. Math. Soc. (3) **68**(1), 145–160 (1994)

99. Wulan, H., Zhu, K.: Lacunary series in Q_K spaces. Studia Math. **178**(3), 217–230 (2007)

100. Xiao, J.: Composition operators: $\mathcal{N}_\alpha$ to the Bloch space to $\mathcal{Q}_\beta$. Studia Math. **139**(3), 245–260 (2000)

101. Xiao, J.: The Q_p corona theorem. Pac. J. Math. **194**(2), 491–509 (2000)

102. Xiao, J.: Holomorphic Q Classes. Lecture Notes in Mathematics, vol. 1767. Springer, Berlin (2001)

103. Yamashita, S.: Gap series and α-Bloch functions. Yokohama Math. J. **28**(1–2), 31–36 (1980)

104. Yang, C., Xu, W.: Spaces with normal weights and Hadamard gap series. Arch. Math. (Basel) **96**(2), 151–160 (2011)

105. Zhao, R.: On a general family of function spaces. Ann. Acad. Sci. Fenn. Math. Diss. **105**, 56 (1996)

106. Zhao, R.: On $F(p, q, s)$ spaces. Acta Math. Sci. Ser. B (Engl. Ed.) **41**(6), 1985–2020 (2021)

107. Zhao, R., Zhu, K.: Theory of Bergman spaces in the unit ball of $\mathbb{C}^n$. Mém. Soc. Math. Fr. (N.S.) **115**, vi+103 (2008)

108. Zhu, K.: Spaces of Holomorphic Functions in the Unit Ball. Graduate Texts in Mathematics, vol. 226. Springer, New York (2005)

109. Zhu, K.: A class of Möbius invariant function spaces. Illinois J. Math. **51**(3), 977–1002 (2007)

110. Zhu, K.H.: Operator Theory in Function Spaces. Monographs and Textbooks in Pure and Applied Mathematics, vol. 139. Marcel Dekker, Inc., New York (1990)

111. Zhu, K.H.: Duality of Bloch spaces and norm convergence of Taylor series. Michigan Math. J. **38**(1), 89–101 (1991)

112. Zhu, K.H.: Bloch type spaces of analytic functions. Rocky Mountain J. Math. **23**(3), 1143–1177 (1993)

113. Zhu, K.H.: Distances and Banach spaces of holomorphic functions on complex domains. J. Lond. Math. Soc. (2) **49**(1), 163–182 (1994)

114. Zygmund, A.: Trigonometric Series. Vol. I, II, 3rd edn. Cambridge Mathematical Library. Cambridge University Press, Cambridge (2002). With a foreword by Robert A. Fefferman

Index

Printed in the USA
CPSIA information can be obtained
at www.ICGtesting.com
LVHW010601061123
763144LV00006B/427

9 783031 397035